ADVANCES IN PRODUCTION MANAGEMENT SYSTEMS

IFIP – The International Federation for Information Processing

IFIP was founded in 1960 under the auspices of UNESCO, following the First World Computer Congress held in Paris the previous year. An umbrella organization for societies working in information processing, IFIP's aim is two-fold: to support information processing within its member countries and to encourage technology transfer to developing nations. As its mission statement clearly states,

> IFIP's mission is to be the leading, truly international, apolitical organization which encourages and assists in the development, exploitation and application of information technology for the benefit of all people.

IFIP is a non-profitmaking organization, run almost solely by 2500 volunteers. It operates through a number of technical committees, which organize events and publications. IFIP's events range from an international congress to local seminars, but the most important are:

• The IFIP World Computer Congress, held every second year;
• Open conferences;
• Working conferences.

The flagship event is the IFIP World Computer Congress, at which both invited and contributed papers are presented. Contributed papers are rigorously refereed and the rejection rate is high.

As with the Congress, participation in the open conferences is open to all and papers may be invited or submitted. Again, submitted papers are stringently refereed.

The working conferences are structured differently. They are usually run by a working group and attendance is small and by invitation only. Their purpose is to create an atmosphere conducive to innovation and development. Refereeing is less rigorous and papers are subjected to extensive group discussion.

Publications arising from IFIP events vary. The papers presented at the IFIP World Computer Congress and at open conferences are published as conference proceedings, while the results of the working conferences are often published as collections of selected and edited papers.

Any national society whose primary activity is in information may apply to become a full member of IFIP, although full membership is restricted to one society per country. Full members are entitled to vote at the annual General Assembly, National societies preferring a less committed involvement may apply for associate or corresponding membership. Associate members enjoy the same benefits as full members, but without voting rights. Corresponding members are not represented in IFIP bodies. Affiliated membership is open to non-national societies, and individual and honorary membership schemes are also offered.

ADVANCES IN PRODUCTION MANAGEMENT SYSTEMS

International IFIP TC 5, WG 5.7 Conference on Advances in Production Management Systems (APMS 2007), September 17-19, Linköping, Sweden

Edited by

Jan Olhager
Linköping University, Sweden

Fredrik Persson
Linköping University, Sweden

 Springer

Library of Congress Control Number: 2007932610

Advances in Production Management Systems

Edited by J. Olhager and F. Persson

p. cm. (IFIP International Federation for Information Processing, a Springer Series in Computer Science)

ISSN: 1571-5736 / 1861-2288 (Internet)
ISBN: 13: 978-0-387-74156-7
eISBN: 13: 978-0-387-74157-4

Printed on acid-free paper

9 8 7 6 5 4 3 2 1
springer.com

CONTENTS

Part II - Strategic Operations Management

Part III – IS/IT Applications in the Value Chain

Part IV – Modelling and Simulation

Part V – Improving Operations

PREFACE

The competitive environment is becoming increasingly more complex and intense. In order to cope, business decisions related to various areas tend to become more interrelated. Firms need to couple their operations strategies to the marketing strategies to best support the competition of their products in the marketplace. The perspectives on production management systems are getting more strategic. A more integrated approach is thus called for, bringing together the various perspectives on production management systems and operations strategy. This relationship is important in any type of operation, perhaps more so in supply chains, production networks and global operations.

This book brings together the latest thinking by leading experts, analysts, academics, researchers, and industrial practitioners from around the world who have worked extensively in the area of production management systems and strategies. In the individual chapters of this book, authors put forward their perspectives, approaches, and tools for use in developing and integrating systems and strategies in production management.

This book is structured in sections to allow chapters which address common themes to be grouped together. In these chapters, the reader will learn about the key issues currently being addressed in production management research and practice throughout the world. This book is composed of five parts, each focused on a specific theme:
- Linking systems and strategies
- Strategic operations management
- IS/IT applications in the value chain
- Modelling and simulation
- Improving operations

The reader will hopefully discover new approaches to integrating systems and strategies in production management and related issues that are relevant for making production into a competitive resource for the firm.

Each of the chapters in this book have been peer reviewed and presented by the authors at the International Conference on Advances in Production Management Systems – APMS 2007, held in Linköping, Sweden on 17-19 September 2007. The conference was supported by the International

Federation of Information Processing and was organized by the Working Group 5.7 in Integrated Production Management, of the Technical Committee 5 from within IFIP. The conference was hosted by Linköping University, Sweden.

We, the editors, would like to thank all of our contributors for taking the time and trouble in preparing and presenting their research and for their willingness to share their insights and ideas. We would also like to thank the members of the IFIP Working Group 5.7 for their support in reviewing and selecting papers.

Jan Olhager, Linköping University
Fredrik Persson, Linköping University

PART I

Linking Systems and Strategies

Supply Chain Redesign Employing Advanced Planning Systems

Jim Andersson and Martin Rudberg
Jim Andersson, Lawson Sweden, Propellergatan 1, SE-211 19 Malmö,
Sweden
Martin Rudberg, Department of Management and Engineering, Linköping
University, SE-581 83 Linköping, Sweden, martin.rudberg@liu.se

Abstract

Higher expectations on supply chain performance force organizations to reinvent themselves in order to cut costs and increase customer service, all to gain competitive advantage. Pursuing the best network of manufacturing, supply and distribution facilities relative the marketplace is therefore on top of many managers "most wanted" list concerning supply chain management. Supply chain planners are thus in need of decision support to be able to establish feasible and sufficient plans. This paper discusses how decision support through advanced planning systems (APS) can assist tactical supply chain planning. A case study is presented showing how APS can act as an enabler in adapting logistics and supply chain principles, as well as reducing costs through streamlining the supply chain. The purpose of this paper is primarily to present findings from a case study regarding supply chain planning with the aid of a master planning APS-module. The case study emphasize that APS in the scope of logistics management have several positive effects on supply chain performance.

Keywords

Supply Chain Management, Supply Chain Planning, Process Industries, Sweden, Production Networks

1 Introduction

Many companies that act on both domestic and global markets experience a growing international competition and recognize the need of supply chain efficiency. This need stems from increasing customer demands on high quality products at a low price plus higher expectations on accurate deliveries and customer service

Please use the following format when citing this chapter:

Andersson, J., Rudberg, M., 2007, in IFIP International Federation for Information Processing, Volume 246, Advances in Production Management Systems, eds. Olhager, J., Persson, F., (Boston: Springer), pp. 3-10.

(Christopher, 2005). Advanced planning systems (APS) can be used as a tool to meet the ever increasing demands on effectiveness that put new pressures on swift and efficient planning and control of the supply chain. APS as a decision support system for production and distribution planning is still a new and fairly unexplored tool (Stadtler and Kilger, 2005; Wu et al, 2000). During the last few years, companies that sell enterprise resource planning (ERP) systems have started developing and implementing APS-modules, which by the aid of sophisticated mathematical algorithms and optimization functionality, supports planning of complex systems such as supply chains (Stadtler, 2005; Wu et al, 2000).

The purpose of this paper is primarily to present findings from a case study regarding supply chain planning with the aid of a master planning APS-module. Unlike traditional ERP systems, APS try to find feasible, near optimal plans across the supply chain as a whole, while potential bottlenecks are considered explicitly (Stadtler and Kilger, 2005). In terms of software, APS means a broad group of software applications developed by various software vendors, such as i2, Manugistics, Oracle, SAP, AspenTech and Lawson. During the last decade, the use of APS for design, integration, and control of supply chain processes have increased. Especially the interest among industrial companies has increased, some have invested in the software, but only few use them in practice on strategic and tactical planning levels. There are also few documented cases showing how standardized APS are used, especially concerning the tactical planning level which is the focus of this paper.

In the following the paper provides a literature review centred on supply chain management and supply chin planning. Thereafter the case study is introduced describing the supply chain redesign, the APS implementation and its results in particular. After an account concerning managerial implications some concluding remarks are also provided.

2   Supply Chain Management

Studies on supply chain management often centre on what has been identified as the two main issues concerning the management of value-adding networks; *configuration* and *coordination* (Rudberg and West, 2008). This paper addresses both issues, *i.e.* the configuration of the network and the coordination of the newly configured network. The ever-present questions of, for example, where to allocate production so as to be most responsive and how to maintain low production and transportation costs, become more important when competition increases. Still, the issue of coordination cannot be left aside and must be included when supply chain redesign is considered. In this paper most focus will be put upon coordination of the newly redesigned supply chain with the aid of a master planning APS module.

2.1   Supply Chain Planning

Considering the complex environment that most companies have to cope with, most decision-support systems advocates a hierarchical distribution of the decision-making processes, where the next upper level coordinates each lower level (Wortmann *et al.*, 1997). Strategic decisions (long horizon and long periods) cannot be based on the same level of detail in the information as is the case for operational decisions (short horizon and short periods). Hence, decisions made at a high hierarchical level are normally based on aggregated information (in terms of product families, factories, *etc.*) and aggregated time periods. Thereafter these high level decisions form the context for the decision-making processes at lower-level decision centres, where decisions are disaggregated into more detailed information and time periods, but the considered horizon is made shorter. Decisions are thus exploded through the hierarchical structure until the lowest level is reached and detailed decisions are executed.

Planning has in the recent years developed to be supported by optimization and simulation tools, especially concerning "higher" planning levels. Complex trade-off analysis can be calculated with the aid of optimization models and solution heuristics in relatively short computing time (de Kok and Graves, 2003; Chopra and Meindl, 2004). The general trade-off in planning is between service, costs, capital expenditure and working capital (Shapiro, 2001). Cost minimization and profit maximization are the two most common ways to control the solution (Stadtler and Kilger, 2005). Many planning related problems in the supply chain are caused by poor communication and coordination throughout the supply chain. APS has been put forward as one solution to these problems.

APS can be defined and explained through different perspectives but commonly APS is viewed as an extension of ERP. On the other hand, standard APS modules stem from the many in-house developed decision support systems (DSS) that aid planners at various levels in the decision hierarchy (De Kok and Graves, 2003). The literature reports on some successful implementations of DSS in either special supply chain planning situations or optimization models regarding the entire chain. Gupta et al (2002), for example, describe a DSS that helps Pfizer to plan their distribution network. The model is useful in both strategic and operational planning situations. Brown et al (2001) presents a large-scale linear programming optimization model used at Kellogg Company to support production and distribution decision making on both strategic and tactical levels. Arntzen et al (1995) comprehensively describe supply chain optimization at Digital Equipment Corporation.

One way to classify standard APS is by categorizing different modules depending on the length of the planning horizon on the one hand, and the supply chain process that the module supports on the other. Figure 1 categorizes the most common standard APS modules according to these two dimensions (Stadtler, 2005; Meyr et al, 2005), which is a module segmentation also used among software vendors. This study focuses on the tactical level, *i.e.* multi-site master planning.

2.2   Supply Chain Master Planning

Master planning (MP) looks for the most efficient way to fulfil demand forecasts and/or customer orders over a mid-term planning interval (see Figure 1), which often covers a full seasonal cycle. Master planning not only balances demand forecasts with available capacities but also assigns demands (production and distribution amounts) to sites in order to avoid bottlenecks (Rohde and Wagner, 2005). Master planning is an important supply chain decision level because to be effective, inputs from throughout the supply chain are required and its results have great impact on the supply chain (Chopra and Meindl, 2004).

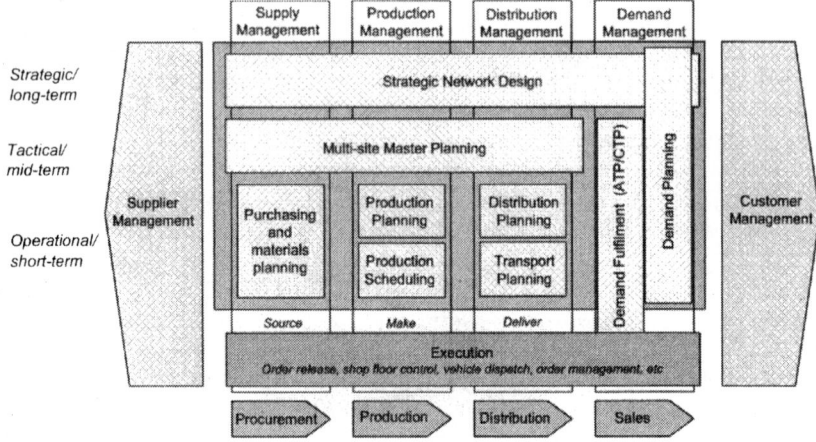

Fig 1:: Typical APS planning structure and categorization of APS modules.

Mid-term planning concerns rough quantities of material supplied, workforce requirements, production quantities and seasonal stock and use of distribution channels. Before operations research (OR) tools like optimization and simulation entered the enterprise planning arena, master planning was often done by traditional infinite MRPII systems, or by simple calculations using spreadsheets without considering capacity limitations (Fleishmann and Meyr, 2003). New OR-related software that conducts master planning using linear programming tries to maximize profit (or minimize costs) while meeting demand (Chopra and Meindl, 2004). To be able to optimize the mid-term supply chain model, production, inventory and distribution must be regarded concurrently. MP uses data on products and material in aggregated product groups. Inputs are demand data and network constraints in terms of a model that defines capacity and dependencies between different processes. The MP results in a common supply chain plan regarding production, distribution, inventory, procurement and materials requirements (Rohde and Wagner, 2005). The case in the following section will describe how supply chain master planning coordinates procurement, production and distribution on the mid-term planning level.

3   Case: The Farmers Group

Svenska Lantmännen (The Swedish Farmers Supply and Crop Marketing Association) is one of the leading groups within the grocery and agriculture industry in Sweden. It is a producer cooperative that works together in marketing, distribution, sales, processing and supply. Large profit margins are not the goal so much as cost reduction through the entire chain. This is due to that the owners, some 44,000 Swedish farmers, are both suppliers and customers to the central production and distribution function. The Group has some 13,000 employees, markets its products in 19 countries and has a yearly turnover of SEK 32 billion (SEK 100 ≈ EUR 11). In Sweden, the group is organised in 13 geographically separate areas and supplies its customers with seed, fertilizers and feed among other things, and of course process and sells what the farmers produce. Prior to 2001 the farmers acted in local and regional cooperatives but in 2001, Svenska Lantmännen was founded out of merging these cooperatives. Since then the group has suffered from inefficiency and surplus capacity. Several structural changes and reorganizations have been carried out in order to streamline the business.

The study presented in this paper centers around the seed supply chain which belongs to the division with company-wide logistics responsibility. The product (seed) is, according to Fisher's (1997) definition, functional in its characteristics: low profit margins, low product variety and long lead times for requiring customized products (5 to 10 years). Due to the functional characteristics of the product, physical efficiency is the main objective for supply chain design, hence a low cost focus through the entire chain of processes. The demand for seed is highly seasonal, and about 70 % of the volume is sold during a period between December and March. The large volumes and the many suppliers and customers make the business dependent on efficient inventory and distribution management. The planning process is difficult due to the high seasonal fluctuations and the fact that seasonal stock can only be built up in restricted amount.

3.1   Supply Chain Redesign

In 2004, a major restructuring of the seed supply chain was undertaken. Two out of six production plants was shut down and two out of four central warehouses were closed. Restructuring the seed supply chain resulted in less capacity in both production and warehouses, and also put higher requirements on distribution activities. Lower capacity and equal requirements on throughput make the planning process harder. The seed supply chain contains 30,000 farmers who act as both suppliers and customers to four production plants and two distribution centres.

Every plant supplies a restricted number of customers within the nearest geographical regions. Only in a few cases the production is differentiated between the plants. The main distribution strategy is to ship finished products directly from the finished products inventory at the plants. This leads to a large number of distribution relations that put pressure on transportation efficiency. To be able to meet the high seasonal demand peaks surplus capacity is needed. Sales forecasts are conducted by the marketing and sales function and are mostly dependent on historical sales data. The access to raw material is in great extent dependent on weather and other factors that are hard to predict.

3.2   Implementing an APS

The restructuring of the supply chain was complemented by changes in the mid-term supply chain master planning in form of a new centralized planning function. Earlier the mid-term planning (including production and distribution) had been carried out locally/regionally with simple spreadsheets. Hence, the new centralized planning function was in need of software decision support in order to find feasible plans for the entire supply chain. The logistics division investigated the use of decision support through APS in a feasibility study in mid 2004. Two months later the APS module Lawson M3 Supply Chain Planner (M3 SCP) was implemented and in use. The tactical planning process and related APS modules are visualized in Figure 2.

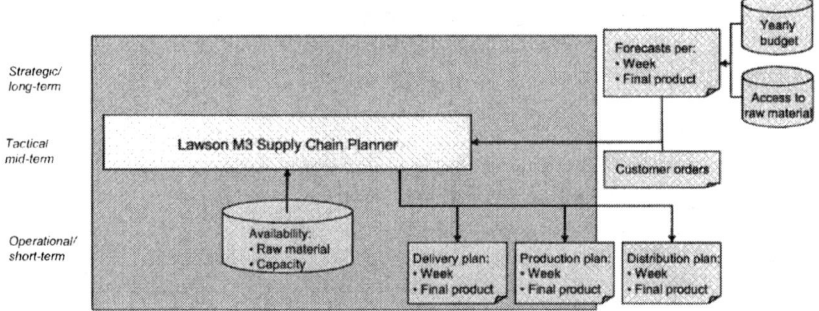

Fig 2. Svenska Lantmännen's tactical planning process and related APS modules

Planning with M3 SCP balances supply and demand for each weekly time bucket during the planning horizon (the remaining season). Forecasts are based on the yearly budgeting process and only updated twice a year. Production and inventory levels are matched with capacity for each period to minimize total supply chain costs, with regard to the four production and two distribution sites. The planning process with M3 SCP is done in an iterative manner where bottlenecks are identified and handled by the planner. Data is extracted from the company's ERP system and used by the M3 SCP module. The solver engine uses linear and mixed-integer programming to solve the planning problem with respect to total cost minimization. Short-term production scheduling as well as planning supply is carried out locally at the plants according to the directions given by the aggregated master plan. Transport planning (routing, loading, etc.) is outsourced to a third party that executes the deliveries within the frames of the distribution plan.

3.3   Results

In a study carried out in 2006 (Andersson, 2006), the effects on total costs regarding the seed production and distribution at Svenska Lantmännen was evaluated with respect to the supply chain redesign. At the same time, the flow of goods and the implementation of the APS at the centralized master planning function were investigated. The results show that the structural changes and the implementation of the APS have streamlined the production and distribution network regarding the flow of goods. Total costs have decreased by some 13 % on a yearly basis, while at the same time the quantity of sold units has increased. This results in a total reduction of cost by some 15 % per tonne. Furthermore, inventory levels in production facilities and warehouses were reduced by almost 50 %, and inventory reductions were realized for raw material, WIP and finished products.

The supply chain planning trade-off have had the following consequences: in general increased production batch sizes, slightly higher transportation costs due to lower fill-rates, decreased production cost and less capital tied up in inventory because of better throughput. The reinvented supply chain planning has reduced the total planning time. Central planning has increased the control of material flows in the chain as well as the cost structure. A higher understanding of the supply chain trade-off makes further development of immediate importance. Optimizing the supply chain has not been the most important objective with the APS implementation. The main focus has been to gain acceptance for central master planning process and to enable communications between functions. In addition to the monetary gains by the restructuring, the new way of planning increased the communication between logistics, manufacturing, marketing and sales functions. The master plans produced by M3 SCP are distributed and discussed between the functions and thereby improved for better fit with actual supply and demand.

4   Concluding Remarks and Managerial Implications

To be able to plan and control complex supply chain structures powerful management decision support is needed. Planning has, therefore, in the recent years found a renaissance in the use of optimization and simulation tools. APS uses such optimization and simulation tools, as it considers the supply chain constraints and produces near optimal solutions (Stadtler and Kilger, 2005).

APS can thereby be used as a tool to meet the ever increasing demands on effectiveness that put new pressures on swift and efficient planning and control of the supply chain. APS as a DSS for production and distribution planning is however still a new and fairly unexplored tool. During the last few years, companies that sell ERP systems have started developing and implementing APS-modules, However, only few cases using standardized APS are documented in the literature, and this paper thereby enhances the understanding of APS and its use in practice through a detailed case study. The paper is therefore of value to both academics and practitioners, since it provides an illustration of how state-of-the-art commercial software is applied in a practical setting. The number of documented cases where standard off-the-shelf APS are used in industry is limited, and this paper contributes with case study findings addressing both implementation issues and the day-to-day operational use of the software.

Finally, this research indicates that APS together with a centralized master planning function contributes to the restructuring of the supply chain with more efficient operations as a result. The conclusions that can be drawn from this study is that positive effects on throughput and inventory levels, and thus supply chain performance, have been realized with the aid of an APS. However, a prerequisite is that the organization employing the APS is operated effectively, in this case through the restructuring of the supply chain.

References

Andersson, J. (2006) *Avancerade planeringssystem som beslutsstöd vid produktions- och distributionsplanering*, Master's thesis, Dept. of Science and Technology, Linköping University, Sweden.

Arntzen, B.C., Brown, G.G., Harrison, T.P. and Trafton, L.L. (1995) Global Supply Chain Management at Digital Equipment Corporation, *Interfaces*, 25(1), 69-93.

Brown, G., Keegan, J., Vigus, B. and Wood, K. (2001) The Kellogg Company Optimizes Production, Inventory and Distribution, *Interfaces*, 31(6), 1-15.

Chopra, S. and Meindl, P. (2004) *Supply Chain Management - Strategy, Planning and Operation*, 2nd edition, Prentice Hall, New Jersey.

Christopher, M. (2005) *Logistics and Supply Chain Management: Creating Value-Adding Networks*, 3rd edition, Pearson Education Limited, King's Lynn.

De Kok, A.G. and Graves, S.C. (Eds) (2003) *Handbook in operations research and management science Vol. 11 – Supply Chain Management: Design, Coordination and Operation*, Elsevier, Amsterdam

Fisher, M.L. (1997) What is the right supply chain for your product? *Harvard Business Review*, March-April, 105-116.

Fleishmann, B. and Meyr, H. (2003) Planning Hierarchy, Modeling and Advanced Planning Systems *In* de Kok, A.G. And Graves, S.C. (Eds) *Handbook in operations research and management science Vol. 11 – Supply Chain Management: Design, Coordination and Operation*, Elsevier, Amsterdam, pp. 457-524.

Gupta, V., Peter, E., Miller, T. and Blyden, K. (2002) Implementing a Distribution-Network Decision-Support System at Pfizer/Warner-Lambert, *Interfaces*, 32(4), 28-45.

Meyr, H., Wagner, M. and Rohde, J. (2005) Structure of Advanced Planning Systems *In* Stadtler and Kilger (red) *Supply Chain Management and Advanced Planning – Concepts, Models, Software and Case Studies*, Springer, Berlin, 109-115.

Rohde, J. and Wagner, M. (2005) Master Planning *In* Stadtler and Kilger (Eds) *Supply Chain Management and Advanced Planning – Concepts, Models, Software and Case Studies*, Springer, Berlin, 159-177.

Rudberg, M. and West, M. (2008) "Global Operations Strategy: Coordinating Manufacturing Networks", *Omega*, 36, 91-106.

Shapiro, J.F. (2001) *Modelling the Supply Chain*, Thomson, Pacific Grove.

Stadtler, H. (2005) Supply chain management and advanced planning – basics, overview and challenges, *European Journal of Operations Research*, 163, 575-588.

Stadtler, H. and Kilger, C. (Eds) (2005) *Supply Chain Management and Advanced Planning – Concepts, Models, Software and Case Studies*, 3rd edition, Springer, Berlin.

Wortmann, J.C., Muntslag, D.R. and Timmermans, P.J.M. (1997) *Customer-driven Manufacturing*, Chapman & Hall, London, UK.

Wu, J., Ulieru, M., Cobzaru, M. and Norrie, D. (2000) Supply Chain Management Systems: state of the art and vision, *International Conference on Management of Innovation and Technology (ICMIT)*.

An Approach for Value Adding Process-Related Performance Analysis of Enterprises within Networked Production Structures

Hendrik Jähn

Chemnitz University of Technology,
Department of Economic Sciences, Professorship BWL VII
09107 Chemnitz, Germany
hendrik.jaehn@wirtschaft.tu-chemnitz.de,
WWW home page: http://www.tu-chemnitz.de/wirtschaft/bwl7/

Abstract

This conceptual paper focuses a methodology for the analysis of performances of enterprises operating in production networks. In order to derive adequate results exclusively the operative perspective of performance analysis is investigated. Operative performance analysis implies the analysis of performances of network members considering a special value-adding-process. The introduced approach is divided into two segments: the value adding process-neutral phases and the value adding process-specific phases. In that context special methodologies for the determination of performance indicators, corresponding parameters, evaluation functions and weightings are focussed. Additionally possible consequences are discussed.

Keywords

Performance Analysis, Network Controlling, Production Network

1 Introduction

In order to successfully stand the competition pressure caused by the large-scale companies as well as the effects of globalisation, especially small and medium-sized enterprises (SME) increasingly form production networks. Thereby, products can be manufactured by several enterprises which remain legally independent and provide their specific core competences. The purpose of the research work predominantly consists in developing an approach for the performance analysis under consideration

Please use the following format when citing this chapter:

Jähn, H., 2007, in IFIP International Federation for Information Processing, Volume 246, Advances in Production Management Systems, eds. Olhager, J., Persson, F., (Boston: Springer), pp. 11-18.

of a high degree of automation. Thereby, the performance analysis does not only comprise processes of performance measurement and evaluation but also the consequences which should be derived from the analysis, for example as sanctions or incentives. For the design and structuring the approach, the individual process phases of the performance analysis were classified as value adding process-neutral and value adding process-related. The first mentioned type particularly includes the processes of determining the performance indicators, corresponding parameters, the evaluation functions as well as the weightings. Those processes just need to be carried out by suitable instances from time to time but not before every analysis process. The value adding process-specific approaches include the processes of measuring and evaluating the performance, weighing the indicators as well as calculating an aggregated performance parameter and finally deriving suitable consequences for the single enterprises. These are processes which need to be carried out for every single value adding process based on pre-defined algorithms.

The introduced approach was realised using an adapted cost benefit analysis under consideration of the claim for a high level of automation of the analysis processes.

2   Value adding process – neutral phases

2.1   Determination of Performance Indicators

The most important task of performance analysis is the determination of the relevant performance indicators which are derived based on an adapted Balanced Scorecard (BSC) [1]. Although in literature network-specific Balanced Scorecards [2, 3] can be found, consecutively it is only applied to identify relevant performance indicators.

At first, it must be clarified which perspectives need to be taken into consideration for the derivation of the performance indicators from the strategy of a production network. The classic BSC disposes of four perspectives. The financial perspective, the customer perspective and the perspective of the internal business processes can be taken over to a large degree. However, due to the dynamic orientation of the network, the learning and development perspective can be neglected in this context and a cooperation perspective can be included instead. All the targets of an enterprise are considered whose achievements contribute to the success of the complete network. The BSC applied in this way thus integrates the enterprises as single organisation units as well as the dynamic production network. This generalisation is efficient because finally only those performance parameters shall be derived which are relevant for the performance analysis. Those performance parameters make possible the analysis of the performance of enterprises but they also take into consideration the network

The different perspectives considered in the BSC including specific targets can be allocated to each other with regard to a hierarchy. Starting from the financial perspective which has the target to achieve profits, connections can be made between the targets for the several perspectives. Finally, basic targets can be identified in the enterprise perspective. Fig. 1 illustrates the described connections graphically.

After determining the performance indicators that need to be considered as well as determining the variables which represent the performance indicators, it is necessary to transform the values of the single performance parameters into an equal metrics in order to allow an aggregation of the results subsequently. This is achieved by determining a value benefit function for every performance parameter.

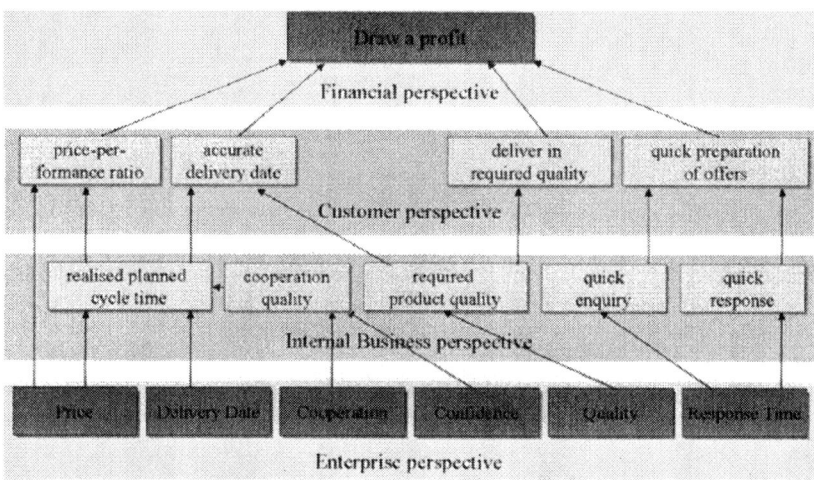

Fig. 1. Methodology for the determination of performance indicators

The value benefit function is of mathematical nature and is used for transforming the results of the measurement process into a credit evaluation. In order to determine such value benefit functions, the preferences of the evaluating person need to be evaluated. Consecutively the preferences of the customers are put on a level with the target value of the corresponding performance parameter, i.e. the target performance.

All criteria have to be illustrated on a 10-credit-scale (10 credits make possible a differentiated evaluation which is easily comprehensible) whereby the value 0 represents the worst and 10 represents the best, the promised, result.

For the six identified performance parameters, in the following chapter a suitable parameter and exemplarily ascertains of the target system and thus of the value benefit function as an evaluation function are introduced. Hereby both hard-facts and social factors (so-called soft-facts [4]) are considered.

2.2   Determination of the Parameters and the Evaluation Function

The performance parameter *price* is represented by the variable price difference. Price difference is defined as the deviation of the realised price from the agreed price in the offer of the enterprise for a value adding process. Thereby, it is important that a possibly cheap price is expected because of the competition in the network in order to increase the chances of an enterprise to be selected. Because the deviation from the agreed price is a big failure, an evaluation function is suggested as a binary relation. This means that the full number of evaluation credits is granted if the offer price was kept. If this is not the case, no evaluation credits are awarded.

The *date of delivery* also is a significant performance parameter which is represented by the variable keeping the delivery date. This is the possible deviation of the realised date of delivery from the promised one within the scope of a value adding process. Because delays in the dates of delivery could also have already been caused by preceding enterprises, a corresponding clearance needs to be made. This means that only enterprises causing a delay can be considered. For deriving an evaluation function, it is necessary to quantify interdependencies between the character and time of the delay. Thus, the full number of credits can be granted if the date of delivery is kept or the goods are even delivered prior to the agreed date while a devaluation is realised if there is a small delay and the granted number of credits approaches to zero with a longer delay. A corresponding mathematical function can be derived by the means of Lagrange interpolation.

During partner selection enquiries are sent to potential enterprises with regard to their core competences during the offer phase by the network broker instance. Because a quick generation of the response is aimed at based on the high level of automation, the *response time* also represents a relevant performance parameter. The appropriate parameter also is called response time that means the time that passes between the enquiry of the broker to the enterprise and the response of the enterprise to the broker instance. The enquiry time has to be cleared for possible times for sub-enquiries. The derivation of an evaluation function can be analogous to the date of delivery. A target response time is assumed which, if kept, leads to the allocation of the full number of credits. If the actual response time is longer than the target, the number of credits is successively reduced up to a certain marginal value starting from which no more credits are allocated. The ascertainment of a function is carried out by the help of Lagrange-interpolation as well.

Quality refers to the quality of the (partial) product which is represented by the parameter with the same name. In the simplest case, a binary relation is chosen: If the product corresponds to the qualitative requirements, the full number of credits is awarded; if not, the enterprise is given zero credits. If a differentiated consideration is aimed at, single quality features can be evaluated with regard to their level of fulfilment and aggregated to a complete value by the means of weightings. This procedure again makes it possible to derive an evaluation function by the means of Lagrange interpolation.

The performance parameter *cooperation* with the parameter cooperation represents a soft-fact which describes the quality of the cooperation in the network. Because soft-facts are primarily described by linguistic terms such as high, medium or few, a quantitative consideration is very difficult. However, there is the possibility to realise a quantification by the help of the Repertory Grid-methodology [5]. That approach is based on the ideas of the Personal Construct Theory [6]. However, further details will not be given here for lack of space. Information concerning soft-facts can be collected by having a corresponding instance ask structured questions to the enterprises. The appropriate levels of fulfilment of the single partial categories in a weighted version can be aggregated consecutively and thus lead to a complete level of fulfilment for the performance parameter cooperation. Then, this complete level of fulfilment can be transformed to an evaluation function by standardisation.

Confidence is an important success factor for networked cooperations and also has to be interpreted as a soft-fact. Thus, there is also the problem of quantification. The confidence climate is used as a parameter. Analogous to the performance parameter cooperation, the Repertory Grid-methodology is also applied for quantification for the factor confidence. Furthermore, a corresponding evaluation function is derived.

2.3   Determination of the Weightings

In order to allow the consideration of the differing standings the performance indicators can be weighted. As the fist step a check of the independence of the performance parameters has to be arranged. In case these preconditions are fulfilled, the determination of the weightings can be started. Therefore, the trade-off-procedure [7] is applied. The application of weightings allows considering the single performance indicators when carrying out the performance analysis dependent on its meaning. Because the approaches for determining the weightings are all quantitative, an automated procedure usually does not face any problems. Detailed questions, however, need to be answered in the forefront by the decision makers.

The value adding process-neutrally ascertained weightings can be applied for the performance analysis for a long term. However, it is recommended to carry out a check or if necessary a correction of the weightings regularely.

3   Value adding process – specific phases

3.1   Measurement of the Performance

The measurement of the completed performance takes place (at the latest) after a value adding process has been finished. It can be carried out by a specific concept of a workflow management system including a monitoring functionality. The support of the enterprises for the coordination of activities in the network is a central task of a workflow management system. Therefore, a distributed workflow management is proposed which consists of several components. The workflow engine as the central coordination instance is the core of the system. The disturbance management, document management, process definition catalogue as well as the monitoring can serve as supporting modules.

Within the scope of monitoring, data are collected during the value adding process and evaluated to a certain extent. Thereby, the evaluation predominantly refers to the disturbance management where it is necessary to react very quickly to deficient developments. The aspect of performance analysis in this step is not the most important issue. However, the data collecting within the scope of monitoring is indispensable and a basic pre-condition for this aspect.

The actual collection of data, i.e. the measurement of the performance without an evaluating element, takes place based on the specific parameters which were introduced earlier. While in the case of the quantitative performance parameters, a high level of automation in the sense of using the ICT can be realised, (pre-defined) repertory grids need to be filled in by the enterprises for the consideration of soft-facts. This can only be done by correspondingly trained human decision makers who represent the enterprise. In order to protect from manipulation, it is desirable to install control mechanisms such as a plausibility check or further decision makers.

Because not all the data which have been collected in the measurement phase by monitoring within the scope of the workflow management are used for the performance evaluation, it is absolutely necessary to make the relevant data accessible as actual values for the ICT. This means that the corresponding data are stored in a suitable form for the subsequent evaluation phase.

3.2 Evaluation of the Performance

The performance evaluation contributes to the realisation of an evaluation of the completed performance of an enterprise under consideration and in comparison with the pre-determined parameters dependent on all performance indicators or their parameters. Thus, the performance evaluation represents a central component of the operative performance analysis.

The evaluation functions of the single performance parameters, which have been determined in paragraph 2.2 gain special importance during the phase of performance evaluation. Thus, clear evaluation functions should be determined for all performance indicators or their parameters. Within the scope of the value adding process-specific performance evaluation, the demarcations of the parameters need to be determined in the corresponding performance parameter-specific kind and subsequently, they need to be included in the evaluation function. In the following, the parameter-specific value benefits can be determined by considering the related degree of fulfilment.

3.3 Interpretation of the realised performance

As a result of the preceding evaluation process, a vector can be formulated which includes the single target fulfilment levels (value benefits) of the parameters for every enterprise. In order to cope with a possible different significance of the single performance parameters, they can be correspondingly weighted. Thereby, the predetermined weightings are applied. The utility of every criterion results by multiplying the value benefits with corresponding weightings. The value benefits represent a partial utility of the total utility. The total utility has to be ascertained during the subsequent phase of the calculation of the aggregated performance parameter.

The additive model can be applied for aggregating the partial utilities to a complete utility of an enterprise to the successful check of the independence of the single indicators. Thereby, the complete utility represents the aggregated performance variable which can be conferred for the evaluation of the performability of an enterprise for a value adding process-specifically configured production network. This evaluation of the complete performance by evaluating the value benefits of the single performance indicators serves as a basis for the determination of sanctions or bonuses within the scope of the further performance analysis.

·The completed performance of an enterprise is represented by the complete value benefit and forms an enterprise-specific parameter. Thereby, it is assumed that the value benefit function considers possible tolerance values with regard to the performance that needs to be completed, in addition to the single variables for the performance parameters and the full number of credits is awarded in the end despite of acceptable deviations.

4   The Performance Analysis Approach

The comprehensive approach for the value adding process-related performance analysis for enterprises in production networks is illustrated in fig. 2.

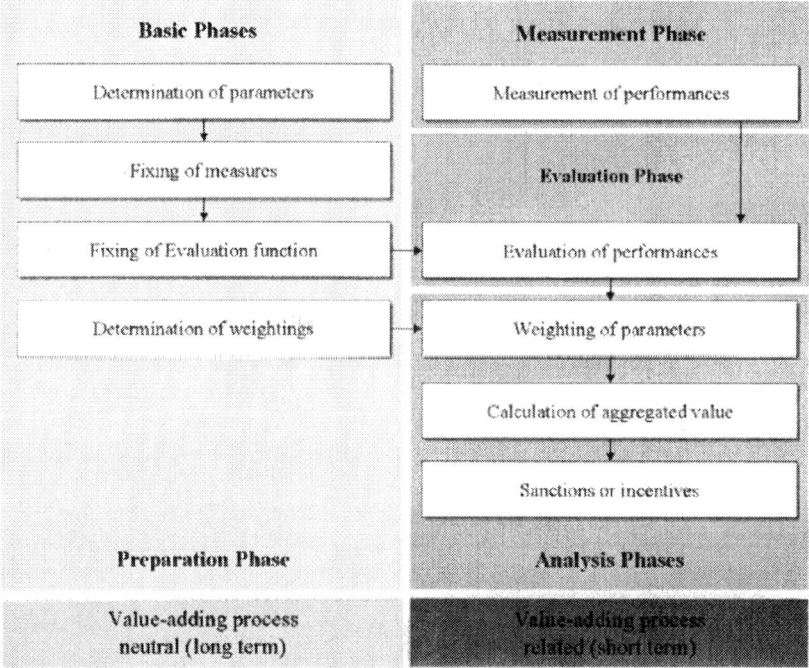

Fig. 2. Approach for value adding process-related performance analysis

From the practical point of view it must be stressed that the application of the introduced approach presupposes a high degree of information which can be guaranteed by the intensive application of the modern ICT. Thereby the analysis data is strictly held confidentially in order to prevent misapplication.

5   Conclusion

The most important finding of the research works is that it could be proved that it is possible to develop an approach for the value adding process-specific performance analysis of single enterprises within networked structures in order to make a largely automated procedure possible. Thereby, the ascertainment of the input information for the soft facts within the scope of the Repertory Grid methodology represents the most challenging task. The approach can be applied generally – however, using an automated network operation and coordination concept is recommended.

The introduced approach focuses on the value adding process-related performance analysis of enterprises within networks and thus closes a gap which established performance measurement systems leave. In most cases approaches for the performance measurement comprise the long term (strategic) view [8, 9]. This means the general economic situation is analysed. Additionally implications of the results are neglected. This approach however analyses the degree of fulfilment of the performance depending on several performance parameters including soft-facts which in order to achieve a comprehensive result, also are considered in addition to hard facts within the scope of the performance analysis. The approach takes into consideration the measurement as well as evaluation and consequences.

By the application of the approach the network controlling will be supported by alterative performance figures and suggestions for consequences in case of a to satisfying performance of single network participants.

6   Literature

1. R.S. Kaplan, D.P.Norton, *The Balanced Scorecard – Measures that drive Performance* (McGraw-Hill, New York, 1992).
2. M. Erdmann, *Supply Chain Performance Measurement* (Eul, Lohmar, 2003).
3. J. Richert, *Performance Measurement in Supply Chains* (Gabler, Wiesbaden, 2006).
4. T. Teich, H. Jähn, M. Zimmermann, *A New Approach for the Analysis of Soft-Facts in Social Networks*, in: Proceeding of the IFIP WG 5.7 Working Conference on Human Aspects in Production Management, edited by G. Zülch (Shaker, Aachen, 2003) pp. 275-281.
5. F. Fransella, D. Bannister, *A Manual for Repertory Grid Technique* (Academic Press, London, 1977).
6. G.A. Kelly, *The Psychology of Personal Constructs* (Routledge, London 1991).
7. F. Eisenführ, M. Weber, *Rationales Entscheiden* (Springer, Berlin, 2004).
8. F. Graser, K. Jansson, J. Eschenbächer, I. Westphal, U. Negretto, *Towards Performance Measurement in Virtual Organizations - Potentials, Needs, and Research Challenges*, in: Collaborative Networks and their breeding Environments, edited by L.M. Camarinha-Matos et al. (Springer, New York, 2005), pp. 301-310.
9. H.I. Kulmala, A. Lonnqvist, Performance measurement of networks: towards a non-financial approach, *International Journal of Networking and Virtual Organisations*, 3, 299-316 (2006).

Implementing and Controlling
an Operations Strategy
in Global Industrial Service Networks

Andreas Nobs[1], André Minkus[2] and Andreas Rummert[3]
1 ETH Zurich, Center for Enterprise Sciences (BWI)
Kreuzplatz 5, 8032 Zurich, Switzerland
anobs@ethz.ch, Tel: +41 44 632 05 31
2 aminkus@ethz.ch, Tel: +41 44 632 05 27
3 Opitzstrasse 31, 22301 Hamburg, Germany
andreas@rummert.com, Tel: +49 40 3612 1258

Abstract

The importance of industrial service implies an increasing need for principles to design and manage the associated operations. This paper focuses on the information exchanges in globally decentralised service networks. It thereby emphasises cross-functional interactions from an operations strategy perspective. A balanced scorecard framework is then proposed in order to facilitate the implementation and controlling of the information operations. It was found that its application fosters the strategic coordination of the operational action in a decentralised service organisation.

Keywords

Industrial Service, Service Operations Strategy, Information Management, Balanced Scorecard

1 Introduction

1.1 Motivation and Problem

Industrial service is the supply of after-sales services (e.g. installing, training, repairing, upgrading and disposing) including tangibles such as spare parts and consumables related to the maintenance of industrial goods [1]. This service is of ever-increasing importance for manufacturers of durable capital goods for several reasons. Due to the concentration on core competencies and the growing complexity of the machinery, there is a rising demand for maintenance and support, since customers use capital goods as a resource in their own operations. Therefore, the

Please use the following format when citing this chapter:

Nobs, A., Minkus, A., Rummert, A., 2007, in IFIP International Federation for Information Processing, Volume 246, Advances in Production Management Systems, eds. Olhager, J., Persson, F., (Boston: Springer), pp. 19-26.

aftermarket of an installed base provides manufacturers with new sources of revenues which tend to have higher margins than the income from the core products [2, 3]. Second, services are more difficult to imitate than mere technical features of the core products and, hence, they offer the providers a differentiation potential and a source of competitive advantage. Finally, the characteristic direct customer involvement permits after-sales service functions to gather market information (e.g. application failures, customer requirements, or competitors' activities). This in turn constitutes an essential input to product and service development, especially in the area of capital goods [4].

In this context, the ability to efficiently manage service process related information from different sources became an important success factor for the performance of the whole business. Chase et al. [5, 6] already pointed to the competitive potentials of the *service factory*, which emphasises cross-functional support from the manufacturing to other functional departments through the exchange of skills, knowledge, and information. Nonetheless, evidence from our case studies shows that the corresponding operations often do not adequately satisfy the increased information requirements, neither concerning the downstream provision nor the upstream feedback of information [7]. In line with the findings of other research in the field, this situation reflects a typical challenge of a manufacturer becoming a decentralised service provider [8-10]. Firstly, it proves a corporate disregard in terms of the growing demand for and the strategic importance of the services. In addition, it demonstrates a specific coordination and controlling problem regarding the information processes in decentralised service organisations (cf. [11], pp. 197 et sqq.).

1.2   Objective

Manufacturers must deliberately optimise the cross-functional information interactions in order to capitalise on the strategic service potentials and to increase their business performance (cf. [12], pp. 535 et sqq.). This is not so much a question of technology, but strategy – the fact that the collaboration in the service network is considered a source of competitive advantage [5]. Companies, therefore, require a new approach facilitating the implementation and controlling of a comprehensive operations strategy in terms of information potentials within their global service. This new approach has to fulfil several requirements. First of all, it has to take into consideration all important competencies and information resources at different functional and organisational company units. Furthermore, since these units usually pursue different objectives according to their conventional division of tasks, it needs to integrate the different views towards a common corporate service perspective. It should provide the possibility to clarify and evaluate the mutual contributions of different company units to their respective performances.

In response to these purposes, the paper firstly integrates the boundary-spanning information interactions into a service operations strategy framework. A balanced scorecard approach is then applied to translate the framework into explicit target areas in order to make transparent the expected operational action based on a common understanding of the relations throughout the organisation.

2   Information Resources in the Distributed Service Organisation

2.1   Organisational Structure and Information Exchanges

The provision of industrial service is a process consisting of different activities, which are carried out in a network of actors belonging to various functions and organisational units [10]. The high degree of customer contact and the intensity of front-line operations require first level service units close to the customer [13]. In connection with the ongoing business internationalisation, the customer proximity necessitates a multi-site expansion of the service organisation [14]. This can be achieved with own local subsidiaries or via independent third party service providers, such as agencies or dealers. Due to the high complexity of the products, the local front-line units are typically backed by central support functions which provide second level assistance [15]. The support centres form the major interface to various other central company and plant departments (e.g. research & development or production & assembly) as well as to suppliers, which all provide third level assistance in case of specific problems that can not directly be solved by the service personnel.

The three level support structure implies complicated information interactions between the different company units (cf. [16]). These relationships may be analysed by means of a *service mapping* methodology [17]. A service map generally facilitates the identification of the patterns and organising principles in the service business processes [18]. This is complemented with an information logic in order to demonstrate the rational connections through information resources within the service systems [7]. On the one hand, this depicts information that run from the central functions "downstream" to the front-line units, building an important resource for the service delivery. This flow comprises, for instance, technical documentation, manuals, and instructions, which are usually prepared by the central service based on the inputs from the preliminary departments, such as design plans from R&D, or inspection sheets from assembly. Other examples are direct technical advice and support as well as education and training provided from second level functions to the local service units. On the other hand, there exists an "upstream" information flow, from the market back to centralised functions. At first, this encompasses requests, inquiries, and all sorts of orders. Moreover, as already introduced above, the market units posses or are at the source of crucial information, both, about the installed base as well as about the customers and how they actually use the products. This information on market and competitive trends needs to be captured, since it serves as one of the most precious inputs for the improvement, redesign, or new development of products and services. Hence, market information, such as for example service histories or front-line personnel's experiences and improvement propositions should flow back to the central company departments.

2.2   Operations Strategy Perspective

In general, an operations strategy is the pattern of decisions which shape the capabilities to meet long-term corporate and business objectives through the reconciliation of market requirements with operations resources [19]. On the one hand, it describes the required operations performance driven by customer needs, competitor activity, and therefore market positioning. On the other hand, it specifies the nature and characteristics of its resources that produce the market performance. Given the market needs in a specific situation, the actual decisions in the operations strategy pertain to the company's resources. The decision categories are usually divided into structural and infrastructural categories (see [20], pp. 31 et seq.). The operations strategy defines the integration of these aspects. Internally, it deals with the fit between structural and infrastructural choices and between different functional areas. External integration aligns the strategic decision areas with the performance objectives [21].

Central to service is the explicit consideration of the nature of customer contact in the so-called service encounter, which takes account of the customer's relationship to and interaction with the service delivery [22]. It implies specific characteristics (see [23], pp. 63 et sqq.) which are considered in a dimension-oriented delineation of a service system by means of potentials, processes, and results. They correspond to the operations strategy aspects and they constitute an operations logic in which the resources and capabilities (potential) are transformed (process) into desired outputs according to customers' needs and expectations (result) (cf. [15], p. 22). This framework is now applied to the service information resources in scope and it thereby explicitly adds the strategic perspective. It builds the basis to identify and specify the information contributions and requirements from different organisational and functional units within a globally decentralised service network.

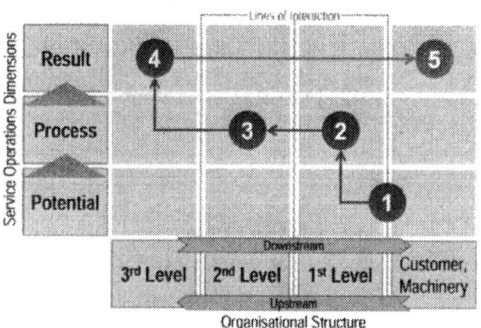

Fig. 1. Information Interaction Matrix

Figure 1 exemplifies the generic integration of the three service operations dimensions with the organisational structure previously outlined. The resulting matrix specifies that the latter is depicted on the horizontal axis and the operations dimensions be arranged consecutively from bottom to top along the vertical axis. Structural relationships between the organisational units are indicated by vertical dotted lines that locate the moments of interaction between organisational units. The explicit consideration of the customer and machinery highlights the fact that it is part of the delivery system. This setup again draws upon the service mapping principles.

The matrix illustrates a simplified upstream information relation. One might think of a structural constructive failure the field technician identifies based on customer interaction or a concrete technical improvement proposal (1). This insight is reported (2) to the central service which analyses and prepares the information about the installed base for the respective functions (3), for instance R&D, where the proposal may lead to a revised technical design (4), and therefore enhances the quality of the product offering (5) that the customer is potentially willing to pay for.

3 Industrial Service Strategy Map

A Balanced Scorecard (BSC) framework is now applied to integrate the information requirements into a comprehensive picture of the service business and to derive operational implementation and controlling measures in the service organisation. Kaplan and Norton [24] introduced the BSC as a performance measurement system to translate an organisation's strategic goals into operational action (i.e. through more tangible goals, actions, and performance measures). The objectives and measures of the scorecard thereby not only focus on a financial perspective, but rather take into account a balanced view of three other performance areas: *customer*, *internal business processes*, and *innovation, learning and growth*. The so-called *strategy map* illustrates the linkages between the four perspectives through a chart of cause-and-effect relationships between different target areas [25]. This central method is adapted to the three service operations dimensions with appropriate target areas in order to identify meaningful objectives as well as to derive manageable measures turning the information requirements to operational action (cf. [26]).

The financial perspective emphasises the overall business performance in terms of tangible goals. The target areas here involve aspects of profitability, revenue growth, or market shares. This level may be viewed part of the result dimension. It is considered and measured in order to control whether the service organisation's operations strategy, its implementation and execution are really contributing to the bottom line, since some of the assumed linkages between operating performance and financial success may be quite uncertain in a particular application. Furthermore, the financial perspective bridges the gap to the business strategy, which finally governs the financial targets. The second perspective focuses on customer satisfaction and retention, which, according to the *service-profit chain* concept [27], are the most important drivers for growth and profitability. It thus directly corresponds to the result dimension and stresses target areas related to the overall availability of goods and services, including choice, quality, delivery time, and the pricing of the offerings. The process perspective highlights critical operations that enable the service organisation to satisfy the strived results and customer needs. They cover the operational assignment of the essential resources in order to develop, improve, and provide the industrial service and the related goods. The industrial service strategy map here covers aspects of information and knowledge processing as well as the development and provision of goods and services. It thereby stresses the horizontal and vertical cooperation within the service network. Finally, the potential perspective includes objectives and measures for the promotion of competencies, assets, and cultural aspects that enable the organisation to effectively perform customer-based and internal processes. This specifically implies to capture the relevant drivers for future innovation and growth in the areas of human, information, and organisation capital [28]. Figure 2 exhibits an example strategy map regarding the upstream report information previously outlined (cf. fig. 1). It defines the strived objectives in the four perspectives in terms of the information interaction and relates them into relationships, similar to causal loop diagrams (cf. [29]).

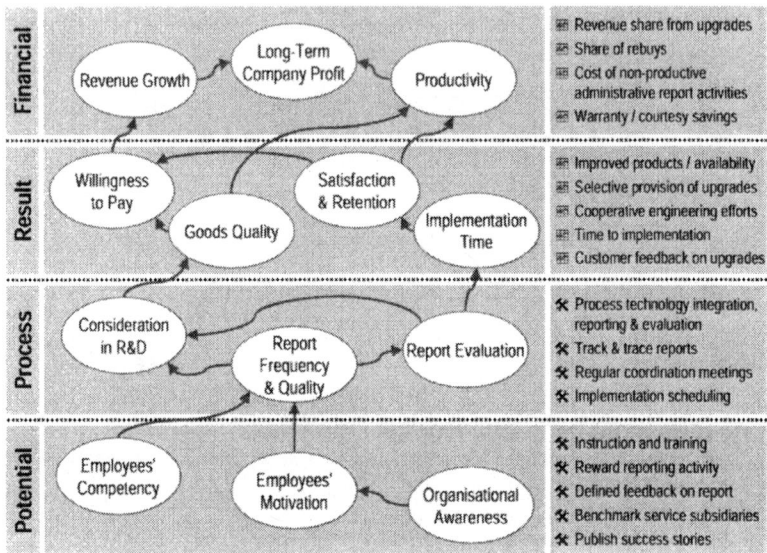

Fig. 2. Industrial Service Strategy Map

This comprehensive picture of objectives is the basis to derive operational activities with corresponding performance indicators (as exemplified on the right hand side). The lower two perspectives thereby mainly address the implementation of measures concerning structural and infrastructural decision categories, whereas the upper perspectives primarily contain aspects to control the performance outcomes.

4   Implications and Benefits

Manufacturers must increasingly adapt their service systems in order to tap the associated economic and competitive potentials. The outlined strategy map provides a suitable concept in consideration of the stated requirements and objectives. It shows the value chain by which a service organisation converts its information resources into desired operational, market, and financial performance. It specifically identifies the driving forces (in terms of potentials and processes) to meet the market needs. This external integration is the necessary condition for a successful implementation of an operations strategy. Moreover, the framework highlights the importance of cooperation in a distributed service organisation. The example in figure 2 demonstrates the relevance of front-line information provided by local service units for the improvement of goods and services, which, in turn, drives the performance of the whole business. This ability to effectively communicate corporate initiatives and operational measures throughout the company is an important benefit of the strategy map. The understanding of the underlying values and hypotheses improves organisational clarity, which in turn facilitates the internal alignment of structural and infrastructural resources across vertically diversified functional company units.

The target areas and objectives are specified from an integrated point of view. The derived operational measures and the performance indicators are then attached to the pertaining organisational units according to the service maps (simplified in fig. 1). This means that the actual implementation maintains the diversification of the operations through measures that are influenced by the decentralised units. The controlling, however, ensures that the operational action and improvement stay in line with the defined overall network objectives. Hence, the performance measurement of the integrative operations should be assigned to the central service function in order to control the success of the implementation and the need for corrective action. To stick with the example, activities to enhance employees' motivation and competency to report market information need to be measured with the frequency and quality of the reports. Such indicators are compared with the implementation in R&D and product improvements, their further impact on customer satisfaction and, finally, sales figures. In addition, this offers the possibility to benchmark the performance of different service units in order to identify particularly successful practices, which in turn allows exchanging these practices horizontally.

The coordinated implementation of the service operations and especially its subsequent controlling and reviewing eventually increases the organisational awareness for and consensus on the strategic service importance. It therefore contributes as a valuable instrument to support the change process towards service system integration in manufacturing companies.

References

1. P. Johansson and J. Olhager, Industrial service profiling: Matching service offerings and processes, *Int. J. Production Economics* **89**, 309-320 (2004).
2. R. Wise and P. Baumgartner, Go Downstream – The New Profit Imperative in Manufacturing, *Harvard Business Rev.*, Sept-Oct, 133-141 (1999).
3. J. Behlke and K. Banki, *VDMA-Kennzahlen Kundendienst* (VDMA, Frankfurt a.M., 2005).
4. T. Levitt, After the sale is over..., *Harvard Business Rev.*, Sept-Oct, 87-93 (1983).
5. R.B. Chase and D.A. Garvin, The Service Factory, *Harvard Business Rev.*, Jul-Aug, 61-69 (1989).
6. R.B. Chase, K.R. Kumar, and W.E. Youngdahl, Service-based Manufacturing: The Service Factory, *Production and Operations Management* **1** (2), 175-184 (1992).

7. A. Nobs, A. Minkus, and T. Pengo, Towards a Process Reference Model for the Management of Industrial Services, in: Exploiting the Knowledge Economy, ed. by P. Cunningham and M. Cunningham (IOS Press, Amsterdam, 2006), pp. 1072-1079.
8. R. Oliva and R. Kallenberg, Managing the transition from products to services, *Int. J. Service Industry Management* **14** (2), 160-172 (2003).
9. S. Brax, A manufacturer becoming service provider – challenges and a paradox, *Managing Service Quality* **15** (2), 142-155 (2005).
10. P. Gaiardelli, N. Saccani, and L. Songini, Performance measurement systems in after-sales service: an integrated approach, *Int. J. Business Performance Management* **9** (2), 145-171 (2007).
11. A.V. Hill, D.A. Collier, C.M. Froehle, J.C. Goodale, R.D. Metters, and R. Verma, Research opportunities in service process design, *J. Operations Management* **20**, 189-202 (2002).
12. A. Picot, R. Reichwald, and R.T. Wigand, *Die grenzenlose Unternehmung: Information, Organisation und Management* (Gabler, Wiesbaden, 2003).
13. M.A. Cohen, N. Agrawal, and V. Agrawal, Winning in the Aftermarket, *Harvard Business Rev.*, May, 129-138 (2006).
14. C.P. McLaughlin and J.A. Fitzsimmons, Strategies for globalizing service operations, *Int. J. Service Industry Management* **7** (4), 43-57 (1996).
15. I. Hartel, *Virtuelle Servicekooperationen: Management von Dienstleistungen in der Investitionsgüterindustrie* (vdf, Zürich, 2004).
16. B.J. Berkley and A. Gupta, Identifying the information requirements to deliver quality service, *Int. J. Service Industry Management* **6** (5), 16-35 (1995).
17. J. Kingman-Brundage, Service Mapping: Back to Basics, in: Understanding Services Management, ed. by W.J. Glynn and J.G. Barnes (John Wiley & Sons, Chichester, 1995), pp. 119-142.
18. J. Kingman-Brundage and W.R. George, Service logic: achieving service system integration, *Int. J. Service Industry Management* **6** (4), 20-39 (1995).
19. N. Slack and M. Lewis, *Operations Strategy* (Pearson Education, Harlow, 2002).
20. M. Rudberg and J. Olhager, Manufacturing networks and supply chains: an operations strategy perspective, *Omega* **31**, 29-39 (2003).
21. A.V. Roth and L.J. Menor, Insights into Service Operations Management: A Research Agenda, *Production and Operations Management* **12** (2), 145-164 (2003).
22. G.L. Shostack, Designing services that deliver, *Harvard Business Rev.*, Jan-Feb, 133-139 (1984).
23. H. Meffert and M. Bruhn, *Dienstleistungsmarketing: Grundlagen – Konzepte – Methoden* (Gabler, Wiesbaden, 2006).
24. R.S. Kaplan and D.P. Norton, The Balanced Scorecard – Measures That Drive Performance, *Harvard Business Rev.*, Jan-Feb, 71-79 (1992).
25. R.S. Kaplan and D.P. Norton, Having Trouble with Your Strategy? Then Map It, *Harvard Business Rev.*, Sept-Oct, 167-176 (2000).
26. A. Assiri, M. Zairi, and R. Eid, How to profit from the balanced scorecard: An implementation roadmap, *Ind. Management & Data Systems* **106** (7), 937-952 (2006).
27. J.L. Heskett, T.O. Jones, G.W. Loveman, W.E. Sasser Jr., and L.A. Schlesinger, Putting the Service-Profit Chain to Work, *Harvard Business Rev.*, Mar-Apr, 164-174 (1994).
28. R.S. Kaplan and D.P. Norton, Measuring the Strategic Readiness of Intangible Assets, *Harvard Business Rev.*, Feb, 52-63 (2004).
29. F. Schoeneborn, Controlling Service Business at a German Printing Press Manufacturer, in: Proc. 19th Int. Conf. System Dynamics Soc., 23-27 Jul 2001, ed. by J.H. Hines, V.G. Diker, R.S. Langer, and J.I. Rowe (2002).

Derivation of Strategic Logistic Measures for Forging Systems

Peter Nyhuis, Felix S. Wriggers and Tim D. Busse
Institute of Production Systems and Logistics,
Gottried Wilhelm Leibniz Universität Hannover,
An der Universität 2, D-30823 Garbsen, Germany
e-mail: wriggers@ifa.uni-hannover.de

Abstract

In order to thrive in global markets, enterprises have to distinguish themselves from their competitors not only by manufacturing high quality products at low costs, but also with superior logistic performance. Logistic Operating Curves (LOC) can be applied to facilitate this as well as to derive strategic measures. This will be demonstrated based on the example of the German forging industry.

Keywords

Logistic Operating Curves, logistic measures

1 Introduction

A superior logistic performance is primarily demonstrated by high delivery reliability and short delivery times [1], [2], [3]. The criticalness of superior logistic performance is particularly noticeable in the German forging industry. Companies in this sector are usually small and medium-sized enterprises (SME) delivering about 70 % of the forged parts directly to automotive OEMs [4], [5]. They thus tend to be integrated in just-in-time and just-in-sequence delivery concepts [6], [7].

Short throughput times and a high level of schedule reliability in production are a prerequisite for accomplishing short delivery times and a high level of delivery reliability. In contrast, low costs require a high output rate [8] whereas the logistic objectives throughput time and output rate, oppose each other: While short throughput times can only be realized with a low work in process (WIP) level, high output rates depend on high WIP levels. This conflict between logistic objectives is commonly known as the 'Dilemma of Operations Planning' [9].

Please use the following format when citing this chapter:

Nyhuis, P., Wriggers, F. S., Busse, T. D., 2007, in IFIP International Federation for Information Processing, Volume 246, Advances in Production Management Systems, eds. Olhager, J., Persson, F., (Boston: Springer), pp. 27-34.

A solution for controlling this dilemma is offered by the Logistic Operating Curve Theory (LOC Theory) developed at the Institute of Production Systems and Logistics, Leibniz University of Hanover [10].

2   Modeling Logistic Processes in the Forging Industry

The starting point for developing all of the sub-models for the Logistic Operating Curve Theory was defining a fundamental production logistics' element – the throughput element. Based on the logistic process levels, the throughput element defines the throughput time of an operation as the time span which an order requires from the ending of the preceding process or from the point in time of the order's input up to the end of the processing in the observed operation. When manufacturing is conducted according to lots, as is common in the forging industry, an order is transported to the following workstation after it is processed or if necessary after a period of waiting. Usually the lot then meets up with orders that are already queued and therefore has to wait until those before it are processed. Initially it is required that the orders are completed according to First In First Out (FIFO). As long as the capacities for processing the orders are available, the station's set-up can be changed and the next lot can be processed. This cycle continues until the order has passed through all the required operations.

2.1   The Funnel Model and Throughput Diagram

With the help of the throughput elements definition, the throughput time, scheduling deviation and their components can be calculated and statistically evaluated. The throughput element also forms the basis for the Funnel Model and the Throughput Diagram derived from it (Figure 1).

With the Funnel Model, similarly to diagrams of flow process methods, the throughput behavior of every random capacity unit in a manufacturer can be completely described through the input, WIP, and output (Figure 1a). Together with those already waiting the lots arriving at the workstation form a store of waiting orders. Once they are processed they flow out the funnel. The funnel opening thus symbolizes the station's output rate, which can vary within the capacity limits.

The funnel's events can be transferred to the so-called 'Throughput Diagram" (Figure 1b). The completed orders are cumulatively plotted with their work content over the completion schedule (output curve). The input curve is developed similarly, in that the incoming orders are plotted with their work content over the input schedule. The start of the input curve is determined by the WIP, which is found on the workstation at the start of the investigation period (initial WIP). At the end of the investigation period, the end WIP can be read from the diagram. Whereas, the mean slope of the input curve corresponds to the mean load, the output curve's mean slope corresponds to the mean output rate.

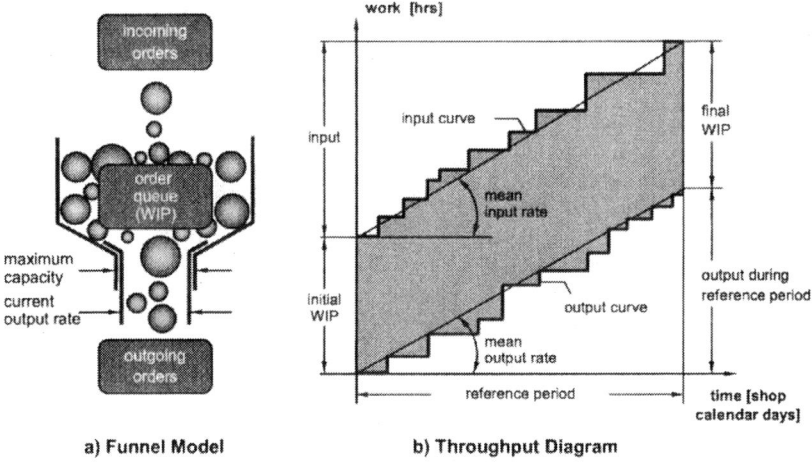

| a) Funnel Model | b) Throughput Diagram |

Fig. 1. The Funnel Model and Throughput Diagram

The Throughput Diagram describes the dynamic system behavior completely quantitatively and exactly with regards to time. It also provides support in identifying causes of plan deviations as well as for deriving control measures. When necessary the Throughput Diagram can be supplemented by diagrams of the throughput time, work content and/or distribution of the scheduling deviations. Fundamental information about the production's processes can thus be gained and analyzed with regards to a variety of problems. Nevertheless, these do not explain or only partially explain the interactions between the logistic parameters.

The following described Logistic Operating Curves (LOC) however, provide support in understanding cause and effect relations between these.

2.2   Logistic Operating Curves

The Funnel Model, Throughput Diagram and figures derived from them each describe a specific relaxed operating state. In the upper part of Figure 2, three basically different operating states are depicted in simplified Throughput Diagrams. These operating states can now be strongly aggregated in the form of Logistic Operating Curves. In order to do so, the relevant values for the output rate and range are plotted as a function of the corresponding WIP level. The Output Rate Operating Curve (OROC) clearly illustrates that a workstation's output rate does not significantly change beyond a specific WIP value. There is a continuous queue of work and thus no WIP related interruptions in the processing. Below this WIP value however there are increasing output setbacks due to a temporary lack of queued orders. The range (and with that the throughput time) though, increases above the critical WIP value for the most part proportionally to the WIP. Nonetheless, when the WIP is reduced both throughput parameters cannot fall below a specific minimum. For the throughput time, this minimum results from the mean operation time for the orders and where applicable the transport time.

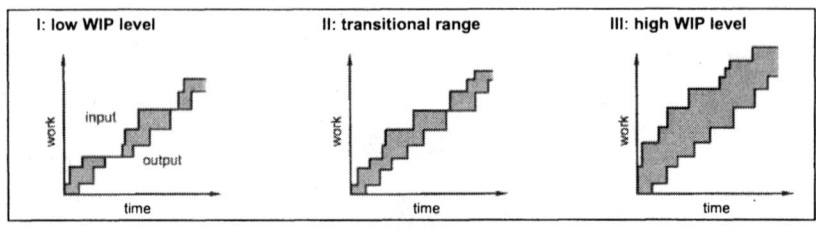

a) typical operating states for a workstation

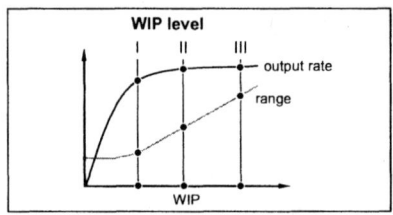

$$WIP_m(t) = WIPI_{min} \cdot (1 - (1 - \sqrt[4]{t}\,)^4) + WIPI_{min} \cdot \alpha_1 \cdot t$$

$$ROUT_m(t) = L_{max} \cdot (1 - (1 - \sqrt[4]{t}\,)^4)$$

$$R_m(t) = \frac{WIP_m}{ROUT_m}$$

$WIP_m(t)$ mean WIP	$WIPI_{min}$	ideal minimum WIP
$ROUT_m(t)$ mean output rate	$ROUT_{max}$	max. output rate
$R_m(t)$ mean range	α_1	stretch factor
	t	running var. $(0 < t < 1)$

b) depiction of operating states in
    Logistic Operating Curves

c) approximation equations for calculating
    the Logistic Operating Curves

Fig. 2. Deriving the Logistic Operating Curves

A momentary state on a workstation always corresponds to only one operating state and thus one operating state point on a LOC. The Logistic Operating Curves represent how the observed system behaves when a different WIP level is set given otherwise stable boundary conditions. They thus characterize the logistic behavior of a production when the WIP changes. Beyond that, it is also possible to develop LOC for changed manufacturing or order structures and to compare these with one another. In this way, the impact of intervening in the production process can be evaluated with regards to logistic aspects.

It can easily be seen that the basic form of the Logistic Operating Curves are applicable to every arbitrary production system: WIP reductions lead to decreased throughput times but also under certain circumstances to breaks in the material flow and thus to utilization losses. However, the specific shape of the LOC for the observed workstations are dependent on different boundary conditions such as the capacity, the orders to be processes (in particular the mean value and their distribution) and how they are incorporated into the system's material flow. A detailed description of the Logistic Operating Curve Theory, the influences and the basic equations can be found in NYHUIS and WIENDAHL [10].

Logistic Operating Curves thus allow the dependencies between the logistic objectives utilization, throughput time, schedule reliability and WIP to be expressed [11]. At the same time, the cost-benefit ratio of applying the LOC is extremely favorable [12]. Due to this they are extremely well suited for increasing an enterprises logistic performance [13] and have therefore been widely received and accepted both in research and on the production floor [10].

3   Fields of Application for Logistic Operating Curves

Logistic Operating Curves represent an ideal basis for developing and monitoring an enterprise's process reliability and capability [13]. LOC can thus be drawn upon for evaluating processes during a production control. They indicate for example which throughput times and WIP levels can be achieved on a workstation with the existing structural conditions, without having to expect appreciable breaks in the material flow and resulting output losses. When applying them within production planning and control (PPC), system parameters can be derived and set which conform with the goals. Depicting the logistic objectives in a diagram makes it possible to decide which attribute should be weighted the most depending on the current market and operational situations as well as the workstation's specific boundary conditions. Simultaneously it can be shown, how changing a parameter impacts the logistic attributes. Instead of searching for an imaginary optimum, a primary, frequently market dependent objective e.g. a desired throughput time, is assumed. The remaining target values such as the output rate and WIP inevitably result from there.

If during the application it turns out that the set target values are not achievable without further intervening measures which change the form of the LOC, then the Logistic Operating Curves can be used to support and evaluate the planning activities corresponding to the imagined possibilities. Alternative planning and control strategies can therefore be evaluated and chosen based on logistic criteria. The LOC Theory can also be directly integrated into PPC such as in methods for determining lot size, scheduling or also for the order release. Moreover, it supports a continual, method based, alignment of the planning and controls with the logistic objectives. During the planning stage of a factory the Logistic Operating Curves can be drawn upon for a logistic oriented evaluation of alternative manufacturing principles or new logistic concepts. Furthermore, it is also possible to evaluate investment decisions (e.g. implementing new transport systems, introducing new production technology) and to extend the modeling of business processes.

Since the LOC express the cause and effect relations between the logistic objectives output rate, throughput time, and WIP both qualitatively and quantitatively, they offer effective support for mastering the 'dilemma of operations planning'. Based on these and depending on the actual operating state – which can also differ from place to place – it can be decided which objective should be assigned the greatest significance.

Prioritizing a goal forms a basis on which the targeted operating points, defined through the WIP, output rate and throughput time can be determined for the individual workstations. Within this process the existing structures for the work content and capacities on the one hand, as well as both the required delivery date, required capacities and the cost structures on the other hand are considered. In the following this procedure will be referred to as a "Logistic Positioning". The Logistic

Positioning is the basis for all of the mentioned applications, it establishes the targets and thus represents a link between all of the individual functions [10].

In addition the determined target values and if necessary the allowable fields of tolerance can be directly checked for consistency. If the throughput time is given for a workstation, the corresponding key figures for the output rate, utilization and WIP result directly from it.

By conducting such a positioning it can be shown whether or not the targeted goals are realistic with the existing boundary conditions. If it is not the case, the target values are not located on the calculated Logistic Operating Curves. Measures can then be taken in order to develop new logistic potential, that is measures which influence the behavior of the Logistic Operating Curves [10].

4   Developing a Method for Choosing Strategic Measures

During the Logistic Positioning, values for the four logistic objectives (WIP, scheduling reliability, utilization and throughput time) are recorded. The actual and target values for these parameters can be formulated as a four dimensional Logistic Position Vector (LPV). The difference between the LPV for the target state and that of the actual state defines the so called Logistic Target Achievement Vector (LTV). Based on the actual state, the Logistic Target Achievement Vector describes the changes required for the production logistic objectives in order to attain the target state.

In order to realize the target state, strategic measures have to be taken. Generally though, there will be a number of both complementary and conflicting possible measures. Decision makers in an enterprise perceive themselves as being confronted with the problem of which measures are best suited for the specific situation. In order to answer this question, a method is currently being developed at the Institute of Production Systems and Logistics, which permits a monetary evaluation of the measures and thus forms the basis of a decision support system (DSS) for choosing strategic measures.

This method requires a description of the functional correlations between the measures and their monetary impact, which can be determined based on positive and negative payment stream parameters. Up until now, the functional correlation between the measures for improving the production logistics and the payment stream parameters was missing. Monetary transfer functions were thus developed. On the one hand, monetary transfer functions produce the correlation between the measures and the revenues and on the other hand, the correlation between the measures and the costs.

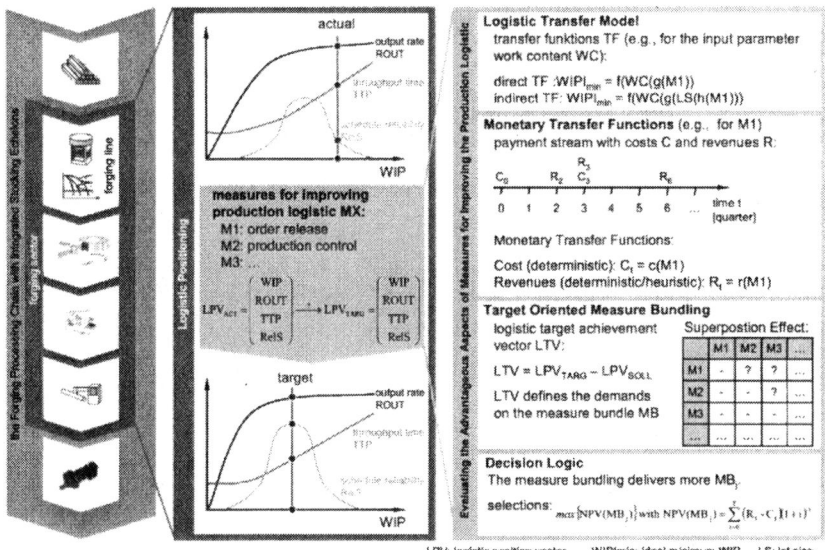

Fig. 3. Expanding the Logistic Operating Curve Theory in order to Derive Strategic Logistic Measures

Evaluating and choosing the measures, should thus occur based on standard economical parameters such as the capital value, the return on investment or the EVA (economic value added).

5 Acknowledgements

The authors would like to thank the German Research Foundation (DFG) for financially supporting the work described in this article as part of the sub-project C2 of the interdisciplinary research center 489 "Processing Chains for the Production of Precision Forged High Performance Components".

6 References

1. Hon, K. K. B., Performance and Evaluation of Manufacturing Systems, *Annals of the CIRP*. Volume 55, Number 2 (2005).
2. Enslow, B., *Best Practices in International Logistic* (Aberdeen Group: Boston 2006).
3. Nyhuis, P., Wriggers, F., Fischer, A., The International Federation of Information Processing (IFIP), Volume 207, *Knowledge Enterprise: Intelligent Strategies in Product Design, Manufacturing, and Management*, eds. K. Wang, Kovacs, G., Wozny, M., Fang, M. (Boston: Springer 2006).
4. Rudau, E., Deutsche Gesenkschmiedeindustrie gerüstet für die Zukunft, *Schmiede-Journal*. (2000).

5. Hirschvogel, M., Schmiedetechnik in Europa und in den USA - technischer oder kultureller Unterschied, *Schmiede-Journal* (2001) pp. 34-36.
6. Wiendahl, H.-P.; Ruta, A., Einsatz der Logistik FMEA, *Schmiede-Journal* (1999), pp. 39-40.
7. Boston Consulting Group: *Steering Carmaking into the 21st Century - From today's best practices to the transformed plants of 2020* (BCG:Boston 2001).
8. Lödding, H., *Verfahren der Fertigungssteuerung - Grundlagen, Beschreibung, Konfiguration* (Springer:Berlin 2005).
9. Gutenberg, E., *Grundlagen der Betriebswirtschaftslehre*, Volume 1 - Die Produktion, (Springer:Berlin 1951).
10. Nyhuis P., Wiendahl H.-P., *Fundamentals of Production Logistics*. Transl. of 3rd ed. (To be published Springer:Berlin 2007).
11. Nyhuis, P., Logistic Production Operating Curves – Basic Model of the Theory of Logistic Operating Curves, *Annals of the CIRP*, Volume 55, Number 1 (2006) pp. 441-444.
12. Nyhuis, P., von Cieminski, G., Fischer, A., Applying Simulation and Analytical Models for Logistic Performance Prediction, *Annals of the CIRP*, Volume 54, Number 1 (2005) pp. 417-422.
13. Wiendahl, H.-P.: Collaborative Supply Chain Planning – A Case Study from the German Cutting Tool Industry, *Collaborative Systems for Production Management* eds. Jagdev, H. S., Wortmann, J. C., Pels, H. J (Kluwer:Norwell 2003).

Strategic Choice of Manufacturing Planning and Control Approaches: Empirical Analysis of Drivers and Performance

Jan Olhager and Erik Selldin
Linköping University, Department of Management and Engineering,
SE-58183 Linköping, Sweden
jan.olhager@liu.se, www.liu.se

Abstract

The design of manufacturing planning and control systems is a strategic decision for manufacturing firms. In this paper we analyze the interrelationships among the choices of planning and control approaches at different hierarchical levels, including sales and operations planning (chase; level), master scheduling (make-to-order; assemble-to-order; make-to-stock), material planning (time-phased; rate-based), and production activity control (MRP-type; JIT-type). We test the relationships with product characteristics and performance. These relationships are explored through a survey of 128 manufacturing plants.

The results show that choice of approaches at the two higher planning and control levels, i.e. sales and operations planning and master scheduling, are strongly interrelated. The choices at the two lower hierarchical levels, i.e. material planning and production activity control, are also strongly interrelated. However, the link between any of the two upper and any of the two lower levels is much weaker. The most significant drivers of the choice of planning and control approach are: (i) product volume and delivery lead time for sales and operations planning, (ii) product variants and delivery lead time for master scheduling, and (iii) production lead time for production activity control, while material planning is not significantly related with any product characteristic. Significant effects on performance are found for sales and operations planning on volume flexibility, master scheduling on product mix flexibility, and material planning and production activity control approaches on delivery speed. This research supports the notion that sales and operations planning is concerned with volume planning and that master scheduling is concerned with mix planning.

Keywords

Empirical analysis, Manufacturing, Performance, Survey research,
Theory testing

Please use the following format when citing this chapter:

Olhager, J., Selldin, E., 2007, in IFIP International Federation for Information Processing, Volume 246, Advances in Production Management Systems, eds. Olhager, J., Persson, F., (Boston: Springer), pp. 35-42.

1   Introduction

In order to achieve manufacturing excellence the planning and control system for manufacturing operations must fit the task relative the market place. The design of a manufacturing planning and control (MPC) system cannot be taken in isolation; it must be based on the specific products produced and their market characteristics. The design of an MPC system is a strategic choice related to the operations strategy. A framework for choosing the appropriate type of planning and control approach relative the market and product characteristics, has been developed by Berry and Hill [1], also presented in [2,3]. They are concerned with three hierarchical planning levels, i.e. master planning, material planning, and shop floor control. The fourth level in an MPC system is sales and operations planning, which has been added to the Berry and Hill framework by Olhager et al. [4], concerned with the choice of planning strategy. The theoretical models are assumingly sound, have been widely referenced, and have not received specific criticism in the literature. Therefore, it is interesting to test these theories on how to design MPC systems related to the market and product characteristics, drawing on the modeling ideas in Berry and Hill [1], with the amendment of Olhager et al. [4]. This paper extends the analysis in Olhager and Selldin [5] by addressing drivers and individual performance measures.

This paper is organized as follows. First, we present the theoretical framework underpinning this research. Then, we discuss the research methodology in terms of the data and measurements used. We then present the survey results, including the interrelationships among the planning and control approaches at different hierarchical levels, the relationships between product characteristics and the choice of planning and control approaches, and the relationships between planning and control system choices and performance outcomes.

2   Theoretical Framework

The models for linking market characteristics to the design of an MPC system can basically be reduced to one, first presented in Berry and Hill [1], and later included in [2,3]. In the Berry and Hill framework there are links and choices at three levels of the MPC system. At each level a set of market requirement attributes are used to make generic choices among a set of level-dependent MPC design variables. At the master scheduling level the choices are reduced to three variables; make-to-order (MTO), assemble-to-order (ATO), or make-to-stock (MTS). ATO covers for similar environments such as build-to-order (BTO), configure-to-order (CTO), and finish-to-order (FTO). At the materials planning level the choices are rate-based or time-phased. Finally, at the shop floor control level the choices are push or pull. The attributes linking market and product characteristics to the MPC approaches are very much the same, namely product type, product range and individual product volume per period, cf. Table 1. Examples of this framework conformity is that firms with high-volume standardized products typically would choose MTS, rate-based, and pull, whereas firms with many low-volume, customized products would choose MTO, time-phased, and push. In ATO environments both sets of MPC choices are applicable to different sections of the value chain upstream versus downstream the customer order decoupling point (CODP). In the case of ATO or similar environments, volumes are typically sufficiently high upstream the CODP to make MTS/rate-based/JIT-type approaches possible, whereas MTO/time-phased/MRP-type approaches are typically required after the CODP due to customized features and low volumes per product. Thus, ATO environments can typically be divided into MTS-type operations upstream and MTO-type operations downstream the CODP.

The three levels used in the Berry and Hill framework are the lower three of four levels, typically found in an MRPII (manufacturing resource planning) hierarchy. The upper level is sales and operations planning (S&OP), which has been added to the Berry and Hill framework by Olhager et al. [4]. A chase strategy would assumingly be used for low-volume, highly customized products, whereas a level strategy would be more suitable for high-volume, standardized products. This addition to the Berry and Hill framework is included in Table 1. Alignment between product characteristics and MPC approaches is assumed when there is a straight vertical line in Table 1.

Table 1. Linking planning and control system to the market and product characteristics, according to Berry and Hill [1] and Olhager et al. [4].

Market and product characteristics:			
Product type [2]	Special	→	Standard
Product range [2]	Wide	→	Narrow
Individual product volume per period [2]	Low	→	High
Planning and control level:			
Sales and operations planning (S&OP) [1]	Chase		Level
Master scheduling (MS) [2]	MTO	ATO	MTS
Material planning (MP) [2]	Time-phased		Rate-based
Production activity control (PAC) [2]	MRP-type		JIT-type

[1] Source: [4]; [2] Source: [1].

While the framework in Table 1 mostly deals with manufacturing operations, Fisher [6] developed a framework for supply chains, linking product characteristics with supply chain design types. Products that are innovative are characterized by demand variability and short product life cycles, they should therefore be transformed through a market-responsive supply chain that has extra capacity, capability of market information processing and that is more flexible. On the other hand, a steady demand pattern, high volumes and long product life cycles characterize products that are functional. A physically efficient supply chain that focuses on cost minimization and high utilization of resources should handle this kind of products. Thus, there is some resemblance with the framework in Table 1. Functional products are basically characterized as standard, narrow range, and high volume, whereas innovative products are special, wide range, and low volume. A physically efficient supply chain would assumingly be able to utilize a level strategy, MTS, rate-based, and JIT-type approaches, whereas a market-responsive supply chain would prefer a chase strategy, MTO, time-phased, and MRP-type approaches. The conceptual model is illustrated in Figure 1.

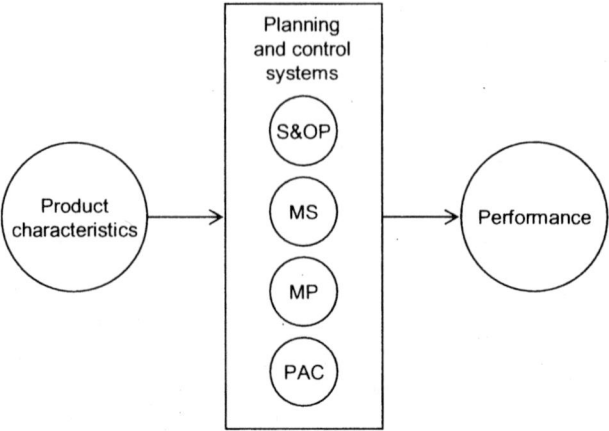

Fig. 1. Illustration of the relationship between product characteristics, planning and control system, and performance.

3   Data and Measurements

The variables for measuring planning and control are related to the different planning levels such as described in [1-4]. Sales and operations planning is measured using a five step Likert scale ranging from a pure chase planning strategy to a pure level strategy. Master scheduling ranging from pure make-to-order to pure make-to-stock, materials planning ranging from pure time-phased to pure rate-based, and production activity control ranging from pure MRP-type to pure JIT-type are all measured using floating scales where the respondents indicate the percentage of planning using each approach.

The product characteristics are based on the Berry and Hill model, with additions by Fisher [6]. All variables are captured using five step Likert scales. Demand volume and number of variants are central product characteristics in Berry and Hill [1], but also in the product-process matrix by Hayes and Wheelwright [7]; a well-known operations strategy model for linking products and processes. In the survey, product volume is captured by a variable ranging from one-of-a-kind product to mass-produced products. The number of variants is captured by a variable ranging from few variants to many variants. Other relevant product characteristics are the effect of seasonal demand (low to high seasonality) and delivery lead time for make-to-order products; cf. Fisher [6]. However, we divide the latter into two variables, delivery lead time and production lead time, to allow not only for make-to-order situations but also for make-to-stock or assemble-to-order situations where there is a difference between production and delivery lead times. Both delivery and production lead times measures range from short to long lead times.

Performance is measured as the respondent's perceptions of the company's performance related to competitors. All variables are measured using five step perceptual Likert scales ranging from "much worse than competitor" to "much better than competitor". The performance is measured by operational performance in terms of quality, delivery dependability, delivery speed, cost efficiency, volume flexibility, and product mix flexibility.

4   Survey

The questionnaire was mailed to 511 manufacturing firms, and 128 usable responses were received. This represents a response rate of 25.0 percent. All respondents are members of PLAN, the Swedish Production and Inventory Management Society, wherefore they all assumingly possess good knowledge about MPC approaches. Single-item measures are used in order to describe the components of the theoretical framework. The unit of analysis in this study is the main product line at the plant level, and the corresponding MPC system. The responding plants represent the whole spectrum of the five generic process choices: project manufacturing, job shop, flow shop, production line, and continuous processing. Both make-to-order and make-to-stock production environments are represented, as well as both consumer goods manufacturers and industrial goods manufacturers. The distribution of different kinds of environments indicates that the sample is representative for the manufacturing industry.

5   Results and Analysis

In this section we investigate the relationships between different planning and control approaches at different hierarchical levels, i.e. at the S&OP, MP, MS and PAC levels. First, we check the consistency among the approaches chosen at different hierarchical levels. According to theory (see section 2), there should be positive relationships among on the one hand a chase, make-to-order, time-phased, and MRP-type approaches, and on the other a level, make-to-stock, rate-based, and JIT-type approaches. Secondly we test the relationships with product characteristics to identify possible differences among which characteristics that drive the choice of MPC approaches at different planning levels. Finally, we analyze the relationship with performance. The significant relationships are summarized in Figure 2.

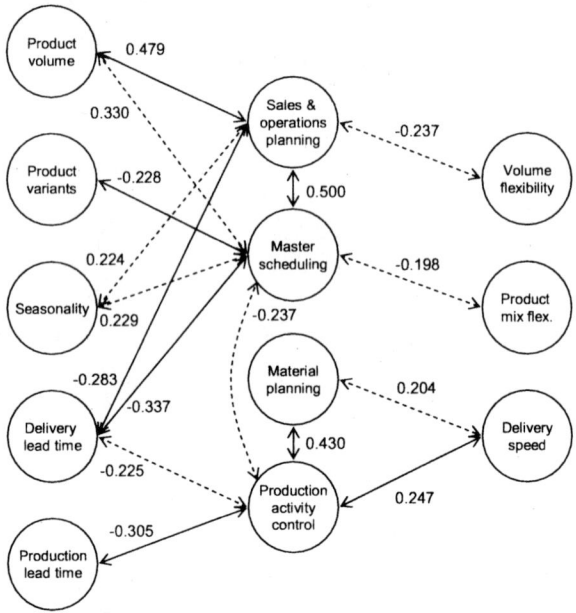

Fig. 2. The total model indicating all significant relationships between product characteristics, MPC approaches, and performance outcomes (a full line indicates significance at the 0.01 level, and a dotted line indicates significance at the 0.05 level).

5.1   Consistency among MPC Approaches

The results show that the relationship between the two higher planning and control levels is highly significant. Firms using a level planning strategy for sales and operations planning predominantly use a make-to-stock approach, whereas a chase strategy tends to be used by make-to-order firms. The relationship between the two lower hierarchical levels is also highly significant. Rate-based material planning is linked to JIT-type production activity control, and time-phased approaches tend to be combined with MRP-type approaches. However, the link between upper and lower levels is weak. There is actually a negative relationship (significant at the 0.05 level) between master scheduling and production activity control approaches, which was not expected. This means that make-to-stock firms tend to use MRP-type approaches to a larger extent that JIT-type approaches. Also, make-to-order firms use JIT-type approaches to a larger extent than MRP-type approaches.

5.2   Relationship between Product Characteristics and MPC Approaches

The theory-based model suggests that the product volume, number of variants and other market and product characteristics drive the choice of MPC approaches at all hierarchical levels. However, the result concerning consistency among MPC approaches suggests that there may be different drivers of the choice at different hierarchical levels.

Product volume is significantly related with the sales and operations planning approach (0.01 level) and with master scheduling approach (0.05 level), but not with the two lower MPC levels. With higher production volumes, firms tend to use a level strategy and apply make-to-stock, which is expected. The number of product variants is significantly correlated (0.01 level) with master scheduling approach. A firm with many variants typically operates on a make-to-order basis, which is expected. Seasonal effects is significantly correlated (0.05 level) with the sales and operations planning approach as well as with the master scheduling approach. Products with high seasonal variations often have a level planning approach and are produced to stock. Assuming that the goods can be stored, the firm can take advantage of a level strategy for make-to-stock operations. Delivery lead time is significantly related with the S&OP approach (0.01 level), with the master scheduling approach (0.01 level), and with the production activity control approach (0.05 level). Short delivery lead times are related to a level planning strategy, make-to-stock, and JIT-type execution, whereas long lead times are associated with a chase planning strategy, make-to-order, and MRP-type execution. Production lead time is significantly correlated (0.01 level) with the production activity control approach. Products with long production lead-times are found to be associated with a push-type approach, while short production lead-times are associated with a pull-type approach.

In summary, the results strongly suggest that the end product characteristics that are related to the marketplace only dictates the choice of MPC approaches for S&OP and master scheduling, i.e. the two planning levels that explicitly deal with end products and product groups (of end products), while the two lower levels are concerned with the detailed planning and execution of individual items at lower levels in the product structure (bill of material). Only delivery lead time showed a significant relationship to both upper and lower MPC levels. Product volume, the number of variants, and seasonality were only associated with upper MPC levels, while production lead time was only associated with one of the two lower MPC levels. The material planning approach did not show any significantly relationship with any of the product characteristics.

5.3   Relationship between MPC Approaches and Performance Outcomes

The performance variables are measured using a five-step Likert scale ranging from "much worse than competitors" to "much better than competitors". Only four significant correlations were found. The S&OP approach is negatively correlated with volume flexibility such that high volume flexibility is linked with a chase planning strategy at the sales and operations planning level. At the next lower level, the master scheduling approach is negatively correlated with product mix flexibility, indicating that make-to-order is linked with high product mix flexibility. The two lower levels, material planning and production activity control are positively correlated with delivery speed, such that high delivery speed is linked with both rate-based and pull-type approaches. Thus, three different performance variables are associated with four different levels of the planning and control hierarchy. The linkages reflected by all four significant correlations are in line with theoretical expectations.

6 Discussion

In this paper we have investigated the relationships among planning and control approaches at different hierarchical levels (sales and operations planning, master scheduling, material planning, production activity control), product characteristics as potentials drivers of the choice of approach, and the relationships with operational performance through a survey with data from 128 manufacturing firms. The contribution to the existing literature from this study is the theory testing of models in the literature about how the choices of planning and control approaches are interrelated and how they are linked to market and product characteristics, and the possible effect on performance. The study supports the conceptual model [1-4]; especially for the sales and operations planning and master scheduling levels, which can be characterized as volume and mix planning, respectively, with respect to the significant relationships with product characteristics as well as performance outcomes.

The implications for managers is that the results show that the choice of MPC approaches at the two upper levels are driven by end product characteristics such as volume, variants, and seasonality, while the two lower levels are driven by the lead time for production and delivery. This suggests that the two lower levels are more related to specific item characteristics rather than end product characteristics.

References

1. W.L. Berry, and T. Hill, Linking systems to strategy, *International Journal of Operations and Production Management*, **12**(10), 3-15 (1992).
2. T. Hill, *Manufacturing Strategy – Text and Cases*, 2nd ed. (Palgrave, Houndsmills, Hampshire, 2000).
3. T.E. Vollmann, W.L. Berry, D.C. Whybark, and F.R. Jacobs, *Manufacturing Planning and Control for Supply Chain Management*, 5th ed. (McGraw-Hill, New York, 2005).
4. J. Olhager, M. Rudberg, and J. Wikner, Long-term capacity management: linking the perspectives from manufacturing strategy and sales and operations planning, *International Journal of Production Economics*, **69**(2), 215-225 (2001).
5. J. Olhager, and E. Selldin, Manufacturing planning and control approaches: environmental alignment and performance, *International Journal of Production Research*, **45**(6), 1469-1484 (2007).
6. M. Fisher, What is the right supply chain for your product?, *Harvard Business Review*, **75**(2), 105-116 (1997).
7. R.H. Hayes, and S.C. Wheelwright, Link manufacturing processes and product life cycles, *Harvard Business Review*, **57**(1), 133-140 (1979).

Structuring Goals and Measures for Information Management

André Minkus, Andreas Nobs and Sören Günther
ETH Zurich, Center for Enterprise Sciences (BWI)
Kreuzplatz 5, 8032 Zurich, Switzerland
aminkus@ethz.ch, anobs@ethz.ch, sguenther@ethz.ch
WWW home page: http://www.lim.ethz.ch

Abstract

Handling the complexity of Information Management (IM) is a very challenging issue – especially in global organizations. Such organizations consist of several entities with different strengths and weaknesses. They have to decide on common goals and agree on coordinated measures for improving their IM activities. However, many organizations lack the abilities to define and operationalize relevant goals. In addition, the complexity of existing measures makes it difficult for them to decide on those measures that are best suited for meeting their IM goals. This paper presents a methodology to structure goals and measures for improving IM. It starts on a strategic level and operationalizes both, goals and measures down to an application level. By considering human, organizational and technical aspects all relevant factors are included and interdependencies are highlighted. Organizations can benefit from the methodology since it supports the design of their IM activities and takes their specific strategies into consideration

Keywords

Information Management, Axiomatic Design Framework

1 Introduction

In recent years the importance of information management (IM) has increased significantly for many companies. IM refers to all activities connected to the acquisition, processing, transfer and provision of information [1]. It is seen as one of the most critical factors for business success. The growing number and complexity of the offered products and services have increased the amount of information to be exchanged. Up to now many companies solely rely on long-established forms of informal communication. This informal communication has proven to be very

Please use the following format when citing this chapter:

Minkus, A., Nobs, A., Günther, S., 2007, in IFIP International Federation for Information Processing, Volume 246, Advances in Production Management Systems, eds. Olhager, J., Persson, F., (Boston: Springer), pp. 43-50.

effective in local organizations. However, a growing globalization requires for more balanced channels of information exchange (IE). In many organizations different branches situated in different countries with different languages, cultures or time zones have to communicate with each other.

1.1. Practitioners Problems and Requirements

Typical examples for organizations like this are global service organizations of equipment manufacturing companies. They have a fast growing international customer base requiring them to establish new subsidiaries in new markets. These subsidiaries have to be supported with all information necessary for offering complex service products. In addition, they have to provide the companies headquarter with information about the performance of the products or about customer requirements.

In global organizations as presented above, information has to be exchanged on a just-in-time basis: at the right time, in a suitable format, at the right place, in the right quality and to the right person. For achieving this, organizations have to operationalize their IM goals and they have to define suitable measures. This requires to coordinate the different elements of the service organization (e.g. subsidiaries, partners) and to consider potential interactions between the different goals and measures. For many companies this imposes a major challenge.

For instance, in one company analyzed by the authors a lack of coordination has led to additional expenses and an increase of time necessary for exchanging information. In this organization, one major subsidiary had identified the need for supporting its IE by information and communication technology (ICT). An ICT solution was developed taking all needs of this subsidiary into consideration. In the same time, a similar solution was developed at one production location. Since these two solutions were developed independently they could only be used locally. Important information had to be transferred manually between them.

Another example illustrates the effects resulting from a lack of considering relevant interactions between different goals and measures. A central service division of a company decided to implement a system supporting different service technicians in their work at the customers' site. This ICT offered a variety of new functionalities. However, the characteristics of the process to be supported were not considered. For instance, using the solution required to enter a lot of data (e.g. equipment numbers) which was not always available at the customers' site. In addition, searching for information was complex and time consuming. This prevented a use in front of a waiting customer. Consequently, the technicians are still calling the hotline if they need any information. The ICT solution did not pay off.

1.2. Overview of Relevant Literature

A review on research literature shows a variety of existing concepts. For instance, there is literature focusing on strategies and goals of IM (e.g.[2]), addressing aspects of process design in information acquisition and provision (e.g.[3]) or presenting methods for supporting information exchange activities (e.g.[4]). While all these types of literature are crucial for the design of IM activities, there are some aspects missing. There exists no concept which structures IM objectives. Usually, the existing literature focuses either on strategic or operative levels but misses to link them. In addition, an analysis of interdependencies between different technical, organizational and human measures is missing. A description of those dependencies could reduce difficulties in selecting suitable measures by identifying necessary requirements and possible impacts of an implementation.

1.3. Objective of the Paper

Motivated by both, the practitioners requirements and the illustrated gap in literature this paper focuses on the development of an Information Management Factor Catalog (IMFC). This catalog structures the goals and measures relevant for improving IM in industrial service organizations and analysis their interactions. It aims to support companies in planning and setting up new structures, processes and methods for IM. Moreover, it can be used as a tool for identifying current weaknesses and potential for improvement.

2   Methodology Used

The research methodology used for the development of the IMFC bases on the principals of action research [5]. It involves the collaboration with four equipment manufacturing companies and their service organizations. For the design of the IMFC elements of the Axiomatic Design Framework (ADF) are used [6]. ADF was developed as a method for designing products and systems. As a fundamental characteristic ADF differentiates between goals ("what we want to achieve") and measures ("how we want to achieve it") [6]. To arrange goals and measures, it provides an illustration in form of a hierarchical tree. Two design axioms provide basic principles for generating a good design. Firstly, the independence axiom is stating that the independence of the different goals must be maintained. Secondly, the information axiom requires that the information content must be minimized and hence the design should be as simple as possible.

Using the ADF to develop the IMFC requires adapting one of its principles. Complete independence as stated in the first axiom can not be achieved because of the complex interdependencies between different organizational, technical and human aspects. Instead, the IMFC tries to reduce the dependencies. By arranging different goals and measures in a certain sequence, their dependencies can be reduced and handled.

3   Setup of the IMFC

The IMFC is structured in a tree diagram consisting of six levels [Fig. 1]. This chapter provides an overview on the proceeding when structuring goals and measures into the six levels. Since a complete explanation of the IMFC is too voluminous, this paper will focus on showing some examples.

3.1. 1st Level: Offering Competitive Products and Services as the Major Goal

The first level refers to the most strategic goal influenced by IM activities. This goal is in line with the overall objectives of companies. It focuses on *supporting the organization to offer superior products and services* or to *create a cost advantage* [7]. One suitable measure helping to fulfill this goal is to provide the organization with all needed information at the right time, to the right person, in the right quality and a suitable format.

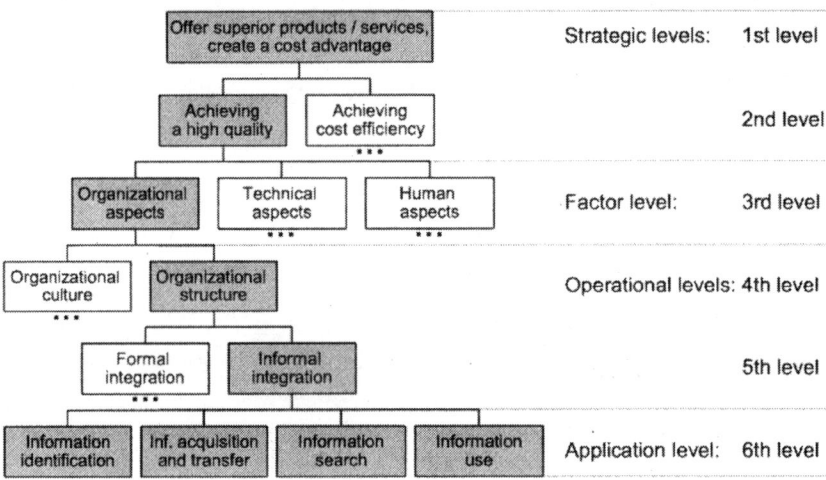

Fig. 1. Example showing the decomposition of the six levels of the IMC

3.2. 2nd Level: Achieving Quality and Cost Efficiency as Strategic Goals

The first goal is decomposed in the second level by considering the different target areas of IM. Since exchanging information is a part of the logistics processes of an organization, the four target areas of logistics – quality, time, flexibility and costs [8] – can be used. However, in contrast to tangible products the quality of information includes the dimensions of time and flexibility [9]. This is evident, since tools needed for ensuring a high quality during the gathering of information can often not be separated from tools supporting a timely delivery during the transfer of information. Consequently, the second level only makes a distinction between the objective of *achieving a high quality in IE* and of *achieving cost efficiency of the IE activities*. To achieve a high quality, the organization has to reduce existing variations in quality. For achieving cost efficiency, an adequate mean is to minimize the assets needed for IE activities and to reduce all kinds of existing waste. For establishing a sequence between the two goals a methodology from the area of production management is used. The so called Sand Cone Model shows that a company can only gain lasting capabilities by starting with a focus on quality. Cost efficiency is then a depended parameter of quality [10].

3.3.   3rd Level: Considering Organizational, Technical and Human Aspects

The decomposition taking place on the third level considers the three major factors relevant for IE: organizational, technical, and human aspects [11]. Understanding their interdependencies is crucial for designing complex systems [11]. Organizational aspects refer to leadership principles and to the design of the structure and the workflows of an organization. The objective is to create *organizational conditions supporting all IE activities* [12]. Technical aspects include tools and methods necessary for supporting the organizational structure and its workflows. The objective is to *provide a suitable infrastructure needed for IE* [4]. Finally, issues closely related to the employees of the organization are summarized as human aspects [13]. This objective aims to *assure the effectiveness of the employees*. For establishing a sequence of the three goals, the numerous interdependencies between the aspects have to be taken into consideration. A possible example of a sequence is illustrated in a concept focusing on the development of knowledge management systems [14]. This concept identifies four pillars in the following order: leadership, organization, technology, and training. Since leadership is defined as a part of the organizational aspects, they will be the first in the sequence followed by technical factors. This is also in line with different findings in literature stating the role of technical aspects like ICT as a support to processes and larger systems (e.g.[15]). According to the mentioned concept [14], training follows the technical aspects. However, a practical validation shows that depending on the qualification of the employees, certain structures can be realized and techniques can be used. Conversely, the development and training of employees is influenced by the organizational structure and its workflows and by the used techniques. Consequently, human aspects would have to be mentioned at the first and last position. But since the selection and training of employees is usually not requiring or inflicting a change in the organization, the placement of human aspects is suitable to the most right.

3.4.   4th and 5th Level: Operationalization of the Different Aspects

On the 4th and 5th level, an operationalization of the different organizational, technical and human aspects takes place. This provides a guideline for implementing or controlling measures and best practices. Figure 1 only shows the operationalization of some organizational aspects.

Operationalization of organizational aspects
An analysis of relevant organizational goals shows organizational culture and organizational structure as two critical success factors for all IE activities [16]. Establishing *an organizational culture to support the IE* is regarded as the most elementary factor to be improved in any organization [16, 12]. It is therefore set as the first goal. A further decomposition on the 5th level splits this goal into two

elements to be achieved. The first is the integration of IM objectives into the strategy of the organization. This includes the communication of the importance of IE to the success of the company [17]. The second refers to the creation of common values in the whole organization by establishing norms for cooperation. This aims at building trust necessary for IE by establishing transparency, and by supporting an integrative collaboration throughout the organization [18].

To provide a suitable *organizational structure supporting the IE* is the second goal on the 4th level. Designing the organizational structure requires to consider the existing organizational culture [12]. A decomposition of this goal on the 5th level shows two major objectives: a formal and an informal integration of IE activities. A formal integration can be achieved by integrating IE tasks into the core and support processes of the company [16]. An informal integration aims at providing opportunities for an informal exchange of information between peoples in the organization [19]. Although no direct sequence between the two tasks can be defined, an excess of formalization might result in rigid processes reducing opportunities for informal IE [20]. Therefore, the formal integration is named first. It includes an analysis of existing processes, the identification of information sources and needs, the definition of responsibility, and the integration into daily work routines. Setting up informal relationships requires reducing contact barriers by creating possibilities for interactions. For instance, work rotation or temporary employee exchanges are suitable measures for communicating the needs and conditions of different partners.

Operationalization of the technical aspects

Technical aspects have to *provide tools or infrastructure necessary for the IE* activities. The most typical example is the provision of an ICT. From a quality perspective, such an ICT has to provide a high perceived functional usefulness, a high perceived reliability and a high perceived ease of use. These three aspects can be defined as goals for the 5th level. A high functionality is the most fundamental aspect essential for the effectiveness of the ICT [21]. It has to support the user to exchange the right information at right process step and in the right format. In addition, it has to minimize the efforts of the user by matching relevant IE tasks with his workflow. This especially includes a reduction of idle times. One way to achieve a fast and process oriented IE is the use of a suitable ICT architecture [22]. A high perceived reliability can be realized by fulfilling the requirements for accessibility at different locations and times – keeping different time zones in mind. This includes both, guaranteeing availability of the ICT and providing all necessary access authorizations. For achieving high perceived ease of use the ICT has to be adjusted to the characteristics of the supported processes [23]. Finally, easy user interfaces have to be provided considering ergonomic norms like established in DIN EN ISO 9241 [24].

Operationalization of the human aspects

Human aspects include two major goals: assuring the *capabilities* and the *commitment* of the employees [13]. Ensuring the capabilities is done by hiring suitable employees, by providing training and by reducing employee turnover [18]. In addition, external support can compensate a lack of certain capabilities, e.g. for structuring, articulating, searching, and using information [25]. Four types of competences can be distinguished and are arranged on the 5th level: the provision of professional, organizational, technical and methodical competences.

For assuring the commitment of the employees, it is relevant to consider intrinsic and extrinsic types of motivation [26]. While intrinsic motivation is a result of all measures mentioned in this paper (operational, technical and human), extrinsic motivation can be achieved by selecting, combining and balancing suitable forms of compensation. However, offering extrinsic motivation can have a negative effect on the intrinsic motivation of the employees (crowding-out effect [26]). It is therefore crucial to identify those types of IE activities, which can be supported by extrinsic motivation without showing "side effects". On the 5th level, three different forms of compensation can be distinguished: social, organizational and financial compensation. They can be used to motivate individual employees or whole organizational units (like subsidiaries).

3.5.   6th level: Application of the Goals and Measures to the IE Process

On the 6th level, all goals are applied to the different steps of the IE process: identification, gathering, transfer, search, and use of information. These steps are related to the lifecycle view of information [1]. A comprehensive description of this level can not be discussed within the context of this paper since it contains more than 40 possible goals and measures. An operationalization at the level of specific IE processes also allows for defining performance indicators which can easily be collected and used for benchmarking and for identifying existing weaknesses. In addition, it shows the impact of different best practices on the IE processes.

4   Business Benefits and Conclusion

The presented IMFC and its underlying methodology support companies in identifying goals and measures for improving IE. The IFC differs from other knowledge- and information management methodologies by connecting strategic, operational and application levels of IE and by showing the dependencies between the different goals and measures of those levels.

A discussion of the approach with the participating companies has shown that the application of this approach is quite intuitive and can support industrial service organizations in designing their IM activities. Firstly, it helps to identify the major goals of IE. After analyzing the current status of the IE activities, the approach supports the prioritization of goals and the balancing of appropriate measures. The goals can be made operational in a systematic way. The hierarchical structure allows to align all operational goals with the top-level objectives and the business strategy of the companies. As a final result, a set of mostly organizational and technical measures van be identified. However, discussions have shown that more specific information is needed about best practices and certain measures. Especially the extend, to which certain measures have been applied in different companies was of a special interest. Therefore, further examinations in this area are necessary. A survey conducted with different global organizations could help to find answers to some of those questions.

References

1. J. Schwarze, *Informationsmanagement* (NWB, Berlin, 1998).
2. F. Caldwell, The new enterprise knowledge management, *VINE: The J. Information and Knowledge Management Systems*. **36**(2), 182-185 (2006).
3. P. Schuelke, *Informationsmanagement im Service zur Verbesserung der Kommunikation mit den Unternehmensbereichen* (FBK, Kaiserslautern, 2001).
4. H. Krcmar, *Informationsmanagement* (Springer, Berlin, 2005).
5. D.J. Greenwood, M. Levin, *Action Research* (Sage, Thousand Oaks, 1998).
6. N.P. Suh, *Axiomatic Design* (Oxford Uneversity Press, New York, 2001).
7. M.E. Porter, What Is Strategy?, *Harvard Business Review*, **74**(6), (1996).
8. P. Schoensleben, *Integral Logistics Management* (Auerbach Publications, Boca Raton, 2007).
9. M. Eppler, *Managing Information Quality* (Springer, Berlin, 2006).
10. K. Ferdows, A. De Meyer, Lasting improvements in the manufacturing performance, *J. Operations Management*, **9**(2), 168-184 (1990).
11. E. Ulich, *Arbeitspsychologie* (vdf, Zurich, 1994).
12. S.L. Pan, H. Scarbrough, A Socio-Technical View of Knowledge-Sharing at Buckman Laboratories, *J. Knowledge Management*; **2**(1), 55 - 66 (1998).
13. L. Argote, B. McEvily, R. Reagans, Managing knowledge in organizations, *Management Science*, **49**(4), 571-582 (2003).
14. C. Baldanza, M. Stankosky, Knowledge management: an evolutionary architecture toward enterprise engineering, *Proceedings of INCOSE* (1999).
15. M. Mohamed, M. Stankosky, Knowledge management and information technology, *J. Knowledge Management*, **10**(3), 103-116 (2006).
16. T. H. Davenport, L. Prusak, *Das Praxisbuch zum Wissensmanagement* (moderne industrie, Landsberg / Lech, 1998).
17. V. Decoene, W. Bruggeman, Strategic alignment and middle-level managers' motivation in a balanced scorecard setting, *Int. J. Operations & Production Management*, **26**(4), 429-448 (2006).
18. H. A. Artail, Application of KM measures to the impact of a specialized groupware system on corporate productivity and operations, *Information & Management*, **43**(4), 551-564 (2006).
19. P. Garg, R. Rastogi, New model of job design: motivating employees' performance, *Journal of Management Development*, **25**(6), 572-587 (2005).
20. M. Hansen, The search-transfer problem, *Admin. Sci. Quart.*, **44**(1), 82-111 (1999).
21. G.-W. Bock, A. Kankanhalli, S. Sharma, Are norms enough?, *European J. Information Systems*, **15**(4), 357 – 367 (2006).
22. M. Kirchmer, A.-W. Scheer, *Business Process Automation*, edited by A.-W. Scheer, F. Abolhassan, W. Jost, M. Kirchmer (Springer, Berlin, 2004).
23. H. Carter, Information Architecture, *Work Study*, **48**(5), 182-185 (1999).
24. L. Bräutigam, W. Schneider, *Projektleitfaden Software-Ergonomie* (TSH, Wiesbaden, 2003).
25. J. Nadler, L. Thompson, L. Van Boven, Learning negotiation skills, *Management Science*, **49**(4), 529-540 (2003).
26. E. L. Deci, *Intrinsic Motivation* (Plenum Press, New York, 1975).

Ensuring the Consistency of Competitive Strategy and Logistic Performance Management

Gregor von Cieminski and Peter Nyhuis
Institute of Production Systems and Logistics (IFA),
Gottfried Wilhelm Leibniz Universität Hannover,
An der Universität 2, D-30823 Garbsen, Germany
Email: {cieminski, nyhuis}@ifa.uni-hannover.de

Abstract

Manufacturing companies often merely attain reduced levels of logistic performance due to inconsistencies between their logistic objectives, logistic performance targets and logistic performance management actions. Qualitative influence models represent the inter-dependencies between these factors on two levels: firstly between logistic objectives, secondly between logistic performance measures and logistic planning and control parameters. Expert knowledge gained from modelling the dynamic behaviour of logistic systems facilitates the interpretation and evaluation of the qualitative influence models. On this basis, consistent logistic objectives can be defined and objective-oriented planning and control actions can be identified. The modelling insights are integrated into a logistic performance diagnosis procedure and tool that improve logisticians' understanding of the consequences of their management decisions and enhance the closed-loop management of logistic performance.

Keywords

Logistic performance, performance management, qualitative model

1 Introduction

Manufacturing companies compete with each other in terms of product functionality, product and service flexibility, price, product quality and logistic performance [1]. Competitive advantages in terms of the first two factors originate from the superior strategic design of products or production systems, respectively. Advantages in terms of the latter three factors firstly require the setting of adequate strategic objectives. Secondly, they also depend on superior operational performance. In order to attain the performance levels required, manufacturing companies have to ensure that their

Please use the following format when citing this chapter:

von Cieminski, G., Nyhuis, P., 2007, in IFIP International Federation for Information Processing, Volume 246, Advances in Production Management Systems, eds. Olhager, J., Persson, F., (Boston: Springer), pp. 51-58.

competitive strategies and their operational performance management actions are consistent [2]. This especially applies to the context of logistic performance: Wiendahl et al. [3] name inconsistencies between logistic performance objectives and operational planning and control parameters as one of the root causes of logistic performance deficits. As logistic performance has become an important customer purchase criterion [4] and developed into a proven success factor for manufacturing companies [5], the companies have to be equipped with performance management approaches that support them in avoiding these inconsistencies.

As a solution, this paper proposes qualitative logistic influence models and an approach for logistic performance diagnosis. It first provides a general overview of the consequences that strategic logistics management decisions have for operational logistic performance management. Second, it introduces two types of qualitative influence models that represent the inter-dependencies between logistic objectives as well as between logistic performance measures and logistic planning and control parameters. General insights gained from the interpretation and evaluation of these models are summarised. Third, as practical applications of the influence models, the paper presents a diagnostic procedure and software tool for factors affecting the operational logistic performance of manufacturing companies.

2   Linking Strategic and Operational Logistics Management

Given the market developments described in section 1, many manufacturing companies adopt competitive strategies based on their logistic performance; their logistic strategies become competitive strategies. Strategic logistics management has two main tasks: the formulation of strategic logistic objectives and the design of logistic and production systems. Both of these tasks have repercussions for operational logistics management. The primary logistic objectives of a manufacturing company depend on its product-delivery strategy, i.e. whether it satisfies customer demands from a finished-products store or whether it fulfils specific customer orders directly from production. Accordingly, the company's operational performance objectives and the ensuing planning and control actions have to be derived from the primary objectives "high service level" and "low delivery delay" or "high delivery reliability" and "short lead times", respectively.

The product-delivery strategy also influences a design aspect of manufacturing companies' logistic systems: the inclusion or omission of the distribution-logistics process. In the more detailed design of the three logistic processes of manufacturing companies – procurement, production and distribution logistics – it is standard practice to follow the prescriptions of logistic reference models. These models typically provide a choice of several process models as design templates for each of the logistic processes. An analysis of the process reference models developed by the Siemens concern [6] on the basis of the supply-chain operations reference model [7] led to the derivation of so-called logistic configurations of manufacturing companies [8] (see fig. 1a and 1b). These configurations indicate those logistic processes that play an active part in the fulfilment of the logistics function. Obviously, only logistically-active processes pursue explicit logistic objectives, carry out logistic planning and control actions, and therefore have an impact on the logistic performance of a manufacturing company. Hence, the specific logistic configuration of a company shapes the basic structure of the relevant logistic inter-dependencies (see fig. 1d). In turn, this structure determines the exact operational performance objectives and planning and control actions whose consistency has to be achieved.

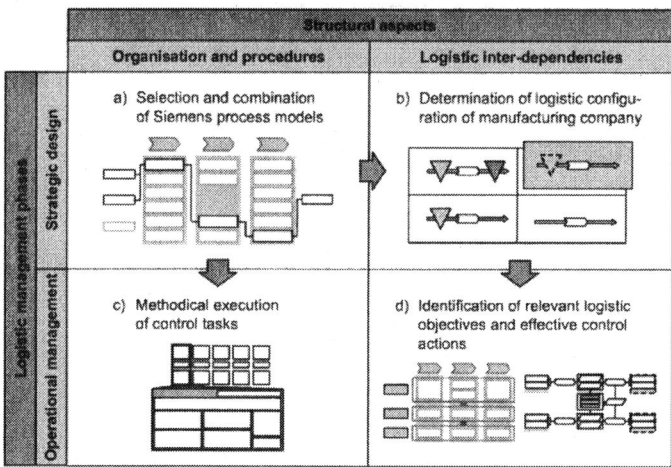

Fig. 1. Configurations of industrial-enterprise logistics determine the basic structure of the logistic inter-dependencies

To ensure the general applicability of the analysis, this paper considers a "make-to-stock" configuration, in which all processes affect the logistic performance.

3 Modelling and Analysing Logistic Inter-dependencies

Generally, analytical models of logistic processes are used to describe and analyse logistic inter-dependencies. These models often fall short of clearly establishing the relationships between logistic performance measures and the operational planning and control actions that the logistic processes can take. Moreover, inter-dependencies between separate logistic processes are not comprehensively taken into account. To overcome these shortcomings, qualitative influence models that holistically identify logistic inter-dependencies within and across process boundaries were compiled [8]. The models are based on the Theory of Logistic Operating Curves developed at the Institute of Production Systems and Logistics (IFA) [9]. This theory includes a range of analytical models of the logistic behaviour of inventories and production systems. Using the mathematical relationships of the models, it is possible to establish logistic inter-dependencies on two levels –between logistic objectives (see section 3.1) and between logistic performance measures and parameters that are manipulated through planning and control actions (see section 3.2).

3.1   Inter-dependencies between Logistic Performance Objectives

IFA has developed sets of logistic objectives for both production and storage processes. For isolated logistic processes, a qualitative characterisation of the inter-dependencies between the objectives is available [9]. As an extension, fig. 2 shows the combination of the sets of objectives of the three logistic processes of manufacturing companies. As for the separate processes, the logistic objectives can be assigned to the two directions logistic performance and logistic costs. The latter can further be differentiated into process costs and inventory costs.

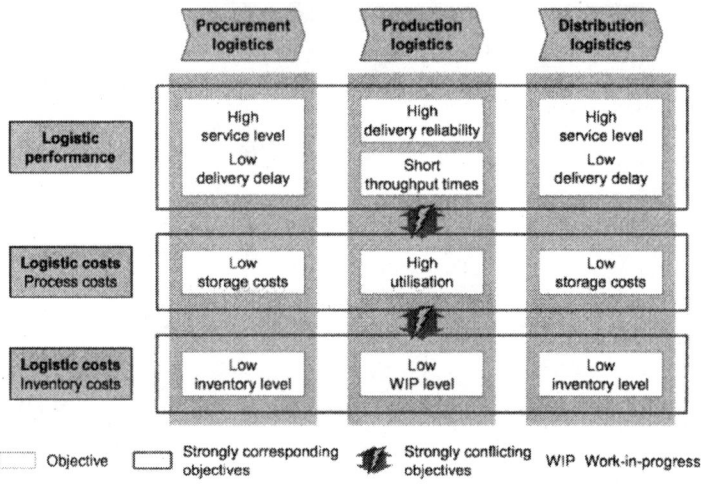

Fig. 2. Overview of inter-dependencies between logistic objectives

The figure also summarises the results of a first qualitative analysis of the inter-dependencies carried out by means of deductive reasoning. The main insight was that the objectives of each of the three directions correspond to each other across all three logistic processes. Conversely, marked conflicts were found to exist between the objectives of the directions logistic performance and process costs as well as process costs and inventory costs. Accordingly, fig. 2 categorises the inter-dependencies between the objectives into strong correspondences and strong conflicts. (More detailed results of the analysis are provided in [8].)

3.2   Inter-dependencies between Planning and Control Actions and Logistic Performance Measures

The effects of operational planning and control actions on the logistic performance of manufacturing companies are modelled through an analogy between technical control processes and logistic performance management. Accordingly, logistic inter-dependencies are illustrated in qualitative influence models that are comparable to the block diagrams known from control-systems theory. An example is shown in fig. 3, the influence model of the production-logistics process adapted from [10].

The qualitative influence models depict the cause-and-effect chains linking the logistic performance measures to the parameters modified by logistic planning and control actions, the so-called determinants. Referring to an example from fig. 3, the performance measures of production-logistics processes "throughput time" and "utilisation" are modelled as functions of the controlled variable "WIP level". The latter is calculated as the difference between the two manipulated variables "actual input" and "actual output". The values of the manipulated variables are in turn controlled by determinants: the actual input by the determinants "actual work content released" and "actual release date", the actual output by the determinants "actual capacity provided" and "date of capacity provision". Obviously, the first pair of determinants is parameterised as part of the order release, while the second pair depends on capacity-control decisions.

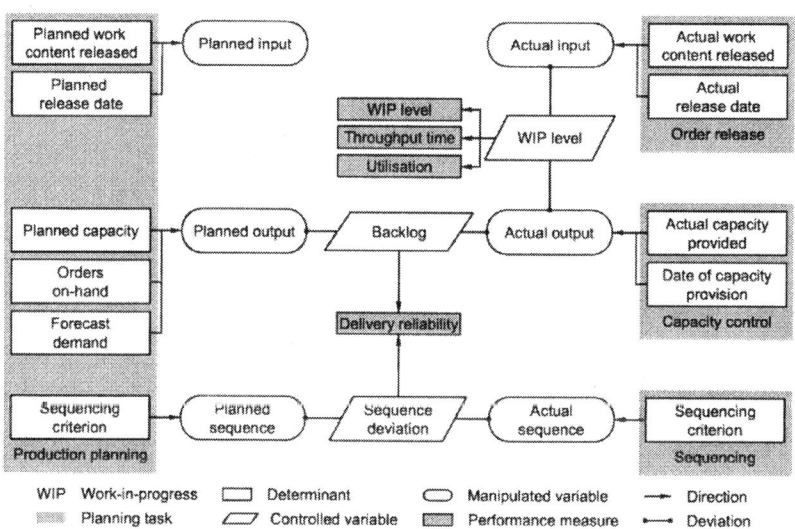

Fig. 3. Qualitative influence model of production planning and control (adapted from [10])

For the inventory management of procurement and distribution logistics processes, a similar influence model has been developed. Additionally, the coupling variables of the logistic processes have been identified so that cause-and-effect chains that extend across process boundaries may be traced as well. Through the use of these models, logisticians can identify all planning and control actions affecting a certain performance measure. Vice versa, they can locate all performance measures that are affected by a certain planning and control action.

3.3   Qualitative Evaluation of Logistic Inter-dependencies

The Theory of Logistic Operating Curves not only determines the structure of the qualitative influence models presented above. IFA's expertise in the interpretation of the quantitative models of the theory also facilitates a qualitative evaluation of the logistic inter-dependencies [8]. This process results in two types of guidelines for logistic performance management: guidelines for the consistent definition of logistic objectives and guidelines for the identification of objective-oriented logistic planning and control actions. Examples for both types are given below:

- The logistic processes of manufacturing companies should jointly pursue corresponding objectives in the directions shown in fig. 2. Conversely, the combined pursuit of objectives from conflicting directions should be avoided.
- Certain target values of logistic processes constitute values of planning parameters in other processes. The value settings for both have to be consistent.
- A logistic process's own planning and control actions are most effective for improving its logistic performance, both in magnitude and responsiveness.
- For maximizing logistic performance, manufacturing companies have to achieve objective-orientation and consistency of planning and control actions across all logistic processes. This is also essential for achieving very good performance levels at low logistic costs.
- Planning and control actions manipulating the magnitude and timing of inputs and outputs of production resources and stores are most effective in improving their logistic performance.
- Sequencing decisions and dedicated capacity adjustments are the most effective planning and control actions from the perspective of single customer orders.

On the premise that the logisticians responsible adhere to these guidelines, manufacturing companies can ensure the application of logistic performance management practices that are both consistent and effective.

4   Using Qualitative Models for Logistic Performance Diagnosis

In order for manufacturing companies to be able to directly apply the qualitative influence models and the guidelines for logistic performance management, IFA developed a logistic performance diagnosis procedure and tool on their basis.

A knowledge base containing the insights gained from modelling and analysing the logistic inter-dependencies lies at the core of the diagnostic procedure. On the one hand, this encompasses the underlying structure of the logistic inter-dependencies. For example, the cause-and-effect chains leading from the logistic performance measure "throughput time" via the corresponding controlled and manipulated variables to determinants such as "actual capacity provided" or "date of capacity provision" are represented (see fig. 3). On the other hand, the knowledge base includes the guidelines for logistic performance measurement and other results of the evaluation of the inter-dependencies. These are represented by logical statements such as: "The actual capacity provided is the second most effective determinant to influence the performance measure delivery reliability."

In the diagnostic tool, the contents of knowledge base are translated into relational databases. Their development and maintenance is facilitated through the expert interface shown at the bottom of fig. 4. Logistics experts are able to modify the structure of the databases and to enter new information into the database tables.

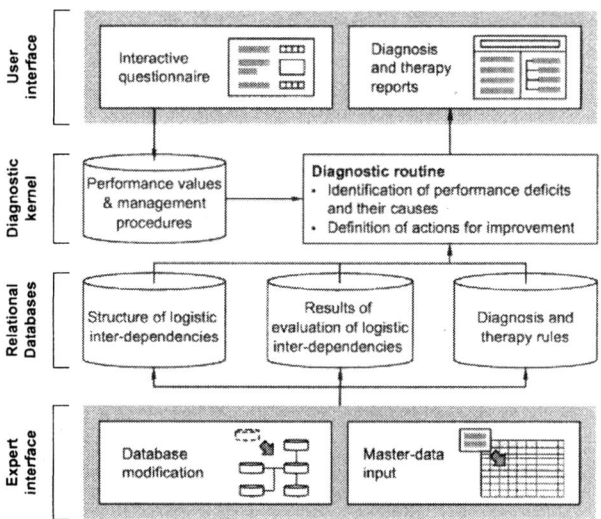

Fig. 4. Structure of the diagnostic software prototype

The application of the diagnostic procedure and tool comprises three phases: performance appraisal, diagnostic analysis and therapy definition. In the first phase, performance data of the manufacturing company under consideration are collected. This is facilitated through the input section of the user interface of the diagnostic tool shown at the top of fig. 4. An interactive questionnaire provides a structure for the required information on the logistic performance. The logistic configuration, target and actual performance values as well as the logistic planning and control practices applied are recorded. At this stage, logistic performance deficits become apparent.

In the second phase, the performance data are analysed using the databases of the diagnostic software tool. A diagnostic routine systematically determines the causes of logistic performance deficits. The routine identifies inconsistencies between logistic objectives as well as inconsistencies between logistic performance measures and logistic planning and control actions. The diagnosis also points out instances, in which the planning and control practices applied do not maximise logistic performance.

In the third phase, therapy rules contained in the database are utilised to ensure the consistency of logistics management practices by defining actions for improvement of the current logistic performance. The causes of logistic performance deficits and the recommendations of actions for improvement are highlighted by reports which can by accessed through the user interface (see top of fig. 4).

5   Conclusions

The qualitative logistic influence models presented in this paper are a practical means to ensure the consistency of logistic performance-based competitive strategies and operational logistic performance management. Most importantly, the models help manufacturing companies identify those operational planning and control actions, with which they can effectively improve specific logistic performance measures. The fact that the models consider inter-dependencies between logistic processes – as well as those within the processes – is an added advantage. Thus, consistent strategic and operational logistics management is facilitated across all logistic processes of a manufacturing company.

Transcending the current state of development, two important enhancements of the qualitative influence models are possible. First, the production processes may be disaggregated into combinations of fabrication and assembly processes with variable numbers of production stages and shifting customer-order decoupling points [6, 11]. The development and analysis of the qualitative models should be extended to the resulting more detailed logistic configurations of manufacturing companies. Second, the qualitative models may serve as a basis for the development of additional quantitative logistic models. For example, it is feasible to extend the Theory of Logistic Operating Curves by establishing and verifying the mathematical relationships between the operational planning and control parameters and the logistic performance measures.

References

1. B.H. Maskell, *Performance Measurement for World Class Manufacturing: A Model for American Companies* (Productivity Press, Cambridge, 1991).
2. J. Mills, K. Platts, A. Neely, H. Richards, and M. Bourne, *Strategy and Performance: Creating a Winning Business Formula* (Cambridge University Press, Cambridge, 2002).
3. H.-H. Wiendahl, G. von Cieminski and H.-P. Wiendahl, Stumbling blocks of PPC: towards the holistic configuration of PPC systems, *Production Planning and Control*, 7 (16) 634-651 (2005).
4. A. Maurer and W.A. Stark, *Steering Carmaking into the 21st Century: From Today's Best Practices to the Transformed Plants of 2020* (Boston Consulting Group, Boston, 2001).
5. G. Stalk and T.M. Hout, *Competing against Time: How Time-Based Competition Is Reshaping Global Markets* (2nd edition, Free Press, New York, 2003).
6. P. Nyhuis and C. Wolter, Quantifying the Rationalization Potential in Logistics through Supply Chain Design, in: Performance Measurement for Increased Competitiveness, edited by H.-P. Wiendahl and G. von Cieminski (Leibniz Universität Hannover, Hanover, 2002), pp. 14-23.
7. Supply Chain Council, *Supply-chain Operations Reference-model Version 8.0* (May 15, 2007), http://www.supply-chain.org.
8. G. von Cieminski, *Einsatz qualitativer Wirkmodelle zur Lenkung der industriellen Unternehmenslogistik* (PhD thesis, Leibniz Universität Hannover, Hanover, 2007).
9. P. Nyhuis and H.-P. Wiendahl, *Fundamentals of Production Logistics* (Springer, Berlin et al., 2007, in press).
10. H. Lödding, *Verfahren der Fertigungssteuerung* (Springer, Berlin et al., 2005).
11. J. Olhager, Strategic positioning of the order penetration point, *International Journal of Production Economics*, 3 (85) 319-329 (2003).

Proposal and Validity of Global Intelligence Partnering Model for Corporate Strategy, "GIPM-CS"

Manabu Yamaji and Kakuro Amasaka
Aoyama Gakuin University,
5-10-1 Fuchinobe Sagamihara Kanagawa, Japan
{yamaji, kakuro_amasaka}@ise.aoyama.ac.jp,
http://web.ise.aoyama.ac.jp/newjit/

Abstract

One of the requisites for winning corporate competitions today is success in the "global marketing" for quickly offering high-quality, latest model products in response to customer needs. For manufacturers to advance "manufacturing" that precisely meets the customers' preferences, it is vital that their affairs and management sections also share the global view and become a core of corporate management and strategy. More specifically, the key to success in "global production" lies in full functionalization of "partnering," in which forefront divisions of technology, production, and sales as well as the affairs and management sections collaborate in a cooperative strategic scheme to realize "global quality and optimal production." This study proposes Global Intelligence Partnering Model for Corporate Strategy, "GIPM-CS" mainly in connection with the administration. Further, the effectiveness of this model is verified at the successful companies.

Keywords

Affairs and Management sections, GIPM-CS, Global Partnering

1. Introduction

In recent years, in the context of quality management, the concept of "quality" has been expanded from "product quality" to "quality in the business process," and then even to "quality in corporate management" [1-3]. Against this background, in order to create products which respond to customer needs and, by extension, even customer wants, it is important to manage the realization of the mission of each division involved in the management of technology through "partnering [2, 3]" the management section, which organically links the technology, production and sales

Please use the following format when citing this chapter:

Yamaji, M., Amasaka, K., 2007, in IFIP International Federation for Information Processing, Volume 246, Advances in Production Management Systems, eds. Olhager, J., Persson, F., (Boston: Springer), pp. 59-67.

divisions and the affairs section, which activates members in all divisions to revitalize the organization.

Having said the above, this study strategically deploys the "next generation quality management technology, *Science TQM* [2, 4]," which has been proposed by the authors and verified as to its effectiveness. The authors then propose the Global Intelligence Partnering Model for Corporate Strategy, "GIPM-CS" which improves the intellectual productivity of the affairs and management sections. Then, the effectiveness of GIPM-CS will be verified by going over the application results observed at Toyota Motor Corporation.

2. Management issue

2.1 Global Production Strategy

The management values shared by so-called "winning companies" are shifting from emphasis on materials to human resources. The companies have amassed human resources, materials, and finances. It is easy to procure materials as well as finances. However, human resources take time to develop, and therefore, is not as easy as the foregoing to procure. The companies are endeavoring to grasp the information on human resources, take hold of the work, and formulate the vision in order to compete at a higher level.

It has been increasingly difficult to differentiate companies only in terms of "high product quality," "cost performance," and "superiority in the business process." It is imperative therefore to improve the value of human resources, but only a few companies have actually constructed a mechanism for improving human resources. Up until now, each department has acquired information on human resources from the personnel affairs division, and systems for offering such information have been insufficient. By improving the system for sharing information on human resources, similar to sharing information on materials and finances, the business assets of companies can be effectively utilized.

Moreover, the information of in-company systems has not been completely updated until the end of fiscal terms. Therefore, the accuracy of the information provided has been inferior, and the judgments regarding management tend to rely on personal experience or inspiration. In other words, information sharing has not been speedy enough. From now on, it is necessary to offer information with high precision based on the PDCA cycle so that decision-making on management matters will be based on facts. To that end, a system capable of analyzing all data, including human resources, from a variety of angles must be prepared. Under such a circumstance, there are many studies abroad for globalization [5-7] and TQM [8, 9]

2.2 Issues of the Affairs and Management Sections

Having said the above, in order to effectively utilize the role and function of each division, it is vital to recognize the issues related to the affairs and management sections, and to work on solving them. The authors find that the issues involving the affairs and management sections which need to be kept in mind when implementing strategic quality management are as follows:

1) The information is kept on a personal basis in many cases, not efficiently shared within the division.
2) They are lagging behind the other divisions, which are directly undertaking manufacturing operations, in the development of information technology (inter-division and intra-division information sharing)
3) Due to lack of communication with those outside the company.
4) They do not have systems for mid-term and long-term human resource development.

3. Improvement in Intellectual Productivity of the Affairs and Management Sections

It has been increasingly important for the affairs section to further advance the corporate management by grasping the environmental changes inside and outside the company in order to reinforce the internal and external management. For this reason, human resource development must be positioned at the core of management planning more emphatically than before so that the "reliability of company, organization, and human resources" can be enhanced. In the stage for utilizing (training) human resources, the function of improving intellectual productivity must be strengthened by partnering with the management section.

Likewise, the management section needs to function at the core of management technology and control. Therefore, it is deemed indispensable for them to reinforce the function of implementing JIT in the business flows of human resources, technology, and materials, through utilizing intellectual information as well as cooperating with the affairs section so that the onsite manufacturing divisions can be well managed internally.

The authors have verified the effectiveness of TIS (Total Intelligence Management System, Fig.1) for the management section, and TJS (Total Job Quality Management System, Fig.2) for the affairs section [10]. In order to activate the role and function of the affairs and management sections, in addition to practical application of the above core technologies, the authors believe that the role and function of the affairs and management sections can be integrally linked to business (high cycle-ization of the business process), as well as with the core technologies of onsite manufacturing divisions, such as TMS (Total Marketing System), TDS (Total Development System), and TPS (Total Production System) to implement strategic JIT.

In this connection, what is required of the affairs and management sections is to have the will and initiative to take responsibility in solving various management issues in cooperation with onsite manufacturing divisions. The authors think that the sections are "expected to play the role" of taking a proactive leadership role in putting together all the related divisions.

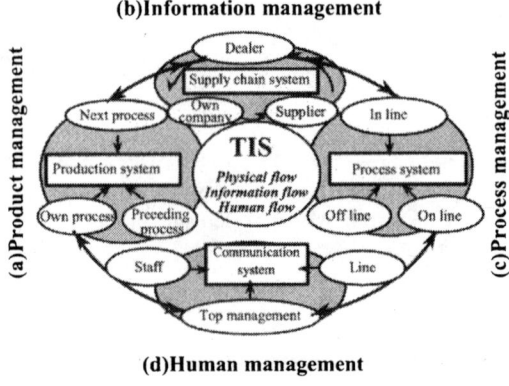

Fig.1 Concept of TIS

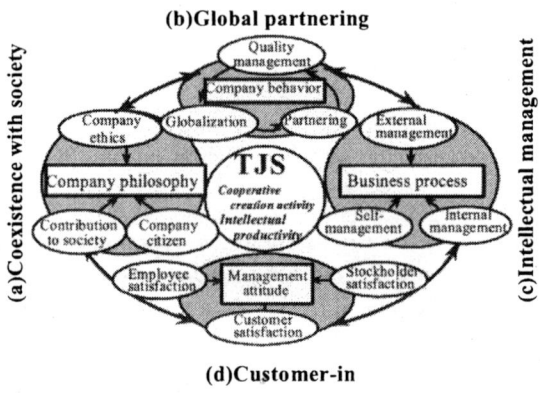

Fig.2 Concept of TJS

4. Proposal of Global Intelligence Partnering Model for Corporate Strategy, "GIPM-CS"

In recent years, Japanese companies have introduced the Western-style division of labor. Such a western labor division system is designed for easy replacement of labor forces and is achievement-oriented, it evaluates the degree of personal contribution to target achievement. Therefore, when a problem arises, it is dealt with not as a personal problem, but only within their responsibility range. Consequently, many activities and actions based on such a principle have been criticized by society. On the other hand, in the former Japanese system, problems were handled by all members across departments, which was a strong point of the Japanese way.

Based on the discussion in the previous chapter, the authors established the Global Intelligence Partnering Model for Corporate Strategy, "GIPM-CS" as shown in Fig. 3. The following are the functions of the affairs and management sections as corporate environment factors for succeeding in "global marketing," customer-first, 1) CS, ES, SS, for 2) high quality product and as a strategic factor to realize it, in order to 3) product reliability and corporation reliability, and success in 4) Intellectual productivity and human resource development. 5) Global production is realized by these. So the same quality and production at optimal locations are achieved. For that purpose, the highest priority was given to the "i) Intellectual information sharing" and "ii) Strategic Co-Creative Action" so that the "Strategic intelligence Application System" and "Business Process High Linkage System" can effectively function.

Fig. 3. GIPM-CS

When implementing a high cycle-ization of a business flow that consists of the setup of management policy, creation of a business plan, budget establishment, business deployment, optimal workforce distribution, task management, and evaluation, the relevant information needs to be shared among many departments and they need to also grasp the numerical values that show company-wide trends. In that way, upon confirming abnormal numerical indications, the problem can be identified and solved at an early stage. For example, if overtime labor cost shows an unusual figure, the project manager should find out the cause by checking which division, as well as which position or which process, is showing such a trend. In the personnel division, consideration needs to be given to possible deterioration of the work environment or lack of labor force, and in the sales division, care must be exercised to not delay delivery. Each division exchanges information with the other divisions to solve the problem as a whole company. The solutions shall be evaluated so that the information sharing of both problems and solutions can become a preventive measure and food for thought for the next plan.

Such a "partnership" among divisions, involves the creation of a system to visualize the information flow as well as its effective and practical application. To that end, a leader who links human resources is indispensable, and therefore, the cornerstone of corporate management is to foster leaders who have the understanding of the vision of management directors, a broad view of world trends, and communication skills to create a network of personnel inside the company. Simply put, the cultivation of an entrepreneurial mind or professional mind is what the authors are intending by proposing this "business model of intellectual productivity improvement for the affairs and management sections." In the following chapter, the effectiveness of the proposed "GIPM-CS" is verified through application cases at Toyota.

5. Application

5.1 Intelligent Quality Control System *"TPS-QAS"* by Utilizing *"T-QCIS"* and *"T-ARIM"*

The application line of the *TPS-QAS* (-Quality Assurance System) [11] using Toyota's Quality Control Information System (*T-QCIS*) is the automated assembly line, which assembles a part that transmits engine-driven power to the tires. This software shows the necessary control characteristics hierarchically specified as Item, Detailed Items, and Extraction Conditions to improve operability and provide an expansion function.

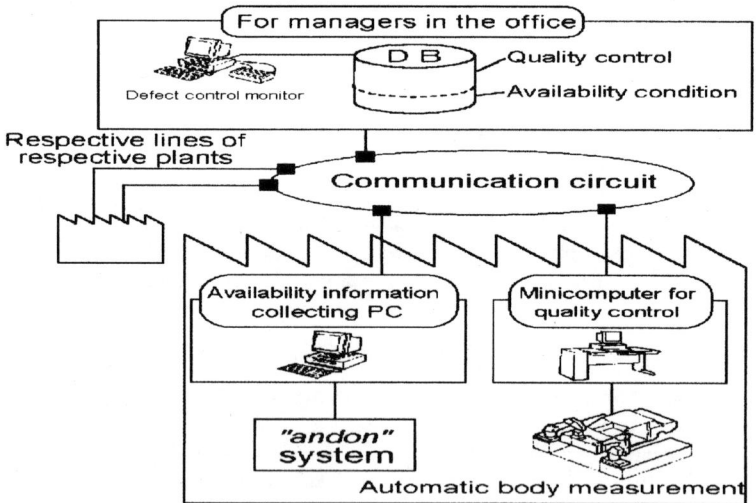

Fig. 4 Example of *T-ARIM* Hardware System

As the support functions for making diagnosis of a process and taking necessary measures according to the findings obtained from the control chart, a scroll function and raw data indication function are newly added. These functions help improve the process by processing and indicating group data according to the purpose of analysis. Regarding the hierarchical factorial analysis, data is allocated after being stratified to trace the causal relation of the factors. Past know-how is utilized by developing the database of the history of improvement, and a warning of abnormal process is generated automatically by using IT to extract processes with and without abnormalities at random. The abnormal diagnosis function makes efficient operation possible. Regarding data link with other application software, real-time factorial analysis becomes possible using the SQC (Statistical Quality Control) analysis software. Introduction of *T-QCIS* has improved the Cm (machine capability) and Cp (process capability) of the overall vehicle production line as expected [12].

To ensure the effectiveness of this *T-QCIS*, the authors have constructed the "Toyota's High-Reliability Production System, Networking of Availability and Reliability Information Manufacturing System (*T-ARIM*)" [11] as shown in Fig.4, a network system constructed by the production engineering control division in coordination with the manufacturing division for the purpose of controlling production line operations and the reliability and maintainability of the lines. By implementing Inline-Online SQC [12], this system collects and processes Inline in-process data Inline in real-time to control the process. It shares this intellectual information with related divisions, both domestic and/or overseas, to maintain or improve the processes scientifically using causal analysis, etc. The results are reflected in prompt improvement of the operating ratio of newly constructed lines. In practice, a system is then established for checking and following up on the quantitative improvement effects by analyzing changes in the failure mode before and after measures against failures, using Weibull chart analysis and other appropriate SQC tools.

5.2 Improvement in Painting Quality of Chassis Parts of Automobiles
Cooperative Activities with *Affiliated* and *Non-Affiliated* Companies

This case is simultaneous achievement of QCD for upgrading of automobile quality. Anti-corrosion has been an important issue for chassis parts of automobiles, and it has been a formidable challenge to resolve as it requires a comprehensive solution which encompasses materials, pre-treatment, paint, painting, and logistics. Conventionally, paint manufacturers and automobile manufacturers prepare paints on their own, and generally, the selection of paints was made after evaluation. This method, however, cannot keep abreast with the sophisticated, accelerating market needs. In order to simultaneously achieve quality, cost and delivery (QCD) across different industries and segmented organizations, it is necessary for paint manufacturers and automobile manufacturers to promote project activity in cooperation all the way from paint designing to building up paint quality.

As an example of this, Toyota carried out a project through formation of a task team with its *affiliated* company, Aisin Chemical Co., Ltd. as well as *non-affiliated* company Tokyo Paint Co., Ltd. Quality Assurance Div. collected quality data in the market. TQM Promotion Div. analyzes that data. The Fig.5 shows the development process of paints used for the rear axel assay of automobiles in each improvement process for cost and quality. Up until now, styrene-altered alkyd resin paint has been used. To prevent the generation of initial rust, which resulted from applying an anti-freezing agent mainly used in overseas markets, phenol-altered alkyd resin paint was adopted (Improvement 1), but as shown in the fig.5, the painting cost increased. The painting cost means not only the price of paints but includes the total cost involved, such as painting operation cost, energy consumption, facility maintenance and cleaning cost, inspection cost, etc. per vehicle.

Against this background, an extensive project was launched in which the ever-increasing quality level required by the market is predicted and the target value is set in several steps for implementation in a planned manner. The first few years were spent improving the paint quality (Improvements 1 to 7) and after that, efforts were made in parallel to improve the painting facilities (Improvements 4 to 8). As a result, a type of paint with anticorrosive properties 10 times the conventional type was developed and the painting cost was reduced by more than 30%, a considerable improvement [3].

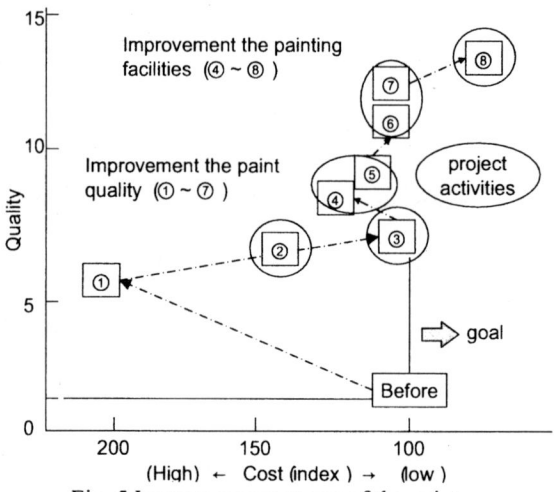

Fig. 5 Improvement process of the paint

6   Conclusions

The authors proposed the Global Intelligence Partnering Model for Corporate Strategy, "GIPM-CS" which improves the intellectual productivity of the affairs and management sections. The effectiveness of GIPM-CS was verified by going over the application results observed at Toyota Motor Corporation.

References

1. Amasaka, K., *New JIT*, A New Management Technology Principle at Toyota, *International Journal of Production Economics*, Vol. 80, pp.135-144, 2002.
2. Amasaka, K., Development of "Science TQM", A New Principle of Quality Management: Effectiveness of Strategic Stratified Task Team at Toyota-, *International Journal of Production Research*, Vol.42, No.17, pp.3691-3706, 2003.
3. Amasaka, K., Applying *New JIT*—A Management Technology Strategy Model at Toyota - Strategic QCD Studies with Affiliated and Non-affiliated Suppliers , *Proc. of the 2nd World Conference on POM, Cancun, Mexico*, pp.1-22, 2006 .
4. Amasaka, K., New Japan Model - Science TQM: Theory and Practice for Strategic Quality Management, *Maruzen*, 2007. *(in Japanese)*
5. Yamaji, M. and Amasaka, K., *New Japan Quality Management Model*, Hyper-cycle model "QA & TQM Dual System", *Proc. of the International Manufacturing Leaders Forum, Taipei, Taiwan*, pp.1-6, 2006.
6. Lagrosen, S., Quality management in global firms, *The TQM Magazine*, Vol. 16, No. 6, pp. 396-402, 2004.
7. Ljungström, M., A model for starting up and implementing continuous improvements and work development in practice, *The TQM Magazine*, Vol. 17, No. 5, pp. 385-405, 2005.
8. Burke, R.J. et al., Effects of reengineering on the employee satisfaction-customer satisfaction relationship, *The TQM Magazine*, Vol. 17, No. 4, pp. 358-363, 2005
9. Evans J. R. and Lindsay W. M., The Management & Control of Quality, *South-Western*, 1995.
10. Amasaka, K *Advanced Science TQM*, A New Japan Quality Management Model: Development of Global Production Strategy in Toyota, *The 18th International Conference on Production Research, Univ. of Salerno, Italy*, pp.1-6 (CD-ROM), 2006.
11. Amasaka K. and Sakai, H., TPS-QAS, new production quality management model: Key to New JIT, *International Journal of Manufacturing Technology and Management*, 2006. *(decided to be published)*.
12. Amasaka, K. and Sakai, H., Availability and Reliability Information Administration System "ARIM-BL" by Methodology in "Inline-Online SQC", *The International Journal of Reliability & Safety Engineering*, Vol.5, No.1, pp.55-63, 1998.

Supply Chain Operations Planning with Setup Times and Multi Period Capacity Consumption

H. Zolfi [1], S.M.T. Fatemi Ghomi [1] and B. Karimi [1]

1 Department of Industrial Engineering
Amirkabir University of Technology, 424 Hafez Avenue, Tehran, Iran,
Tel.: +98 21 66413034; Fax +98 21 6641 3025
Home page: http://www.aut.ac.ir

Abstract

Developing an efficient heuristic algorithm to solve a *supply chain operations planning* model is the main purpose of this paper. The model considers multi period supply chain planning with capacitated resources. The concept of *multi period capacity consumption* has been developed recently at the context of supply chain management that realizes resource planning at a supply chain. Because of considering setup times and costs, the model contains binary variables. Since the mixed integer model is strongly NP-hard problem and finding a feasible solution is NP-complete, developing an efficient algorithm is remarkable. In this paper a heuristic algorithm is developed to solve this complicated model. Two reasons encouraged the authors to solve this complex problem. First, the model is an advanced and applicable operations planning model at the supply chain environment. Second, this model is strongly NP-hard. So it is of important task to develop a solution for the problem to be capable of feasible and efficient.

1 Introduction

De Kok and Fransoo (2003) define supply chain operations planning (SCOP) as coordinating material and resource release decisions in the supply chain such that predefined customer service levels are met at minimal cost. Extensive discussion about supply chain operations planning models can be found at De Kok and Fransoo (2003). A few researches in the past explained the concept of multi period capacity consumption. As a first work, Negnman (2000) described this concept as a new additional property for planned lead time. According to multi period capacity consumption, capacities of resources are allowed to consume at any internal periods during the lead time. A supply chain model with multi period capacity consumption and its concept can be comprehended clearly at Spitter et al. (2005). Their model did

Please use the following format when citing this chapter:

Zolfi, H., Ghomi, S. M. T. F., Karimi, B., 2007, in IFIP International Federation for Information Processing, Volume 246, Advances in Production Management Systems, eds. Olhager, J., Persson, F., (Boston: Springer), pp. 69-76.

not consider the setup times and costs. So it was formulated as two LP[1] model. We develop our new model strongly based on the first model of Spitter et al. (2005). New model considers non zero setup times and costs. Considering non zero setup times and costs make it possible to be used at supply chains that setup times and costs are important. For example a part manufacturer factory can be considered as a good real example for our model. Consider that this factory is supplier of special parts as a member of Car Company's supply chain. It is producing some special parts for car manufacturing companies using a lot of bending, drilling and pressing machines. This factory must produce wide variety kinds of products. Since it is required to have none zero times and costs to changeover the setup of each machine, therefore setup times and costs are not negligible. It is apparent that all of machines are capacity restricted. At this situation the model of this paper is very suitable to determine lot sizes of each part on each machine. Additionally because of considering multi period capacity consumption, capacity planning of machines is more realistic and simultaneous planning of capacities and orders will be possible.

Models of supply chain planning problems and lot sizing models are very similar in many ways (see Voβ and Woodruff (2000)). So we can consider this new supply chain operations model as a new Multi Level Capacitated Lot Sizing Problem (MLCLSP) with setup times and costs. From this viewpoint the concept of multi period capacity consumption is considered at an MLCLSP with setup times. Multi period capacity consumption realizes the coordination of the release of materials and resources at the MLCLSP model.

From the aspect of solving the model, our model is complicated. Because of setup variables, the model will be an MIP[2] model (contrary to Spitter's model) and it is more complicated than the previous lot sizing models too. The most complex lot sizing model that has been considered yet is the MLCLSP with setup times with general assembly structure. This problem is NP-hard problem (Dellaert et al. (2000)) and finding a feasible solution is NP-complete (Maes et al. (1991)). As we know just two papers have discussed this model up to now. Tempelmeier and Derstroff (1996) solved this problem using Lagrange relaxation of capacity and inventory balance constraints and Katok et al. (1998) have solved the problem using LP based approach. The model of this paper will be more complicated considering the concept of multi period capacity consumption. Therefore developing a good feasible heuristic algorithm will be so important. The developed algorithm in this paper is based on decomposition the major model into two simpler models at a hierarchical structure and solving the model of each level efficiently. Also developed algorithm for these conditions can be used for models with zero setup times and costs. Therefore this model can be considered as an advanced and applicable supply chain planning model. Now the problem is how these operational situations can be modeled as a mathematical model and how the model can be solved that is strongly NP-hard.

---

[1] Linear Programming
[2] Mixed Integer Programming

2   The Problem formulation

To formulate the problem, it is required to define the notations which are very close to Spitter et al. (2005). i, u, t and s are indices of products, resources, order release times and resource release times respectively. n, k and T are number of items, resources and time horizon respectively. τ_i is the planned lead time of item i.
h_{ij} indicates the number of units of item i used to produce one unit of item j. α_{it}, β_{it} and SC_{ius} are holding, backordering and setup costs respectively. The exogenously determined demand of item i for period t is shown by D_{it}. ST_{ius} is the setup time of item i on resource u at period t. C_{us} is the maximum available capacity of resource u at period t. R_{it} is the planned order of item i at time t, Z_{iuts} is the part of R_{it} produced on resource u at period s so that $s = t+1,...,T+\tau_i \,|\, s \leq T$. Amount of R_{it} and Z_{iuts} are given for past periods $(t = -\tau_i+1,...,-1)$ as $\overline{R}_{it}$ and $\overline{Z}_{iuts}$. The binary setup variable is δ_{ius} that is 1 if setup of item i occurs on resource u at period s and 0 otherwise. Considering the concept of multi period capacity consumption and above notations, the problem is modeled as follows:
Problem A

$$Min \ \sum_{t=1}^{T}\sum_{i=1}^{n}(\alpha_{it} I_{it} + \beta_{it} B_{it}) + \sum_{u=1}^{k}\sum_{i=1}^{n}\sum_{s=1}^{T}\delta_{ius} SC_{ius} \tag{1}$$

s.t:

$$R_{it-\tau_i} + I_{it-1} - I_{it} - D_{it} - \sum_{j=1}^{n} h_{ij} R_{jt} + B_{it} - B_{it-1} = 0 \qquad \forall t=1,...,T, i=1,...,n \tag{2}$$

$$R_{it} = \sum_{s=t+1}^{t+\tau_i}\sum_{u=1}^{k} Z_{iuts} \qquad \forall i=1,...,n \quad ,t=-\tau_i+1,...,T-1 \atop s<-T \tag{3}$$

$$\sum_{i=1}^{n}\sum_{t=s-\tau_i}^{s-1} Z_{iuts} + \sum_{i=1}^{n}\delta_{ius} ST_{ius} \leq C_{us} \qquad \forall u=1,...,k \quad ,s=1,...,T \tag{4}$$

$$\delta_{ius} \geq \frac{\sum_{t=s-\tau_i}^{s-1} Z_{iuts}}{M(=C_{us})} \qquad \forall i=1,...,n \quad ,u=1,...,k \quad ,s=1,...,T \tag{5}$$

$$B_{it} - B_{it-1} \leq D_{it} \qquad \forall i=1,...n \quad ,t=1,...,T \tag{6}$$

$$R_{it} = \overline{R}_{it} \qquad \forall i=1,...,n \quad ,t=-\tau_i+1,...,-1 \tag{7}$$

$$Z_{iuts} = \overline{Z}_{iuts} \qquad \forall i=1,...,n \quad ,u=1,...,k \quad ,t=-\tau_i+1,...,-1 \quad ,s=t+1,...,0 \tag{8}$$

$$R_{it}, B_{it}, I_{it} \geq 0 \qquad \forall i=1,...,n \quad ,t=1,...,T \tag{9}$$

$$Z_{iuts} \geq 0 \qquad i=1,...,n \quad ,u=1,...,k \quad ,t=-\tau_i+1,...,T-1 \quad ,s=t+1,...,t+\tau_i \,|\, s \leq T \tag{10}$$

$$\delta_{ius} \in \{0,1\} \qquad \forall i=1,...,n \quad ,u=1,...,k \,, s=1,...,T \tag{11}$$

The objective function minimizes the total inventory holding, backorder and setup costs. Constraint set (2) implies the inventory balance equations. Equation (3) presents the relation between orders and their feasible productions according to multi period capacity consumption. Constraint set (4) presents capacity feasibility on each resource and at each period. The constraints set (5) mean if there is at least one positive Z_{iuts} ($t \in [s-\tau_i, s-1]$), setup of item i occurs on resource u at time slot s. C_{us} is set as an upper bound for $\sum_{t=s-\tau_i}^{s-1} Z_{iuts}$. Equation (7) enforces that backorders are allowed only for end items and amount of backordering is less than the independent exogenous demand. At equations (7) and (8) the effect of past orders and productions which still have influence on the future is considered. Equations (9) and (10) are sign restrictions and equation (11) claims δ_{ius} is binary variable.

3   The heuristic algorithm

The above MIP model is very hard to solve. The major factors of complexity are:
1) Relation between orders at different levels in the inventory balance equations.
2) Capacity constraints.
3) Existence of setup times in the capacity constraints.
4) Different summation bounds on the production variables in the various equations.

We tackle the complexity of this problem by hierarchical planning. The main model is decomposed into two new related models. At the first level we try to overcome the first factor of complexity. Therefore at the second level there will be a **single level**, multi item, multi period and multi resource problem. By Lagrange relaxing the equation (3) at the model of second level, several single item, single resource and single period MIP models are obtained and thus remaining factors of complexity are controlled. In this way the first model is engaged to plan orders, based on the exact holding and backordering costs and approximate setup costs. At the second level, planed orders are split up to the lot sizes; based on setup times and costs. Considering the cited points, problem A is disintegrated as follows:

Problem B1: a linear model for the first level of hierarchy

$$Min \;\; \sum_{t=1}^{T}\sum_{i=1}^{n}(\alpha_{it} I_{it} + \beta_{it} B_{it}) + \sum_{u=1}^{k}\sum_{i=1}^{n}\sum_{s=1}^{T} Ic_{ius}\, SC_{ius} \tag{12}$$

s.t:

$$R_{it-\tau_i} + I_{it-1} - I_{it} - D_{it} - \sum_{j=1}^{n} h_{ij}\, R_{jt} + B_{it} - B_{it-1} = 0 \qquad \forall\, t=1,...,T \;\;, i=1,...,n \tag{13}$$

$$R_{it} = \sum_{\substack{s=t+1 \\ s<=T}}^{t+\tau_i}\sum_{u=1}^{k} Z_{iuts} \qquad \forall\, i=1,...,n \;\;, t=-\tau_i +1,...,T-1 \tag{14}$$

$$\sum_{t=s-\tau_i}^{s-1} Z_{iuts} \le V_{ius} \qquad \forall\, i=1,..,n \;\;, u=1,...,k \;\;, s=1,...,T \tag{15}$$

$$\sum_{i=1}^{n} V_{ius} \le C_{us} \qquad \forall\, u=1,...,k \;\;, s=1,...,T \tag{16}$$

$$B_{it} - B_{it-1} \leq D_{it} \qquad \forall i = 1,...,n \quad ,t = 1,...,T \tag{17}$$

$$R_{it} = \overline{R}_{it} \qquad \forall i = 1,...,n \quad ,t = -\tau_i + 1,...,-1 \tag{18}$$

$$Z_{iuts} = \overline{Z}_{iuts} \qquad \forall i = 1,...,n \quad ,u = 1,...,k \quad ,t = \tau + 1,...,-1 \quad ,s = t+1,...,0 \tag{19}$$

$$R_{it}, B_{it}, I_{it} \geq 0 \qquad \forall i = 1,...,n \quad ,t = 1,...,T \tag{20}$$

$$Z_{iuts} \geq 0 \qquad i = 1,...,n \quad ,u = 1,...,k \quad ,t = -\tau_i + 1,...,T-1 \quad ,s = t+1,...,t+\tau_i \,|\, s \leq T \tag{21}$$

$$V_{ius} \geq 0 \qquad \forall i = 1,..,n \quad ,u = 1,...,k \qquad ,s = 1,...,T \tag{22}$$

Problem B2: a single level MIP model for the second level of hierarchy

$$Min \; \sum_{u=1}^{k} \sum_{i=1}^{n} \sum_{s=1}^{T} \delta_{ius} SC_{ius} \tag{23}$$

$$R_{it} = \sum_{s=t+1}^{t+\tau_i} \sum_{u=1}^{k} Z_{iuts} \qquad \forall i = 1,...,n \quad ,t = -\tau_i + 1,...,T-1 \tag{24}$$
$$\scriptstyle s <= T$$

$$\sum_{t=s-\tau_i}^{s-1} Z_{iuts} + \delta_{ius} ST_{ius} \leq V_{ius} \qquad \forall i = 1,...n \quad ,u = 1,...,k \quad ,s = 1,...,T \tag{25}$$

$$\delta_{ius} \geq \frac{\displaystyle\sum_{t=s-\tau_i}^{s-1} Z_{iuts}}{V_{ius} - ST_{ius}} \qquad \forall i = 1,...,n \quad ,u = 1,...,k \quad ,s = 1,...,T \tag{26}$$

$$Z_{iuts} = \overline{Z}_{iuts} \qquad \forall i = 1,...,n \quad ,u = 1,...,k \quad ,t = \tau + 1,...,-1 \quad ,s = t+1,...,0 \tag{27}$$

$$Z_{iuts} \geq 0 \qquad i = 1,...,n \quad ,u = 1,...,k \quad ,t = -\tau_i + 1,...,T-1 \quad ,s = t+1,...,t+\tau_i \,|\, s \leq T \tag{28}$$

$$\delta_{ius} \in \{0,1\} \qquad \forall i = 1,...,n \quad ,u = 1,...,k \quad ,s = 1,...,T \tag{29}$$

Simultaneous consideration of constraints (15) and (16) implies constraint (4) without setup times. Setup times are studied at the problem B2 exactly. At the objective function of problem B1, δ_{ius} is replaced with linear estimator called lc_{ius}. Based on *fixed charge problems* literature (e.g. see Taha (1975)), the definition of lc_{ius} can be stated as follows:

$$lc^{q+1}_{ius} = \begin{cases} \dfrac{\displaystyle\sum_{t=s-\tau_i}^{s-1} Z^{q+1}_{iuts}}{\displaystyle\sum_{t=s-\tau_i}^{s-1} Z^{q}_{iuts}} & if \; \displaystyle\sum_{t=s-\tau_i}^{s-1} Z^{q}_{iuts} > 0 \\[4mm] 0 & if \; \displaystyle\sum_{t=s-\tau_i}^{s-1} Z^{q}_{iuts} = 0 \end{cases} \tag{30}$$

lc_{ius} behaves as δ_{ius} implicitly. δ_{ius} is replaced with lc_{ius}, because it is desired to have an LP problem at the first level. lc_{ius} is determined in the iterative way. At iteration $q+1$ ($q \geq 0$), Z^{q+1}_{iuts} is the variables of current iteration and Z^{q}_{iuts} is the solution of problem B1 at the previous iteration. After problem B1 is solved the initial solution (not necessarily feasible for problem A) is gained and initial values for I_{it}, B_{it}, R_{it}, Z_{iuts} and V_{ius} are obtained. If no feasible solution exists for problem B1, the problem A is infeasible. Otherwise at problem B2 we try to find a feasible good solution restricted to obtained R_{it}. Fixing R_{it} at problem B2 implies fixed I_{it} and B_{it}, and fixing these variables at the first level makes the second level independent from inventory balance equations. At the next step the algorithm tries to find a feasible solution.

- Consider $\sum_{t=s-\tau_i}^{s-1} Z_{iuts}$ and define δ_{ius} according to the following equation:

$$\begin{cases} if \ \sum_{t=s-\tau_i}^{s-1} Z_{iuts} > 0 \ \Rightarrow \ \delta_{ius} = 1 \\ if \ \sum_{t=s-\tau_i}^{s-1} Z_{iuts} = 0 \ \Rightarrow \ \delta_{ius} = 0 \end{cases} \tag{31}$$

- Correct V_{ius} to obtain feasible equation (25) as follows:

$$\begin{cases} if \ \sum_{t=s-\tau_i}^{s-1} Z_{iuts} > 0 \ \Rightarrow \ V_{ius} = \sum_{t=s-\tau_i}^{s-1} Z_{iuts} + ST_{ius} \\ if \ \sum_{t=s-\tau_i}^{s-1} Z_{iuts} = 0 \ \Rightarrow \ V_{ius} = 0 \end{cases} \tag{32}$$

By changing V_{ius} as above, problem B2 becomes feasible and all constraints of problem B1 remain feasible except constraint (16) that may violate feasibility. In this step the algorithm tries to satisfy constraint (16) through decreasing V_{ius} so that other constraints remain feasible. Decreasing V_{ius} may just break up the feasibility of constraint (25). To prevent Z_{iuts} infeasibility of constraint (25), if V_{ius} decreases for special i, u and s, some related Z_{iuts} are shifted to other periods or on the other resources that called *the destination point of transition* and indicated by "*". The indices of selected Z_{iuts} to shift is called as *the origin point of transition* and signed by " ¯ ". The destination point is selected according to some conditions such that R_{it} is not allowed to change. Following pseudo code consists of detailed description of this step:

1) *for* $\bar{s} = 1$ *to T*

2) *for* $\bar{u} = 1$ *to k*

3) *if constraint (16) is infeasible*

4) *select the smallest positive* $\sum_{t=\bar{s}-\tau_i}^{\bar{s}-1} Z_{i\bar{u}t\bar{s}}$ *and indicate related i with* $\bar{i}$

5) *for* $\bar{t} = \bar{s} - \tau_{\bar{i}}$ *to* $\bar{s} - 1$

6)   *if* $Z_{\bar{i}\bar{u}\bar{t}\bar{s}} > 0$

7)   *for* $u^* = 1$ *to k*

8)   *for* $s^* = \bar{t}+1$ *to* $\min(T, \bar{t}+\tau_{\bar{i}})$

9)   *if* $C_{u^*s^*} - \sum_{i=1}^{n} V_{iu^*s^*} > 0$

10)   *if* $\sum_{t=\bar{s}-\tau_{i^*}}^{s^*-1} Z_{i^*u^*t s^*} > 0$

11) $\begin{cases} i^* = \bar{i}, \ t^* = \bar{t} \\ transfer = \min(C_{u^*s^*} - \sum_{i=1}^{n} V_{iu^*s^*}, \ Z_{\bar{i}\bar{u}\bar{t}\bar{s}}) \\ Z_{i^*u^*t^*s^*} = Z_{i^*u^*t^*s^*} + transfer \\ Z_{\bar{i}\bar{u}\bar{t}\bar{s}} = Z_{\bar{i}\bar{u}\bar{t}\bar{s}} - transfer \\ V_{i^*u^*s^*} = V_{i^*u^*s^*} + transfer \\ V_{\bar{i}\bar{u}\bar{s}} = V_{\bar{i}\bar{u}\bar{s}} - transfer \end{cases}$

After a feasible solution is found at previous step, the algorithm is going to find a good near optimal solution for problem B2. To solve problem B2, first it is decomposed using Lagrange relaxation of constraints (24). Lagrange multipliers are defined as follows:

λ_{it} : Lagrange multiplier for constraint $R_{it} = \sum_{s=t+1}^{t+\tau_i} \sum_{u=1}^{k} Z_{iuts}$, $i = 1,...,n$, $t = -\tau_i + 1,...,T-1$

$1 \le s \le T$

the relaxed problem of B2 after rearranging of objective function is:
Problem RB2:

$$\min \ \sum_{i=1}^{n}\sum_{u=1}^{k}\sum_{s=1}^{T}\delta_{ius} SC_{ius} - \sum_{i=1}^{n}\sum_{u=1}^{k}\sum_{s=1}^{T}\sum_{t=s-\tau_i}^{s-1}\lambda_{it} Z_{iuts} \qquad (34)$$

s.t.:

$$\sum_{t=s-\tau_i}^{s-1}Z_{iuts} + \delta_{ius} ST_{ius} \le V_{ius} \qquad \forall i = 1,...n \ , u = 1,...,k \ , s = 1,...,T \qquad (35)$$

$$\delta_{ius} \ge \frac{\sum_{t=s-\tau_i}^{s-1}Z_{iuts}}{V_{ius} - ST_{ius}} \qquad \forall \ i = 1,...,n \ , u = 1,...,k \ , s = 1,...,T \qquad (36)$$

$$Z_{iuts} \ge 0 \qquad i = 1,...,n \ , u = 1,...,k \ , s = 1,...,T \ , t = s - \tau_i,...,s-1 \qquad (37)$$

$$\delta_{ius} \in \{0,1\} \qquad \forall \ i = 1,...,n \ , u = 1,...,k, s = 1,...,T \qquad (38)$$

The problem RB2 can be decomposed into nkT single item, single resource and single period problems. Each of decomposed problems is solved using the following subroutine:

1) if $V - ST \le 0$ then $Z_t = 0$ $\forall t \in [s - \tau_i, s-1]$ and $\delta = 0$.
2) if $V - ST > 0$ and if $\lambda_t \le 0 (\forall t \in [s - \tau_i, s-1])$ then $Z_t = 0$ $\forall t \in [s - \tau_i, s-1]$ and $\delta = 0$
3) if $V - ST > 0$ and if there is at least one $\lambda_t > 0$
4) select the maximum $\lambda_t > 0$ (denoted by λ_{t^*})
5) if $SC - \lambda_{t^*}.(V - ST) < 0$ then $(Z_{t^*} = V - ST$ and $\delta = 1$ and $Z_t = 0$ $(\forall t \in [s - \tau_i, s-1] \ne t^*))$
6) else $Z_t = 0$ $\forall t \in [s - \tau_i, s-1]$ and $\delta = 0$

Problem B2 is solved iteratively by updating Lagrange multipliers at the each iteration. We use subgradiant optimization technique to update Lagrange multipliers. Therefore, at each iteration problem RB2 must be solved to obtain new lower and upper bounds for problem B2. The objective function value of new optimal solution for problem RB2 is a new lower bound for problem B2. At the each iteration of subgradiant optimization method, the best lower bound of problem B2 is updated by selecting the maximum lower bounds obtained up to now.

4   Our Results

The algorithm has been tested using randomly generated problems. The quality of solutions of the heuristic algorithm is evaluated using LINGO optimal solutions for the small size examples. For small size examples (average binary variables is 50), the solutions obtained by heuristic algorithm are, on average, 2.69% worse than the optimal solution. Katuk et al. (1998) have reported average 4% of optimality gap for their algorithm. Comparing with their solutions, our algorithm performs better. For all small examples, the heuristic algorithm could find the feasible solution and, on average, about 25% of solutions were optimum. Running time of the small size problems is not reported because it is negligible (less than 2 seconds).

LINGO can not find the optimum solutions of the medium size (average binary variables is 500) and the large scale examples (average binary variables is 1500) at reasonable time. Therefore, we evaluate the heuristic algorithm using time-truncated LINGO solutions (close to Katok et al. (1998)). For medium size examples the solutions of LINGO, on average, are 3.22% better than the heuristic solutions that are acceptable with respect to the complexity of the problem. For large scale examples at 64.12% of the problems, LINGO could not find a feasible solution at runtime of the heuristic. According to the average deviation, it is clear that the heuristic is very efficient comparing to LINGO solutions, because the heuristic have produced the solutions with -317.18% deviation from LINGO. The number of problems that the heuristic solutions are better than the LINGO solutions is 94.87% that proves high performance of the heuristic. The magnitude of 16.73% is obtained for $\frac{upper\ bound - lower\ bounded}{lower\ bounded}\%$ that is comparable with 16.5% reported by Tempelmeier and Derstroff (1996). The results indicated a good performance for small size and medium size problems, and high performance for large scale problems.

5   References

Dellaert, N., Jeunet, J., Jonard, N., 2000. A genetic algorithm to solve the general multi-level lot-sizing problem with time varying costs. International Journal of Production Economics 68, 241–257.

De Kok, A. G., Fransoo, J. C., 2003. Planning supply chain operations: Definition and comparison of planning concepts. Pages 597–675 of: De Kok, A. G., & Graves, S. C., (eds), Handbook in Operations Research and Management Science, Volume 11: Design and Analysis of Supply Chains. Amsterdam.

Katok, E., Lewis, H.S., Harrison, T.P., 1998. Lot sizing in general assembly systems with setup costs, setup times, and multiple constrained resources. Management Science 44 (6), 859–877.

Maes, J., McClain, J.O., Van Wassenhove, L.N., 1991. Multilevel capacitated lot sizing complexity and LP-based heuristics. European Journal of Operational Research 53, 131–148.

Negenman, E. G., 2000. Material coordination under capacity constraints. Ph.D. thesis, Eindhoven University of Technology, Eindhoven, Netherlands.

Spitter, J. M., Hurkens, C. A. J., de Kok, A. G., Lenstra, J. K., Negenman, E. G., 2005. Linear programming models with planned lead times for supply chain operations planning. European Journal of Operational Research 163, 706–720.

Taha, H. A., 1975. Integer Programming, Theory, Applications and Computations. Academic Press Inc., Florida, USA.

Tempelmeier, H., Derstroff, M., 1996. A Lagrangean-based heuristic for dynamic multi-item multi-level constrained lot sizing with setup times. Management Science 42, 738–757.

Voß, S., Woodruff, D. L., 2003. Introduction to Computational Optimization Models for Production Planning in a Supply Chain. Springer, Berlin.

PART II

Strategic Operations Management

Mass-Customized Production in a SME Network

Dario Antonelli[1], Nicola Pasquino[2] and Agostino Villa[1]

1   Politecnico diTorino, Dept. Sistemi di Produzione ed Economia
dell'Azienda, c.so Duca degli Abruzzi, 24 – 10129 Torino Italy
dario.antonelli@polito.it; agostino.villa@polito.it
2   Università degli Studi di Salerno, Fisciano [SA]
WWW home page: http://www.unisa.it

Abstract

The most promising feature of a manufacturing system oriented to mass customization is to have at disposal a layout and a governance such to allow: a) to include a new product within the family of products under manufacture; b) and to apply the required modification of the manufacturing process in front of the market-requested product innovations, in such a way to minimise the cost for product inclusion. The inclusion of a new product in a SME network implies to approach two complementary problems: 1) a *post-ponement problem*, that means to recognise the new characters of the innovated product such to specify its difference with respect to the set of other products already processed, and to adjoin the new working sequence in the existing processing program (to be possibly modified at least) already applied by the SME network; 2) an *order-fulfillment problem*, that means to include the a-priori estimated production flow required for the new product, within the programmed flows pattern in the existing SME network, by adding the minimum possible innovations to the network itself. The paper will discuss the proposed solution phases, and illustrate a set of integrated procedure to be applied in order to obtain a mass-customisation strategy of practical utilisation for managing SME networks.

Keywords

post-ponement, order fulfilment, mass customisation.

1   Introduction

Manufacturing system oriented to mass customization implies use of layout and governance such to allow an easy and fast inclusion of a new product within the family of products under manufacture, and a simple modification of the

Please use the following format when citing this chapter:

Antonelli, D., Pasquino, N., Villa, A., 2007, in IFIP International Federation for Information Processing, Volume 246, Advances in Production Management Systems, eds. Olhager, J., Persson, F., (Boston: Springer), pp. 79-86.

manufacturing process in front of the market-requested product innovations, such to minimise the cost for product inclusion.

This task is particularly important in a Network of Small-Mid Enterprises (SME), where a large effort is necessary to coordinate

1. the definition of the necessary modifications of the working sequence (which should be applied in the different SMEs of the network, depending on their specialisation and the phases of the product working sequence they are able to implement),
2. the innovation of the existing pattern of production flows among the various SMEs.

Then, the inclusion of a new product in a SME network implies to approach two complementary problems:

(i)   a *post-ponement problem*, i.e. recognise the new characters of the innovated product such to specify its difference with respect to the set of other products already processed, and adjoin the new working sequence in the existing working sequence;

(ii)   an *order-fulfillment problem*, i.e. include estimated production flows of new product within the existing flows pattern but with the minimum possible innovations to the network itself.

A sketch of the two joined problems is illustrated in the following Figure 1.

Planning the inclusion of a new product in the mass-customised SME network can be obtained by the following phases:

1° Phase: Detecting the request of the new product, in terms of expected features;

2° Phase: Specification of the new product in terms of new "product tree":

  1° Step: Recognition of new expected functions, different from those which can be obtained by existing products and production activities of the different SMEs in the network;

  2° Step: Selecting new product components through a comparison with already existing product trees;

3° Phase: Definition of the new working sequence:

  1° Step: Transforming the new tree of components into a working sequence;

  2° Step: Given the new working sequence, selecting the required resources to be included in some SMEs of the network, such to assure the execution of the new working sequence;

4° Phase: Inclusion of the new product into the SME network:

  1° Step: Integrating the layout of the existing SME network and the production capabilities of each SME with the new resources required at the previous step;

  2° Step: Given an a-priori estimation of the new expected demand, assigning production flows on the various SMEs in the network.

The paper will discuss the proposed solution phases, and illustrate a set of integrated procedure to be applied in order to obtain a mass-customisation strategy of practical utilisation for managing SME networks.

The developed solution strategy can be validated by applying real data contained in the web portal developed within the EU-funded CODESNET (COllaborative

Demand & Supply NETwork[1]) project. Owing to space constraints, description of some industrial applications will be done during the APMS'07 Conference.

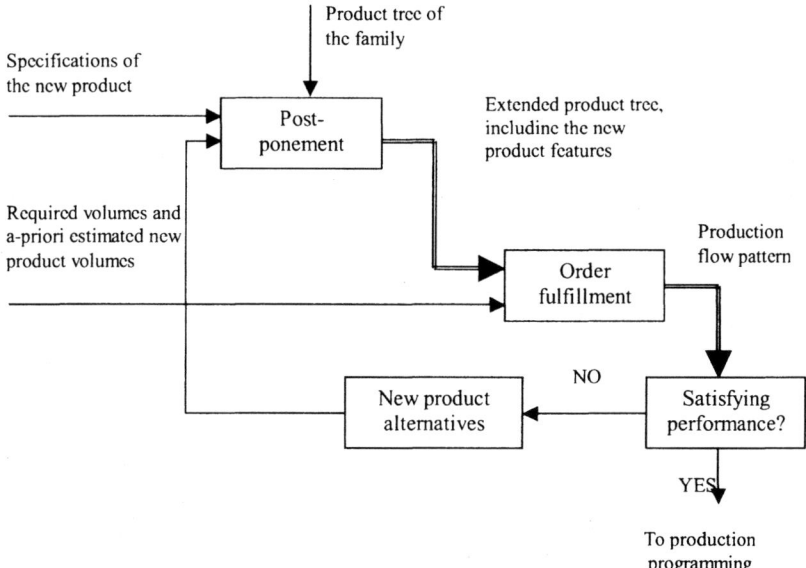

Figure 1. Loop connection of the two joined problems of "post-ponement" and "order fulfillment"

Known Results:
Several research efforts have been dedicated to the two complementary problems of post-ponement and order fulfillment in a mass customization industrial system. Usually, the two problems are separately approached, with the scope of defining procedures sufficiently simple for the following two scopes: for the former, to generate a wide mix of product configurations, possibly by delaying the product differentiation on the final part of the manufacturing lines [1, 2, 3]; for the latter, search for strategies minimizing due date even in case of large variety of products [4]. The integration of both for assuring an assignment of working sequence phases to the line stages maximizing the line utilization (then, minimizing delivery delays) even for a wide mix, has still to be defined and validated.

---

[1] CO-DESNET is the acronym of the Coordination Action (CA) project n° IST-2002-506673 / Joint Call IST-NMP-1 on COllaborative DEmand & Supply NETworks, supported by the European Commission, Information Society Directorate General, Communication Networks, Security and Software Applications, under the coordination of Politecnico di Torino (see the web portal address www.codesnet.polito.it).

2   Practice-oriented problem solution

The approach outline sketched in Figure 1 contains the following organisation of solution phases for the joint problem of post-ponement & order fulfilment in a mass customization system:

1st phase: recognize a new product to be manufactured;

2nd phase: characterize the new product in terms of multi-attribute product
> tree [5, 6]
> 1st step: identify all new functions required to the new product, not yet supplied by existing products;

> 2nd step: select new components through a comparison with existing product trees.

3rd phase: Define the new working sequence [7]:
> 1st step: transform the new product tree into a sequence of manufacturing operations;
> 2nd step: Given the new working sequence, select the resources required to manufacture.

4th phase: Include the new product into the manufacturing system:
> 1st step: integrate the existing manufacturing system with newly required resources (see 2nd step 2nd phase) [7];
> 2nd step: given the expected average demand of the new product, assign production flows to the manufacturing system network [8].

In details, the four phases have to be accomplished according to the following actions.

1st phase: recognize a new product to be manufactured;

Data: the enterprise receives the demand for a new product, whose innovation with respect to the existent mix is identified in terms of: a) product "mission", stated by new functionality; b) product "structure", in terms of new components.

Action: the enterprise detects the demand for new product, compare required functionality with its product mix data base (each product being described by a specific product tree).

Remark: comparisons of several product trees could be time-consuming: then, new procedures based on pattern recognition could be necessary.

2nd phase: characterize the new product in terms of multi-attribute product tree:

Data: the enterprise, detected the innovation characters of the newly requested product, specify it in terms of new product tree.

Actions:
1st step: Identification of all new functions required to the new product, not yet supplied by existing products, will be performed through a comparison of two/many product trees of the same "product family". To this aim, the new product must be recognized as belonging to a given product family according to the following:

Rule 1: Two products belong to the same product family if: a) both have the same "mission"; b) the parameters describing the functionality of one differ from those of the other at most for a given bound; c) the product tree structure of one differs from the other of at most a given number of components.

2ⁿᵈ step: Selection of the new components through a comparison with existing product trees will be done according to the following:

Consequence of Rule 1: Necessary and sufficient condition to assure a correct comparison between two products is that both present the same "mission".

3rd phase: Define the new working sequence:

The scope of this third phase is to design the working sequence for the new product with attention to the as small as possible number of changes from existing sequences.

1ˢᵗ step: Transformation of the new product tree into a sequence of manufacturing operations can be obtained according to the following

Rule 2: The optimal allocation of the manufacturing operations required for a new product is stated as the problem of selecting both the best patterns of product components and the minimum cost working phases, among those characterising the product family.

Note that this is a combinatorial problem of reduced complexity, because the number of feasible alternative in practice does not appear to be high.

2ⁿᵈ step: Given the new working sequence, the selection of the resources required to manufacture is a standard problem of manufacturing/assembling system design, often denoted "machine layout problem", as in [1].

4ᵗʰ phase: Include the new product into the manufacturing system:

This last phase is devoted to estimate at which level new product orders can be satisfied. estimation of this level is based on the a-priori evaluation of the new product demand, in terms either of an average value over a mid-long time span, or a known variable evolution depending on the market demands (e.g., seasonal periodicity). Such an evaluation is the basis of a standard "order fulfilment" problem.

1ˢᵗ step: The integration of the existing manufacturing system with newly required resources (result of the second step of the previous phase) consists of a completion of the graph of manufacturing centers included in the production system, such to make it able to perform all the working operations necessary to process both the old products and the new one. Then, this step is just a completion of the "resource selection" problem above mentioned.

2ⁿᵈ step: Given the expected average demand of the new product, the assignment of production flows to the manufacturing system network aims to estimate the loading conditions of any manufacturing center in the system, such to give an evaluation of the system utilization and of the production costs there involved.

Starting point is the new product tree and the related working sequence, obtained at 1ˢᵗ step of the previous phase.

To the new working sequence and product tree as well as to the set of working sequences of the other products in the mix, a graph of production flows corresponds. Within said graph, both "old" production flows and the ones for processing the new product must coexist. The following Figure 2 shows an example of the mentioned graph.

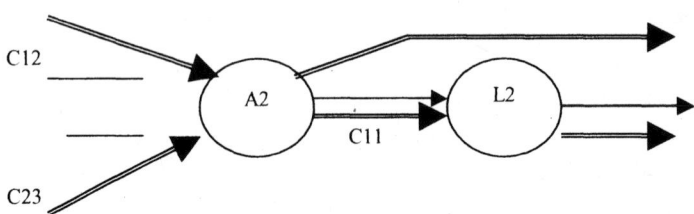

Fig. 2. Graph of production flows in a simple manufacturing system.

    ——▶ denotes a production flow for the new product components;

    ═▶ refers to production flows of the "old" mix in the system.

The following rule has to be applied:

Rule 3: The minimum cost to include a new product demand into an existing system, provided that the required new resources have been put inside, is obtained by solving a standard Aggregate Production Planning [9] problem, i.e. by optimising production flows on the graph of manufacturing centers through minimization of production costs given the centers capacities.

It is a LP problem, whose solution tries to minimise distance between demands and produced volumes, in each time bucket of the planning horizon.

From this flows optimisation problem a further result concerning the overall objective of deciding if to include or not a new product in the production stream, follows:

Rule 4: Decision of either including or not a new product in the previous mix of an enterprise depends on the joint solution of two problems:

  i.   select an alternative of product components and related working operations of minimum cost, among those contained in a given product family;

  ii.   optimise production flows over the network of manufacturing centers such as to minimise production costs.

Remark: Note that the best (minimum cost) decision is that of including a new product if it implies the minimum possible perturbation in the manufacturing system, in terms of a non-congested utilisation of the actual *capacity margin* of the production centers.

This remark should suggest the system manager in selecting heuristic criteria and rules for taking into considerations new product, market pushed, and evaluating the cost involved if they will be included in the actual manufacturing operations.

3   Some concluding remarks

The approach sketched in Figure 1, and detailed through the four phases above, can receive two types of interpretations:

a.   from a theoretical point of view, the four solution phases come from a *decomposition* of a unique complex combinatorial optimisation problem, i.e. that of selecting an organised set of new components (the new product tree) according to which to modify the production flows and loads in such a way to minimise the production costs related to manufacturing all the components for the whole (previous and new) set of products;

b.   from a practical point of view, the four solution phases summarise the sequence of steps that a production manager must apply in order to evaluate the convenience of including or not a new product into the existing mix.

Applications under development concerns the production of bicycles for disabled people (tricycle, with special features for sport or out-road use). New features are sometime requested, to make the cycle easier to use and lighter.

Obtained results in selecting the new features and in identifying how to modify the existing production line, according to the four phases – four rules above presented, seem to be really promising. Evaluation data will be presented at the APMS Conference.

References

1.   D. He, A. Kusiak, and T-L Tseng, Delayed product differentiation: a design and manufacturing perspective, Computer-aided Design, vol. 30, n. 2, pp. 105-113, 1998.
2.   A. Garg and C.S. Tang, On postponement strategies for product families with multiple points of differentiation, IIE Transactions, vol. 29, pp. 641-650, 1997.
3.   R.S. Farrel and T.W. Simpson, Product platform design to improve commonality in custom products, J. Intelligent Manufacturing, vol. 14, pp. 541-556, 2003.
4.   J. Jiao, Q. Ma and M.M. Tseng, Towards high value-added products and services: mass customization and beyond, Technovation, vol. 23, pp. 809-821, 2003.
5.   J. Jiao, M.M. Tseng, V.G. Duffy, and F. Lin, Product family modelling for mass customization, Computers Industrial Engineering, vol. 35, n. 3-4, pp. 495-498, 1998.

6. F. Salvador, C. Forza, and M. Rungtusanatham, Modularity, product variety, production volume, and component sourcing: theorizing beyond generic descriptions, J. Operations Management, vol. 20, pp. 549-575, 2002.
7. A. Villa, *Analisi di Sistemi di Produzione Industriale*, Ed. CLUT, Torino, 2006 (in Italian).
8. J.P. Burbidge, *Production Flow Analysis for Planning Group Technology*, Clarendon Press, Oxford, 1989.
9. P. Brandimarte and A. Villa, *Advanced Models for Manufacturing Systems Management*, CRC Press, Boca Raton, 1995.

Extended Service Integration: Towards "Manufacturing" SLA

Frédérique Biennier, Loubna Ali and Anne Legait
INSA LYON, LIESP. INSA de Lyon, Bat B. Pascal, 69621
Villeurbanne Cédex – FRANCE
{frederique.biennier, loubna.ali, anne.legait}@insa-lyon.fr,
WWW home page: http://liesp.insa-lyon.fr

Abstract

Market constant evolution towards more and more customisation and call for "product/service" increases to need of agile and lean organisations, making an heavy use of information and communication technologies. To bring the necessary openness, interoperability and agility features to the enterprises information systems, one can use fruitfully Service Oriented Architecture. Already used at a business level as a potential interoperable and integrating framework, this technology must be adapted to define manufacturing services and to take into account manufacturing constraints, namely time constraint and security integration. In this paper, we propose a global framework to define a "manufacturing service bus", paying a particular attention to the manufacturing service definition. We also show how mobile agents can be used to set dynamically monitoring systems.

Keywords

SOA, Interoperability, SLA, manufacturing agreements, ESB

1   Context

Due to the economical context involving more and more customisation and "service oriented products", enterprises have to adapt their organisational strategy: while focusing on their core business, outsourcing or collaborative strategies must be set to fit the market requirements (i.e. getting a critical size and being able to provide a high service level to the consumer). These organisational trends rely on an heavy use of information and communication technologies calling for agility, i.e. the ability to answer to structural changes quickly (client requests, technological or activity changes, supplier management…), for reducing waste (leading to lean manufacturing organisation) and for developing a product/service strategy. This involves

Please use the following format when citing this chapter:

Biennier, F., Ali, L., Legait, A., 2007, in IFIP International Federation for Information Processing, Volume 246, Advances in Production Management Systems, eds. Olhager, J., Persson, F., (Boston: Springer), pp. 87-94.

reorganising the enterprise according to the services it can propose and defining "manufacturing services", taking into account industrial constraints (security, time constraints, quality, management strategy consistency...).

To fit the openness and required agility levels involved by the enterprise constant changes according to the market evolution, one can reorganise the information System (IS) according to a Service Oriented Architecture (SOA). SOA organises a component based architecture, namely services, that are mostly business process oriented: thanks to the orchestration level, elementary services are composed according to the business process specification and orchestrated. As services provide well identified interfaces (mostly defined in XML format), they can be fruitfully used to bring the necessary interoperability level whereas the composition and orchestration mechanisms support the IS agility (figure 1).

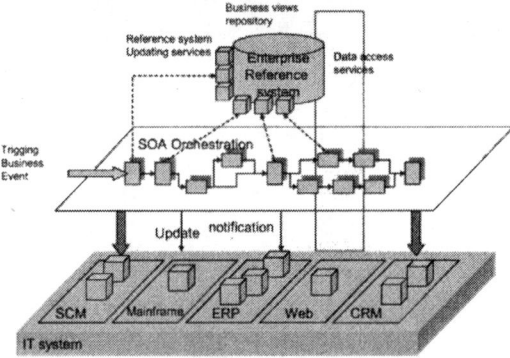

Figure 1: Interconnection of information system thanks to SOA from (Rivard et al. 2005), pp.25 and 39

Implementing a SOA requires both a middleware technology to enable service integration AND a distributed communication services. Focusing on the middleware side, SOA requires first to define service repository, a business service choregrapher (in charge of composition and orchestration) as well as a middleware engine in charge of routing, message transformation, security control... Such functions are commonly implemented in enterprise service bus which provides access to external services (Schmidt et al. 2004) (figure 2).

After the composition process, orchestrating these different services leads to a "service chain" organisation" which may involves different partners. This involves defining contracts for the shared process and interface definition. Service underlying philosophy is that the execution should be environment independent. Such a requirement can not fit the manufacturing requirements where processes are heavily related to the exact hardware resource. Moreover, answering times and other Quality of Service parameters are not taken into account in the service contract. To support the "end to end" contractual relationships including QoS, Service Level Agreements (SLA) are set between "connected services", linking the different partners among the service chain. (figure 3). The required Quality of Service defined in SLA is described thanks to Key Performance Indicators and Key Quality Indicators, taken from patterns associated to the business and network services.

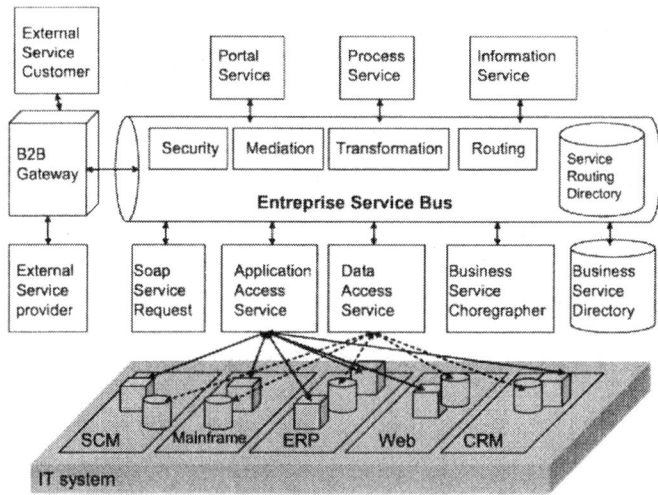

Figure 2: ESB position

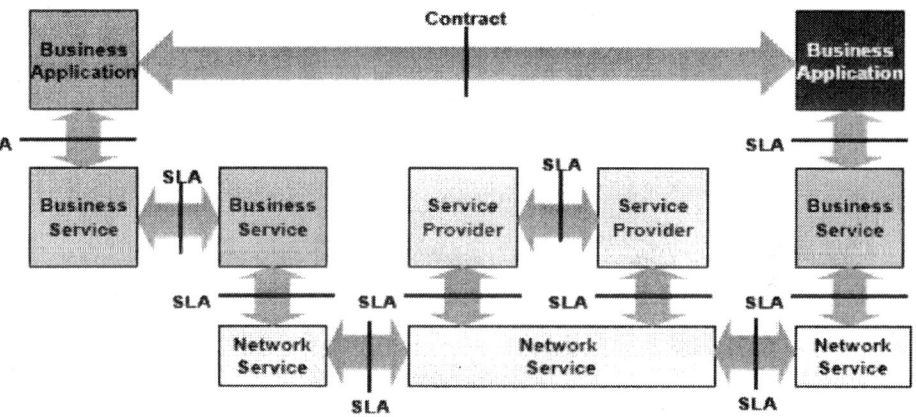

Figure 3: SLA and service combination (taken from (Open group, 2004), p.3)

Despite of the openness provided by the multi-level organisation, such a SLA framework is limited to the ICT infrastructure and does not fit an extended service chain organisation. In order to integrate a product-service dimension, we must first define how service-chain eco-system can be model to derive extended service models and then integrate manufacturing and management constraints in the SLA organisation.

2   Holistic manufacturing chain organisation

To integrate manufacturing constraints, we propose to extend the "service chain" specification by adding a "manufacturing chain" based on a SCOR model (Supply Chain Council, 2000) in order to capture the management properties of the linked objects and to define the information to exchange. Due the plethoric variety of information which can be collected to describe and evaluate a manufacturing chain, we propose to establish a multi-dimensional modelling framework which can be enriched incrementally. We define a **manufacturing chain recursively either as a manufacturing chain which links a set of manufacturing services, or as a simple stand-alone service. Services can be described as a set of competencies** (i.e. what the service can achieve) **whereas their position in the manufacturing chain is defined as a role** (i.e. a set of competencies according to which the service is selected and plays a part). Based on this definition, **we propose to analyze and model the manufacturing services according to an incremental holistic model (gathering IT, industrial, legal, economical, human related... dimensions) associated to four different points of view:**

1. **The "Internal" Service View** describes the service competencies (i.e. what the service can be used for), the service's own organization (i.e. how the service is designed to achieve the works it can do) and the service management agents (including set of indicators/dashboards, acquisition and building processes). The service's own organization and the service management sub-views are conditioned by the service type (i.e. pure human service, semi-IT supported (i.e. human service using allied IT related services) or IT supported (which may include a human part heavily connected to the IT support). To support incremental description as well as to include best practice capitalization, the service model is split among different levels associated to different purposes (security requirements, service competencies, interconnection rules, exchange constraints...).

2. **The Service Chain View** is related to the role a service plays in the chain This view represents the public part of the service model and uses the service interface (i.e. the service competencies) to select the convenient service to set an adapted service chain. This model gathers 2 dual views: the External view which describes the service chain competencies and the way it can be integrated in a higher level chain and the Internal view. **This internal view is based on the composition process description**, i.e. definition of a strategic assembly policy (i.e. description of the different roles and of the criteria used to select the convenient services) and "assembly controllers" used to conveniently propagate the different constraints after selecting the services and assembling them in a service chain. Of course, after achieving a composition process, the current service chain described as a "service workflow" is stored in the model. To support management and evaluation services, this internal service-chain sub-model is also related to service chain management agents, used to evaluate the added-value creation / propagation among the service chain and to evaluate the chain component satisfaction, describing the alignment of each service goal within the service chain's global goal.

3. **Environmental View** is used to describe the service chain selection policy (namely, goals and service chain roles, management criteria, strategic selection criteria as partner trust chain, localisation...). This view is related to the "exposition" of the service chain external view. Nevertheless, the environmental model is not reduced to a simple service chain exposition; it also includes the environmental impact of the service chain (mostly structuring impact). Then, management agents are used to orchestrate the service chain and service management sub-models: these environmental agents are used to select the convenient dashboard patterns, collect environmental information used to define composition criteria and aggregate performance indicators. According to a mid-term exploitation of the collected data, this management system will be used to capitalize emerging behaviour to guide further composition selection criteria.

Gathering these different views leads to a multi-interface extended service model:

-   **"Conceptual service"**: this interface is used to the service "semantic, i.e. the service "competencies" (what is achieved by this service)
-   **Manufacturing interface**: this interface describes HOW the manufacturing service will be achieved. It includes
    o   *Product view:* this interface gathers information on material and products (bill of material, management policy, quality indicator...) as well as on the production management strategy.
    o   *Process view:* this includes both the routing specification, potential resource constraints (this interface is used to couple the manufacturing service to the resources), process qualification according to the CMMI classification (Williams and Wegerson 2002). For this last point we use the information stored in the different parts of the information system to asset the process maturity level (i.e. the manufacturing process can be described; its results are evaluated in a dashboard, documented...).
-   **SLA control interface:** Real-time constraints are used to describe global QoS constraints to be included in SLAs whereas safety constraints are used to define pre-emptive or not pre-emptive services. This interface is coupled to the dynamic monitoring system to define dynamically the adapted QoS monitoring features

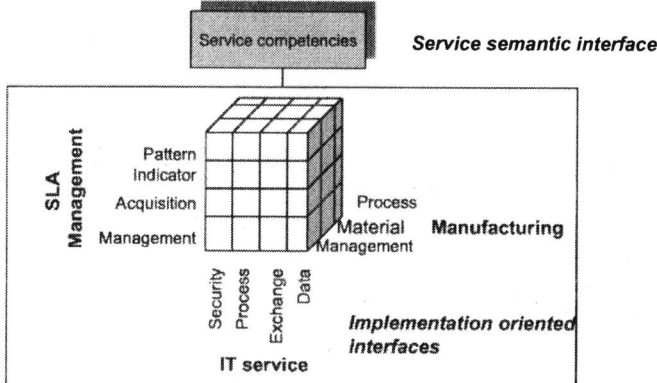

Figure 4: Extended service model

- **IT View :** this interface describes the connection with the IT support services / systems and addresses mostly information transformation (thanks to standards as B2MML or FDML), security services

3   Towards a Manufacturing service bus

To fit such an extended service, the underlying middleware organisation must be adapted in order to support information transformation (from the business to the manufacturing level), QoS and safety requirements. This leads to a **manufacturing service bus (MSB)** organisation able to process functional (service semantic, management strategy...) and non functional (namely security and QoS) interfaces. The MSB is based on the enterprise service bus organisation but, as it must integrate strong manufacturing constraints (i.e. service execution is related to the resources it is "run" on, QoS constraints, safety constraints leading to "pre-emptive" services...), the MSB integrates the manufacturing service composition system and the service orchestrator as well as a dynamic monitoring system:

- The service composition process enriches the service traditional semantic selection: our multi-criteria service selection uses information from the different interfaces: first services are selected "semantically" according the "competency oriented interface" and the implementation constraints are processes in the following order: first manufacturing constraints are processed (i.e. management strategy, process maturity level...) in order to define consistent manufacturing chains and then QoS constraints are taken into account by comparing the current context to the required performance level. IT service interface is not used to select or reject a manufacturing service: this view is lastly used to select the convenient IT services (information transformation, security components) to add in the composed service chain.
- Safety requirements (i.e. pre-empting resources in emergency cases) is implemented by adding execution priorities (standard or emergency) to the services while they are processed by the orchestrator and by defining" roll-back" processes associated to the different service in order to get back after an emergency stop.
- Service mediation and transformation includes is based on B2MML (Emerson 2006) to implement information interface services towards business service providers and FDML to implement interfaces towards manufacturing resources
- Real-time constraints are taken into account by adding a QoS monitoring system.

This QoS monitoring system is implemented in a distributed way thanks to mobile agents. Based on the Aglet platform (Lange and Oshima 1998), we develop a set of different agents built incrementally to monitor the service chain (figure 6). The mobile agent architecture we use to implement the distributed management system consists in different agents types that can be assembled to set new management functions so that a dynamic management system is set. Three main types of agents are used:

- **Itinerary agents** are in charge of the "mobility management", i.e. visit the different nodes belonging to a given itinerary to collect, configure… information before going back to the management node which has created them.
- **Code agents** are in charge of running the indicator building programs to set more complex indicators. They can be used in a distributed way in conjunction with itinerary agents.
- **Action agents** are used to orchestrate the area management, setting the convenient itinerary and action agents. These agents are organised into several categories(security management, QoS management, service management…).

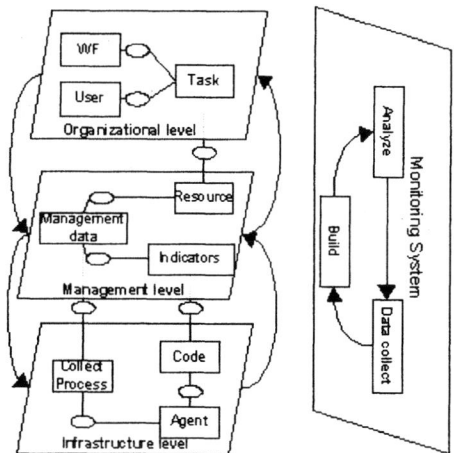

Figure 7: Monitoring system organisation

The management system is organised hierarchically and consists in three interacting monitoring algorithms running on the different nodes:

- **The server node** is. in charge of information analysis and of notifying contextual information to the manager nodes. It manages the monitoring area information base, computes indicators by starting convenient code agents after receiving raw information via itinerary agents, extracts and sends back monitoring information. It can also launch an information acquisition process according to the information call received from manager nodes.
- **The manager node** is in charge of context analysis (either by analysing the information sent by the information server or by processing the notification sent by another area manager), identifies the convenient actions to achieve and set the itinerary agent (it define the visited node sequence and the convenient action to be launched on each node).
- **Managed nodes algorithm** is simpler: these nodes only wait for itinerary agents, and then identify the action part. Then, it has to get and install the code agent associated to the management action and set properly parameters. Lastly, at the end of the code agent process, a simple itinerary agent is started to send back the action result to the manager node.

4   Conclusion and further work

To fit collaborative manufacturing environment, we proposed to adapt the business oriented service architecture to the manufacturing context. This leads us to define extended service, including different interfaces, as well as a "manufacturing service bus". This new middleware concept is based on the enterprise service bus but it also includes a manufacturing service composition process as well as a dynamic QoS monitoring system to fit real time constraints. We focus first on the real-time constraints, experiment various kinds of agents to collect information and set convenient configuration as well as to build more complex indicator. Nevertheless, we do not integrate by now the semantic associated to the different SCOR objects.

Acknowledgements: this work is supported by the INSA BQR project "Enterprise urbanism" and Rhone Alpes area council through the GOSPI Cluster project INTERPROD.

References

Lange D.B. and Oshima M., 1998. Programming and deploying java mobile agents with aglets, Addison-Wesley

Mathieu H., Ali L., Biennier F., 2006. A distributed management system: towards pro-active information system management in virtual enterprise. INCOM'06 Proceedings.

Open group, 2004. SLA Management handbook volume 4: Enterprise perspective, available at http://www.opengroup.org/pubs/catalog/g045.htm

Rivard F., Brendel C., Buche N., Delayre S., Mocaërr A., Nevers J., 2005. SOA et urbanisme: le rôle des architectures orientées service dans l'alignement des systèèmes d'information. Unilog White paper, available at http://admin.kermeet.com/Data/kmreed_informatique/event/F_374d0882e6f2c44d80bb816 b5c87600d42ba63161965c.pdf, 109p.

Schmidt M.T., Hutchinson B., Lambros P., Phippen R., 2005. The enterprise service bus. : Making service oriented architecture real. IBM System Journals, vol. 44, n° 4, pp.781-797

Supply Chain Council, 2000. Supply Chain operations reference Model – SCOR Version 5.0, 255 pages

Williams R., Wegerson P., 2002. MINI CMMI(SM) (SE/SW/IPPD/SS Ver 1.1) Staged Representation. Cooliemon

Designing Manufacturing Networks
– An Empirical Study

Andreas Feldmann, Jan Olhager and Fredrik Persson
Linköping University, Department of Management and Engineering,
SE-58183 Linköping, Sweden
andreas.feldmann@liu.se, jan.olhager@liu.se, fredrik.persson@liu.se
www.liu.se

Abstract
The design of the manufacturing network for a firm is an important factor for its competitive position. By manufacturing network we mean the plant or plants of the manufacturing firm and the relationships with external suppliers. The way that these operate together is central to the entire supply system supports the competition of the products in the marketplace. The decisions are typically categorised as related to facilities and vertical integration, two decision categories in an operations strategy. This paper presents the results of a survey of 84 Swedish manufacturing plants. The results show that competitive priorities such as quality and price play different roles in the networks, and that there is a significant difference in terms of how internal and external suppliers are selected.

Keywords
Empirical analysis, External suppliers, Internal suppliers, Manufacturing networks, Market characteristics, Survey research.

1 Introduction

The competitive positions of manufacturing firms stem from the design of the entire manufacturing network that needs to be in alignment with the market opportunities. The literature contains few models that help managers to design and manage plant networks. Colotla et al. [1] and Shi and Gregory [2] view a manufacturing network as a factory network with matrix connections, where each node (*i.e.* plant) affects the other nodes and hence cannot be managed in isolation. Rudberg and Olhager [3] analyze manufacturing networks and supply chains from an operations strategy perspective. They relate the manufacturing network to the decision categories of vertical integration and facilities, concerning both configuration and coordination.

Please use the following format when citing this chapter:

Feldmann, A., Olhager, J., Persson, F. 2007, in IFIP International Federation for Information Processing, Volume 246, Advances in Production Management Systems, eds. Olhager, J., Persson, F., (Boston: Springer), pp. 95-102.

The issues of configuration and coordination dominate the research agenda on manufacturing networks. The former has its origins in the multi-plant research, and location-based criteria dominate (see *e.g.* DeBois et al. [4] and Ferdows [5,6]). The latter is mainly concerned with technology transfer and diffusion, as well as within-network learning (see *e.g.* Gailbraith [7] and Flaherty [8]). However, in some instances attempts have been made to integrate the two issues to get an overall view of the manufacturing network (see *e.g.* Shi and Gregory [2] and Porter [9]).

This paper is based on a mail questionnaire survey to Swedish manufacturing firms, in order to capture the plant level perspectives of designing manufacturing networks in Swedish manufacturing firms. 500 manufacturing firms were contacted in January 2007. 84 useable responses were received, with a response rate of 16.8 percent. The survey is carried out at the plant level, providing the plant perspective of the manufacturing network. We report on the link to environmental factors such as market and product characteristics, including qualifiers and order winners. We explicitly treat the role and design of plants in the network and the principles for network design, including the reasons for location. We distinguish between internal and external suppliers and study the reasons for selecting a specific type of supplier. Using the survey data this paper gives an overview of the issues of importance for manufacturing network design in Swedish manufacturing firms. This is to our knowledge the first large scale empirical study on manufacturing network design.

In this paper we first present the research methodology. Then, we present and discuss the results concerning the market, plant, and supply characteristics.

2   Research Methodology

The questionnaire was designed with respect to the guidelines and recommendations presented in Dillman [10] and Forza [11]. As noted in [11] it is not possible for a company or a plant itself to answer any questions, this has to be done by human respondents. Therefore, people working with manufacturing and logistics in manufacturing companies were contacted and asked to respond in the survey. The respondents were all upper level managers related to production or logistics, and thus expectedly knowledgeable about the survey questions. The largest group of respondents was logistics/supply chain managers (32.6 percent of the responses), followed by production managers (30.1 percent), plant managers (6.0 percent) and presidents or vice presidents (3.6 percent). Other respondents include e.g. supply managers and logistics project leaders. The sample includes smaller, medium-sized as well as larger manufacturing plants, based on number of employees and sales turnover; see Table 1. The sample includes manufacturers of industrial goods (86.9 percent), consumer goods (10.7 percent) and both types (2.4 percent). All types of customer order decoupling point position are included in the sample; engineer-to-order, make-to-order, assemble-to-order, make-to-stock, and finally make and distribute to stock. The last position refers to holding finished goods inventory in the distribution system, beyond the plant inventory. All these characteristics suggest that the responding firms in the sample are a good representation of the population.

Table 1. Firm characteristics.

Characteristic	Distribution
Number of employees:	
– 99	10.7 %
100 – 199	17.9 %
200 – 499	32.1 %
500 – 999	14.3 %
1000 –	24.0 %
Sales turnover (MSEK; 1€=9.3 SEK, May 2007):	
– 99	4.0 %
100 – 499	36.5 %
500 – 999	17.6 %
1 000 – 4 999	32.4 %
5 000 –	9.5 %

3 Survey Results

The survey results are presented first in terms of market characteristics to describe the manufacturing environment. Then, we present the results concerning plant roles and responsibilities and the reasons for location. Finally, we present the results concerning the design of the supplier network; both internal and external.

3.1 Market Characteristics

In the survey market characteristics are presented along two dimensions, geographical distribution and order winning criteria; see e.g. Hill [12]. The geographical distribution was measured by letting the respondents assign share of sales they had in nine different regions. In recent years the Baltic region has attracted attention as a region for internal or external suppliers to Swedish manufacturing firms, i.e. a near-shore low-cost country. Table 2 shows that the major markets for Swedish firms can be found in Europe, including Sweden, and North America. The "weighted average" is an average of all companies with presence. The relative large difference between the overall average and the weighted average in some regions indicates that once a company is establishing itself on a market it is for a substantial share. Taking North America as an example the overall average is 11.2 %. However only half of the companies (44 out of 84) have markets in North America and the ones that do, have on average 21.1 % of sales there.

Each respondent rated 13 different competitive priorities on a 7-point Likert scale, ranging from "low importance" to "decisive importance" with respect to the relevance as an order-winning criterion. The results are presented in Table 3. The results reveal that quality comes out as the highest ranking order winner, with both the highest mean and lowest variance, and price can be found surprisingly far down the list. This might be explained by that manufacturing companies in Sweden, with traditionally high labor costs, have to find alternative means of competing, thus focusing on quality, product characteristics and delivery precision instead of price. The survey also allowed for other criteria being added, and brand, traceability, and environmentally friendly product all rated as decisive for the companies in question.

Table 2. Geographic distribution of markets.

Market	Overall average	Frequency	Weighted average with respect to presence
Sweden	34.8 %	74	37.1 %
Baltic region	1.1 %	21	4.2 %
Other Europe	37.0 %	72	40.6 %
North America	11.7 %	44	21.1 %
India	0.7 %	18	2.9 %
China	2.7 %	27	7.9 %
Other Asia	7.1 %	41	13.7 %
Other	4.7 %	36	10.3 %
Total	100.0 %	-	-

Table 3. Competitive priorities.

Order winning criteria	Mean	Std.dev.	Median	Mode
Quality (conformance to specification)	6.08	0.92	6	6
Product characteristics	6.00	0.98	6	6
Delivery precision	5.45	1.04	6	6
Product range	4.84	1.64	5	6
Volume flexibility	4.59	1.51	5	6
Delivery speed	4.54	1.36	4	4
After-sales service	4.52	1.66	5	6
Price	4.51	1.37	4	4
Geographical coverage	4.33	1.71	5	6
Offer a range of logistical solutions	4.13	1.48	4	4
Design flexibility	3.89	1.40	4	4
Size of focal company	3.81	1.37	4	4
Geographical proximity	3.34	1.66	3	2

3.2   Plant Characteristics

The plant itself was the focal point of the survey. We investigated the reasons for location as it is perceived currently, as well as the competence and responsibilities of the plant. These issues are based on the issues discussed in Ferdows [5,6] and Veerecke and van Dierdonck [13]. Ferdows find three primary strategic reasons for the site; access to low-cost production, access to skills and knowledge, and proximity to market. Veerecke and van Dierdonck also consider socio-political reasons and competition. Other typical reasons are proximity to raw materials, transport hubs, and cheap energy, wherefore we included all these issues in our survey. The plant location factors were captured using a 7-point Likert scale ranging from "unimportant" to "of the utmost importance" concerning how important the individual factor is for the plant currently. The result is shown in Table 4. The survey allowed for other criteria being added; among these were proximity to the residence of original founders, as well as a combination of history, existing buildings, need for secrecy, and old factories.

Table 4. Reasons for geographical plant location.

Reason for location	Mean	Std.dev.	Median	Mode
Proximity to knowledge	4.09	1.92	4	4
Proximity to transport hubs	3.67	1.63	4	4
Proximity to market	3.40	1.84	3	2,4
Sociopolitical climate	2.95	1.61	3	2
Proximity to cheap labor	2.80	1.45	3	2
Proximity to cheap energy	2.67	1.52	2	2
Proximity to raw materials	2.54	1.61	2	2
Proximity to competition	1.70	0.95	1	2

We find that proximity to knowledge is the primary reason for the site location. This is in line with expectations, since many firms choose to locate new sites near industrial parks nearby universities. Proximity to transport hubs comes second, which can be motivated by the relative low density of the population in Sweden and long transportation distances. Proximity to markets is the third major reason, which most likely is related to locations near major cities. The other potential reasons for site location are of lesser importance.

The other plant characteristics in the survey were concerned with the competences and responsibilities of the plant. Again, we based the set of issues on Ferdows [5,6] and Veerecke and Van Dierdonck [13]. The plant competencies and responsibilities were captured using a 7-point Likert scale ranging from "no local responsibility" to "full local responsibility". The result is show in Table 5.

Table 5. Competences and responsibilities at the plant.

Area of plant responsibility	Mean	Std.dev.	Median	Mode
Production	6.63	1.02	7	7
Production planning	6.43	1.14	7	7
Technical maintenance	6.22	1.42	7	7
Process development	5.83	1.59	6	7
Logistics	5.57	1.66	6	7
Introduction of new process technologies	5.28	1.95	6	7
Sourcing	5.25	2.05	6	7
Supplier development	4.67	2.14	5	7
Supply of global markets	4.41	2.49	5	7
Introduction of new product technologies	4.30	2.37	4,5	7
Product development	4.14	2.19	4	7

Overall, plants seem to possess many competences and responsibilities, indicating that plants typically have higher strategic roles, such as "source", "lead", and "contributor", using the typology by Ferdows [5,6]. All measures of central tendency indicate that the average plant at least share the responsibility or have full local responsibility concerning all competence areas.

3.3   Supply Characteristics

As a final part of the mapping of the manufacturing networks of Swedish manufacturers, the characteristics of the supply are presented. The survey was concerned with both internal and external suppliers, to be able to detect the extent of similarities and differences between them. We checked for geographical distribution and the criteria for choosing suppliers of both types.

The geographical distribution of the supply networks were found to be focused around Europe, including Sweden, and North America; cf. Table 6. The so-called low cost countries have a relatively small proportion of the inbound supply. The "weighted average with respect to presence" is the average share of inbound material from a country for those companies that have suppliers in that particular region. This gives an indication of whether a low average indicates a small number of companies with heavy presence in a region or a large number of companies with small presence.

Table 6. Geographical distribution of internal and external suppliers (all 84 plants have external suppliers, while 45 of these also have internal suppliers).

Geographical distribution of suppliers	Internal (N=45)			External (N=84)		
	Fre-quency	Average	Weighted average wrt presence	Fre-quency	Average	Weighted average wrt presence
Sweden	32	48.1 %	67.6 %	73	55.7 %	57.2 %
Baltic Region	4	3.7 %	42.0 %	23	2.0 %	6.6 %
Other Europe	28	37.4 %	60.0 %	68	29.1 %	32.1 %
North America	10	7.5 %	33.9 %	30	4.7 %	11.8 %
India	4	1.5 %	17.3 %	27	4.2 %	11.7 %
China	1	0.0 %	2.0 %	12	0.5 %	3.4 %
Other Asia	2	0.5 %	11.0 %	19	2.1 %	8.1 %
Other	2	1.2 %	26.5 %	13	1.8 %	10.5 %
Total	-	100.0 %	-	-	100.0 %	-

Perhaps more interesting is the criteria on which suppliers are selected. The respondents were asked to rate the importance of 15 different criteria for choosing suppliers, using a 7-point Likert scale, ranging from "low importance" to "deciding importance". The result is presented in Table 7. Two clear conclusions can be drawn from the comparison between internal and external supplier selection. First the decision to use an internal supplier is based on very few criteria (as opposed to the selection of external suppliers); many criteria has received either a very high (7) or a very low score (1). Second, there is a significant difference in how internal and external suppliers are selected. "Corporate decision" is used significantly higher for internal suppliers, while all other criteria rank higher for selecting external suppliers. This indicates that the choice of an internal supplier is to a large extent based on a single corporate decision, reflecting competence, quality, control and synchronization. Many of the other issues are even significantly more important in the choice of external supplier, indicating a multi-criteria decision process. Here, quality, price and delivery dependability are the top three criteria; all significant at the 0.01 level. A paired t-test was used for testing the differences between criteria for selecting internal and external suppliers, for the 45 companies that has both internal and external suppliers.

Table 7. Criteria for selecting internal and external suppliers ("control and synchronization" was not available as an answer for the section on external suppliers").

Criteria	Internal				External			
	Mean	Std.dev.	Med.	Mode	Mean	Std.dev.	Med.	Mode
Corporate decision***	5.72	1.74	6	7	4.02	1.99	4	4
Exploit/use/keep important competence	4.51	2.28	5	1,6,7	4.67	1.54	5	6
Quality (conformance to specification)***	4.49	2.21	5	6	6.01	1.02	6	6
Control and synchronization	4.17	2.11	4	6	-	-	-	-
Delivery dependability ***	4.07	2.09	4	1	5.33	1.25	6	6
Volume flexibility***	3.87	1.92	4	5	4.59	1.39	5	4
Price***	3.72	2.08	4	1	5.49	1.14	6	6
Delivery speed*	3.60	1.91	4	1	4.28	1.48	4	4
Design flexibility	3.37	2.00	3,5	1	3.77	1.66	4	4
Product range	3.15	1.91	3	1	3.48	1.46	4	4
Geographical proximity**	3.11	2.03	2	1	3.31	1.56	3	5
Logistical solution***	2.78	1.75	2	1	3.85	1.70	4	4
Size of company***	2.74	1.74	2	1	3.90	1.37	4	4
Geographical coverage	2.57	1.87	2	1	3.19	1.65	3	2
After-sales service**	2.42	1.67	2	1	3.27	1.75	3	2

*** Difference between internal and external suppliers is significant at the 0.01 level;
** Significant at the 0.05 level; * Significant at the 0.10 level.

4   Discussion and Concluding Remarks

In this paper we have investigated how manufacturing networks are designed, related to the market and product characteristics, the distribution and roles of plants, and the selection of internal and external suppliers, based on a survey of 84 Swedish manufacturing plants. The role of quality and price is interesting. Quality is perceived as a high priority for competing in the market, for plant location and roles, and for both internal and external suppliers. Thus, there is a strong alignment concerning quality. As for price, the products do not compete on price in the market, the plants do not have a low-cost focus, but cost is an important issue for choosing external suppliers. Thus, this supports the role of cost as a major issue for outsourcing parts of the manufacturing network, that concern items for which cost is important.

The most important lesson to be learned for managers from this research is that companies have a differentiated treatment of internal and external suppliers. Internal suppliers are primarily chosen for quality, competence, control and synchronization related to a corporate decision, while external suppliers have to compete on quality, competence, price and delivery as well as many other issues. Even though this study focuses on Swedish manufacturing networks, the results are most likely representative of many western countries. The design of the manufacturing network including the external suppliers has to be aligned to the market characteristics and take the particular products into consideration.

References

1. I. Colotla, Y. Shi and M. J. Gregory, Operation and performance of international manufacturing networks, *International Journal of Operations & Production Management*, **23**(10), 1184-1206 (2003).
2. Y. Shi and M. Gregory, International manufacturing networks – to develop global competitive capabilities, *Journal of Operations Management*, **16**(2,3), 195-214 (1998).
3. M. Rudberg and J. Olhager, Manufacturing networks and supply chains: an operations strategy perspective, *Omega - The International Journal of Management Science,* **31**(1), 29-39 (2003).
4. F.L. DuBois, B. Toyne and M.D. Oliff, International manufacturing strategies of U.S. multinationals: a conceptual framework based on a four-industry study, *Journal of International Business Studies*, **24**(2), 307-333 (1993).
5. K. Ferdows, Mapping international factory networks, in: Managing International Manufacturing, edited by K. Ferdows (Elsevier Science Publishers B.V, North-Holland, New York, 1989), pp. 3-21.
6. K. Ferdows, Making the most of foreign factories, *Harvard Business Review*, **75**(2), 73-88 (1997).
7. C.S. Gailbraith, Transferring core manufacturing technologies in high-technology firms, *California Management Review,* **32**(4), 56-70 (1990).
8. M.T. Flaherty, *Global Operations Management* (McGraw-Hill, New York, NY, 1996).
9. M.E. Porter (Ed.) *Competition in Global Industries* (Boston: HBS Press, 1986).
10. D.A. Dillman, *Mail and Internet Surveys: The Tailored Design Method* (John Wiley & Sons Inc., New York, 2000).
11. C. Forza, Survey research in operations management, *International Journal of Operations and Production Management*, **22**(2), 152-194 (2002).
12. T. Hill, *Manufacturing Strategy* (Palgrave, New York, 2000)
13. A. Vereecke and R. van Dierdonck, The strategic role of the plant: testing Ferdow's model, *International Journal of Operations & Production Management*, **22**(5), 492-514 (2002).

Methodologies for Dividing Profit in Networked Production Structures

Hendrik Jähn and Joachim Käschel
Chemnitz University of Technology,
Department of Economic Sciences, Professorship BWL VII
09107 Chemnitz, Germany
hendrik.jaehn@wirtschaft.tu-chemnitz.de,
WWW home page: http://www.tu-chemnitz.de/wirtschaft/bwl7/

Abstract

This conceptual paper focuses the problem of dividing the profit earned in a production network to the different network members. In this context different theoretical approaches for profit division are introduced and discussed. It is assumed that a high degree of automation is aimed at by using the modern information and communication technology (ICT) intensively. The task of profit division is integrated in a comprehensive approach considering incentive and sanction mechanisms for harmonising the interests of the network members as well. In that context the basic assumptions of the New Institutional Economics serve as a theoretic basis.

Keywords

Production Network, Network Controlling, Profit Division

1 Introduction

The production of goods within networked organisation structures especially allows small and medium-sized enterprises (SME) to enter new markets. For the operation and coordination of production networks the management concept "Extended Value Chain Management" (EVCM) [1] was developed. This approach operates by automation at a high degree. It consists of nine phases representing the typical phases of a life-cycle of a production network.

For the modelling of the approaches of profit division the methodology of linear optimisation served as a base. Furthermore the microeconomic approach of the New Institutional Economics [2] with its basic assumptions limited rationality,

Please use the following format when citing this chapter:

Jähn, H., Käschel, J., 2007, in IFIP International Federation for Information Processing, Volume 246, Advances in Production Management Systems, eds. Olhager, J., Persson, F., (Boston: Springer), pp. 103-110.

opportunistic behaviour of actors and individual maximisation of utility in connection with asymmetrically distributed information among the economic actors is considered. For the solution of the problem the Principal-Agent-Theory serves as a valuable basis. Hereby the EVCM which represents the network can be interpreted as Principal while the network participants are the Agents. By using instruments for harmonising the interests of both actor groups the utility can be maximised. Options in that context are the introduction of incentive or sanction mechanisms [3] as specific modules of a comprehensive model for profit division. That approach is illustrated in fig. 1.

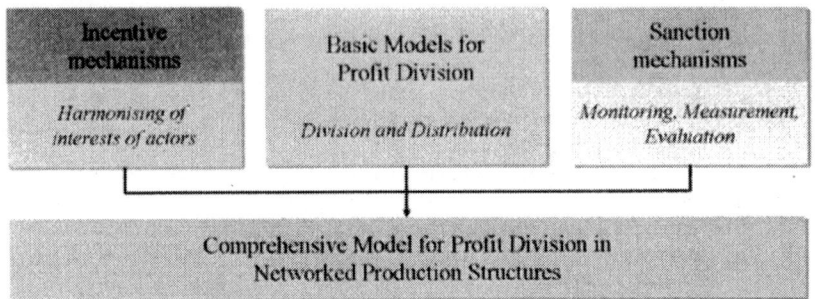

Fig. 1. Comprehensive Approach for profit division in networked production structures

It is obvious that there must be flexible but harmonised interfaces between the partial models. That is realised by the calculation of specific measures which allow a transfer of information.

2 The Comprehensive Model

2.1 Basic Models for Profit Division

In existent literature hardly any comprehensive models focusing that topic can be identified. As an exception a few game-theoretic approaches [4, 5], macroeconomic approaches [6] or approaches requiring a special framework [7, 8] can be identified. However these approaches are highly theoretic and applicable for real-world networks only to a very limited degree what marks the initial point for the development of specific approaches for SME-based production networks.

All profit division models described within this contribution consider three influencing parameters: a fixed share to cover the fixed costs, an added value-dependent share and a profit expectation dependent share. The least mentioned parameter represents the most important variable influencing the ascertainment of the individual profit shares of the single enterprises. Thereby, the individual profit expectation belongs to the input parameters of the model which are assumed and already introduced.

Due to the consideration of the three components, some possibilities of variation occur for the further modelling. Such variations above all refer to the question whether the offer price or offer profit correspond to the sales price or profit or not. Thus, there are several possibilities of component integration for calculating profit shares under consideration of the three components. For example, the profit shares could first of all be calculated via the profit expectation and if a profit share remains, it can be divided under consideration of further components. A further opportunity is the introduction of weightings for each of those three components.

In principle, the calculation of the network profit which is included in the offer is based on the individual profit expectations of the enterprises. However, it is also imaginable that the profit expectation-dependent profit share paid out after the value adding process has been carried through, does not correspond to the total profit amount that can be divided. That so-called non-divided remaining profit is negative if a too high amount has been divided and it has a positive value if a profit share that has to be divided remains. In both cases, measures have to be taken in order to divide the earned profit exactly. Therefore, we suggest either a corresponding revaluation or devaluation by using a standardisation parameter or the consideration of the two remaining components.

2.2 Incentive Mechanisms

Starting from the situation that a customer initiates a request to a network for the production of a good, the network management has to search for suitable network participants. In some cases the situation can arise that not for every value-adding step a suitable enterprise could be found. Several reasons for that situation are imaginable. In case one or more core competences are not represented by network members, external partners need to be acquired. However in case every value-adding step could be realised by at least one enterprise there is a different situation. In that case the "missing" enterprises need to be convinced to participate. One reason can be the lacking profitability of an order.

Economic transaction is fulfilled by two actor groups: principals and agents. In network theory the network management can be interpreted as principal while the network participants (SME) act as agents. Both parties tend to maximise their utility with respect of their economic activities. In case an enterprise does not want to participate in a value-adding process although it has the necessary core competences the decision comes from the tendency for maximisation of utility. In that case incentives can be granted to convince the enterprise to participate. The incentive mechanism will be successful when the enterprise accepts the incentive. That process can be described as harmonisation of interests of principal and agent.

Based on that idea quantitatively orientated incentive mechanisms for the participation of enterprises in networks for the case core competencies are missing for one special value adding process can be applied. For example it can be imagined that the profit of the entire network can be increased in case one specified enterprise can be engaged which under regular conditions would not take part. Therefore it is necessary to grant incentives. In that case monetary incentives seem to be most efficient.

2.3 Sanction Mechanisms

According on the realised performances of an SME, sanction mechanisms represent a further valuable instrument of the network controlling. This mechanism is applied in case the performance agreed upon has not been completed in a satisfying degree considering selected performance indicators. Sanctions reduce the individual profit share and need to be restricted to an extension that is recognised as justified by the enterprises involved.

Sanctions must not lead to the fact that a future cooperation is made harder or does not even take place. For the quantification of sanctions the observation of the behaviour of the enterprises with regard to the fulfilment of the contract is necessary in order to quantify the sanction amount [3]. Within that scope, the actual and target values of important performance attributes are balanced for every enterprise participating at a value-adding process. A "degree of fulfilment" in the sense of a credit evaluation needs to be determined for each of those parameters. The ascertained degree of fulfilment as an actual value is subsequently compared to the target value for every attribute.

The variable "network conformity" is finally ascertained by the means of a weighting of the single parameters. This aggregated variable is considered after finishing the value-adding process in the profit division model. Therefore, a connection is made between the "network conformity" and the payment of enterprise-specific profit shares by means of a function.

3 Approaches for Profit Division

3.1 Conceptual Framework

All subsequently introduced approaches for profit division consider three influence parameters. There are a value-adding-independent (fixed), a value-adding-dependent (variable) part and a profit expectation-dependent component. The profit share of an enterprise which is determined by the last mentioned influence factor is called profit-expectation-dependent profit share. The individual profit expectation of an enterprise (with regard to the individual value-adding-share) can be indicated as an amount or a percentage of the value-adding-share. It is stored in a central data base which is neither accessible to the network members nor to the broker instance. There is no "open book" strategy within the cooperation.

In the further modelling it has to be differentiated whether the product offer price of the network (respectively the offer profit) correspond to the final sales price (respectively the sales profit) to the customer. Thus, several possibilities of component integration arise for the calculation of the profit shares. The following sections will focus that structure. A further approach consists in the introduction of weightings for each component which is not discussed in this contribution.

In principle, the calculation of the network profit calculated in the offer is based on the individual profit expectations of the enterprises. However, it is imaginable that, after the value adding-process has been finished, the divided profit expectation-dependent profit share does not correspond to the complete profit that can be divided among the network members. This so-called undivided remaining profit is negative if a too high amount has been divided already whereas it is positive if a profit share remains and can be divided. In both cases, measures need to be taken in order to exactly divide the earned profit. Therefore, either a corresponding revaluation or devaluation by the help of a standardisation parameter or the consideration of the two remaining components are imaginable. While the first mentioned approach will be discussed in section 3.2.1, the second option will be focussed in section 3.2.2. For the less complicating case that the offer profit and the realised profit correspond to each other, two different approaches are discussed next in the sections 3.3.1 and 3.3.2.

It has to be stressed that the different approaches for profit division presupposes a flat-hierarchical structure of the cooperation with an independent broker unit. The sensible data are stored centrally for the calculation of individual profit shares in an automated way using the modern information and communication technology.

3.2 Models considering no remaining profit

3.2.1 Enterprise-related profit expectation as a basis

The calculation of the offer is based on the individual value-adding-shares of the cooperation members and the corresponding profit expectations an enterprise stated. If the offer including the suggested price is accepted by a customer, the goods are produced and delivered. Usually, the customer then pays the agreed sales price including the profit, which is based on the individual profit expectations and thus corresponds to the offer price. The achieved profit can be completely divided to the participating enterprises of the network. No amount remains undivided. The calculation of the enterprise-related profit shares results from multiplying the enterprise-related profit expectation (in per cent) by the net value-adding of an enterprise.

Basically, enterprises are given their desired profit share without any reductions or supplements. The divided (profit-expectation-dependent) profit results from the sum of individual profit expectation-dependent profit shares of all enterprises. This profit expectation-dependently divided profit corresponds to the offer profit and thus, the complete profit was divided. There is no remaining profit. One problem of this procedure is the initial parameter "individual profit expectation". The enterprises indicate this parameter independent from a certain value adding-process. However, there is the problem of intended incorrect information. For this reason, a parameter should be introduced which eliminates outliers and obviously wrong numbers. This parameter is called the average percentage of the expected profit and it is equal for all enterprises. However, that the divided profit does not necessarily correspond to the profit that can be divided. Thus, further mechanisms need to be applied which are also valid if a profit, that can be divided, was achieved which deviates from the offer profit. These mechanisms are considered in the following section.

3.2.2 Enterprise-related average profit expectations as a basis

For calculating the (weighted) average profit (measured in per cent of the value-adding share), the individual percentage of the profit expectation is weighted by the share of the individual value-adding-process of the complete value-adding-process. Thereby, it has to be considered that this parameter is enterprise-independent. In the next step, it is possible to ascertain the profit expectation-dependent profit share for every enterprise. This variable represents the most important component of the complete profit share of an enterprise. Finally, the complete (profit expectation-dependent) profit, that has so far been paid out, results from the sum of those individual profit expectation-dependent profit shares of all enterprises. Summing up the profit shares per enterprise is necessary in order to determine the further procedure. Thus, a comparison between the already divided profit and the profit after the value adding-process, that can in total be divided, is the next step. By subtracting the already divided profit from the distributable profit, a non-divided remaining profit results, which still can be divided among the enterprises

The total profit could be completely divided based on the (non-weighted) profit expectation (per cent). When applying the profit expectation, that is weighted using the individual value adding-process, it has to be checked subsequently if this condition is fulfilled. Here, it is significant to stress an effect of the application which occurs in case the offer profit corresponds to the distributable profit. Starting from the enterprise-related percentage of the profit expectations (related to the enterprise-related net value-adding-process and unweighted), the profit expectation-dependent profit share is calculated by multiplication. This procedure usually leads to different values. It can be remarked that the distributable profit G will in any case need to be divided among the enterprises without a remaining profit. This happens independently from the kind of calculation. This effect is especially interesting because thus it is made sure that this procedure provides usable results in a simple way despite of the improved modelling. It is guaranteed that this approach does not provide any procedure-specific remaining profit. However, those procedure-specific remaining profits must not be changed with the remaining profits, which occur in case the distributable profit from the sales price is higher than the offer profit in the offer price. This will be the focus of the following sections.

3.3 Models considering a remaining profit

3.3.1 Solution with a standardisation parameter

In case the distributable profit differs from the offer profit the balance is called remaining profit. That amount can be positive (realised profit > offer profit) or negative (realised profit < offer profit). In both cases, a corresponding division mechanism needs to be applied for the remaining profit. One possibility is the application of a standardisation parameter.

In the following, that procedure will be described. It is assumed that a partial amount of the realised profit could be divided based on individual profit expectations. Hereby it is of secondary importance whether the division took place based on the individual profit expectation or on the weighted profit expectation, because of the same character. In case a profit could be realised which exceeds the divided profit an alternative division approach is required. The suggested standardisation parameter is calculated by proportioning the total profit and the profit expectation-dependent profit share that has already been divided. Finally, the profit that has already been divided depended on the expectations of the enterprises is multiplied with the standardisation parameter for calculating the individual profit share of an enterprise. When calculating using concrete figures, it strikes that the profit share of an enterprise of the complete profit corresponds to the value adding-share of an enterprise of the entire value adding-process.

3.3.2 Division by means of a fixed and a variable profit share

A further division variant results by the inclusion of fixed and variable profit shares in addition to the profit expectation-dependent profit share. This approach can again be applied for positive as well as negative remaining profits. The division of that remaining non-divided profit is made subsequently by a fixed and a variable profit share. The undivided remaining profit hereby is divided by the number of active enterprises of the network and multiplied by a parameter which weights the fixed share. The remaining variable profit share is calculated by multiplying the corresponding weighting parameter with the undivided remaining profit and the ration of the individual value adding share to the total value adding of the network. The sum of both weighting parameters must be one.

By applying those interdependencies, a division of the profit to the enterprises that is based on three components is realised. The calculation of the complete profit share of an enterprise results by summing up the profit expectation-dependent profit share, the fixed profit share and the variable profit share.

It becomes clear that this model applies a division parameter. Several possibilities are imaginable for determining that parameter [3]. Thereby, it seems to be probable that, after the division, the complete profit still has not been divided. In that case the standardisation parameter must be applied again.

Because on the one hand, the performance-oriented profit division is favoured, but on the other hand enterprise with a small value-adding share should not be disfavoured, the profit division based on three components and using a variable division parameter is recommended in case the numbers are similar.

3.4 Findings

The necessity of taking remaining profits into consideration predominantly arises when the offer price and the sales price (and thus offer profit and sales profit) do not correspond. It has to be remarked that the application of weighting parameter is only obligatory in selected situations. In principle, several weighting parameters could be applied in this connection. Our example, however, is restricted to the application of an individualised division parameter. All the further models render that parameter dispensable – a fact that is absolutely desirable because the (only) consideration of the enterprise-related profit expectation promises a higher rate of being accepted by end-consumers. However this option will increase the complexity of the process.

4 Conclusion

Selected approaches for the division of profit were introduced in this paper. These methods serve different functions. In order to give a wider comprehension also incentive and sanction mechanisms as well as the fundamental framework were introduced. Transferring the approaches into practice seems efficient and disposes of a high success potential with regard to the structure and operation of promising cooperations. Thereby, it especially has to be stressed that the introduced approaches for the profit division within networked production structures have predominantly been formulated quantitatively and therefore meet the claim of having a high level of automation. The individual maximisation of utility can be considered as well by including an expectation-dependent component. It has to be mentioned that the introduced profit division approaches presuppose a high degree of automation of the network controlling processes. That precondition is fulfilled by the network operation and coordination concept Extended Value Chain Management (EVCM) in an outstanding way. Furthermore all network participants must agree the application of a suitable model. In practice the high degree of automation can lead to a shorter time from request to offer and a shorter process time. Within that framework the application of the EVCM including the profit division approach can serve as a valuable instrument of network controlling helping especially SME to cope with the challenge of global competition and rising process costs.

5 Literature

1. T. Teich, *Extended Value Chain Management (EVCM) – Ein Konzept zur Koordination von Wertschöpfungsnetzen* (Verlag der GUC, Chemnitz, 2003).
2. E.G. Furubotn, R. Richter, *Institutions and Economic Theory: The Contribution of the New Institutional Economics*, 2nd ed., (University of Michigan Press, Ann Arbor, 2005).
3. H. Jähn, M. Fischer, M. Zimmermann, *An Approach for the Ascertainment of Profit Shares for Network Participants*, in: Collaborative Networks and their breeding Environments, edited by L.M. Camarinha-Matos et al. (Springer, New York, 2005, pp. 257-264).
4. B. Fromen, *Faire Aufteilung in Unternehmensnetzwerken* (DUV, Wiesbaden, 2004).
5. E. Sucky, *Koordination in Supply Chains* (DUV, Wiesbaden, 2004).
6. H. Rehkugler, Die Verteilung einzelwirtschaftlicher Wertschöpfung (Disertation, Ludwig-Maximilians-Universität, München, 1972).
7. M.A. Krajewska, H. Kopfer, *Profit Sharing approaches for freight forwarder: An overview*, in: Logistics, Supply Chain Management and Information Technologies, edited by. D. Ivanov, A. Kuhn, V. Lukinskiy (Saint Petersburg Publishing House of the State Polytechnic University, St. Petersburg, 2006).
8. M. Jin, S.D: Wu, *Supplier coalitions in on-line reverse auctions: Validity requirements and profit distribution scheme*, International Journal of Production Economics 100, 183-194, (2006).

Coordinating the Service Process of Two Business Units towards a Joint Customer

Rita Lavikka, Riitta Smeds, Miia Jaatinen and Emmi Valkeapää
SimLab, Department of Computer Science and Engineering, Helsinki
University of Technology (TKK), P.O. Box 9220, FIN-02015 TKK, Finland
http://www.simlab.tkk.fi/contact.htm
Tel. +358 50 384 1662, Fax +358 (0)9 451 4698, rita.lavikka@tkk.fi

Abstract

The paper presents a new theoretical framework for coordinating an inter-unit collaborative service process towards a joint customer. The common service process is itself presented as a central coordination mechanism. It defines how tasks and responsibilities are shared between the collaborating units. The framework presents the factors supporting cooperation between the units, the prerequisites of the common service process, and the ways of coordination suitable for inter-unit cooperation. The framework is developed through a single-company longitudinal, qualitative case study that consists of three action research projects.

Keywords

Coordination, collaboration, service process, case study, action research

1 Introduction

Cooperation between organizations is necessary in the fast developing business world to add value for the end-customer. In practice, added value can be produced by broadening the service offering. According to Grönroos [1], it is a challenge to provide a joint service offering and it requires cooperation between the service providers. Cooperation between companies, but also between business units inside one company, must be supported and coordinated. Thus, a study on how collective human activity should be organized is required [2].

The objective of our research was to study how cooperation between business units of a single company can be supported and coordinated. As the main result of this research we present a new framework for coordinating a collaborative service process to produce better service offering to the joint customer.

Please use the following format when citing this chapter:

Lavikka, R., Smeds, R., Jaatinen, M., Valkeapää, E., 2007, in IFIP International Federation for Information Processing, Volume 246, Advances in Production Management Systems, eds. Olhager, J., Persson, F., (Boston: Springer), pp. 111-119.

The framework was developed in a case study in which cooperation between two business units inside one company was improved and the service processes of the units were integrated and coordinated. The need for coordination arose as the units realized that they needed to serve the common customers better by enlarging the product and service offering, but the units were too decentralized to do this together. The business units also tried to improve their competitive advantage by finding synergies through collaboration. [3]

2 Theoretical Background

2.1 Coordination of Interdependencies

Different units of organizations are usually specialized in certain operations, which differentiates the units from each other. However, the interdependencies between the units need to be managed to produce high-quality products and services together. Thus, coordination, i.e., the integration of different parts of the organization is needed in order to achieve common objectives [4].

The difficulty lies in determining how to coordinate different interdependencies between business units. Hall [5] claims that there is no one way to coordinate, but the suitable coordination mechanisms depend on the environment.

Thompson [6] presents three types of task interdependencies (pooled coupling, sequential coupling, and reciprocal coupling) and three modes of coordination (standardization, planning, and mutual adjustment) to manage these interdependencies. Mintzberg [7] continues Thompson's work in the field of coordination by stating that a fourth mode of coordination, i.e., direct supervision exists. Mintzberg's modes of coordination are intended for different organizational structures.

We combined Thompson's [6] and Mintzberg's [7] work on coordination. Thus, we present four modes of coordination suitable for different kinds of environments: planning, mutual adjustment, direct supervision, and the standardization of work processes, output, or skills needed to accomplish the work. Planning is suitable in dynamic situations where tasks are changing frequently. Mutual adjustment suits in dynamic environments when tasks are interrelated. Direct supervision is suited for a small organization where task interdependencies change. Standardization is suitable for environments that are mainly stable and the work activities are interrelated. [6, 7]

2.2 Supporting Cooperation and Improving Service Quality

Hoeg [8] and Simatupang et al. [9] state that cooperation between organizations needs to be improved in order to together form a product that satisfies customers. Axelsson and Easton [10] present that organizations should be integrated (i.e., become a whole) before the coordination of their operations can be started.

We hypothesize that factors supporting cooperation provide organizations with integration mechanisms. We present five factors supporting cooperation: common will and values [11], common understanding [12, 13], common development projects [11], internal customership [14], and communication [15]. We chose these factors because they suited well in the context of our case study, i.e., two business units trying to cooperate inside one company.

Service quality depends on the quality of both the service process and its end-result. Service process can be defined as a chain of sequential and/or parallel activities that have to work in order to produce a service. [16] According to Grönroos [1], a service process involves both the production and delivery of services. The relationships between quality generating resources (personnel, service idea, systems, and customers) have to be coordinated in order to improve service quality [1].

3 Methodology

This study applies abductive reasoning [17, 18]. First, a preliminary theoretical framework was developed based on theory. Then, the framework was tested using deductive reasoning in empirical research. The findings from the empirical research were generalized inductively to theory as a new framework. The empirical part is based on a qualitative case study [19, 20] in which action research was carried out [21].

The case study of this paper is focused on cooperation between two business units of a Finnish media company. The data was collected during three developmental action research projects following the SimLab™ method. During SimLab action research project, researchers prepare and implement a process simulation together with case companies. This includes setting goals, interviewing relevant parties, modeling the selected business processes, organizing a simulation day, analyzing results, and giving feedback. [22, 23] The simulation day includes a facilitated group discussion in front of a visual process model and group work sessions for developing further the solutions. The first and third writer of this paper acted as facilitators in the first action research project, whereas the first and fourth writer acted as facilitators in the second and third action research projects.

Each of the three consecutive projects lasted around three months. The data was gathered between October 2004 and November 2005. The qualitative data in the preparation and running of the three action research projects included altogether 40 semi-structured interviews, observation, and three questionnaires. Additional sources of information included documents provided by the business units or created during the action research project, such as process charts and management presentations on the case company's business models and objectives.

The semi-structured interviews of this study concentrated on predetermined topics selected by the researchers, but the flow of discussion was free, hence, relevant topics emerging during the interviews could be talked through [24]. The topics of the interviews were related to developing and coordinating the cooperation between the business units. The interviews were digitally recorded and then transcribed into text files word for word. After that, they were analyzed by marking the relevant pieces of the text and classified into themes (i.e., content analyzed).

The observation in the study was holistic in nature [25]. We made observations on the behavior of the participants during the projects and later talked through these observations with other researchers. The participants of the interviews, simulation days, and questionnaires included employees from the management level as well as from the operational level. Three common customers of the units were interviewed and altogether 21 common customers participated in the first and second simulation days.

Three questionnaires were designed to ask the respondents' opinions about ways to manage cooperation and the ways of coordination. The questionnaires were handed to the participants of the simulation immediately after each simulation day. There were open-ended questions and rating questions. The questionnaires were anonymous in order to elicit honest opinions. The answers to the open-ended questions were classified into themes, and the answers to the rating questions were average calculated. In the first project all 27 participants answered. In the second project 21 out of a total of 26 participants answered. In the third project 35 out of a total of 37 participants answered. Some of the respondents participated in all three projects and answered three times.

4 Case Description

The case study involved two business units (product and service unit) within a Finnish, international company operating in media industry. The units served partially the same customers. The product unit provided traditional physical products, whereas the service unit provided an Internet-based service. Both units had their own customers as well as common customers. The units had separate business models. In the beginning of the study, the personnel regarded these two issues as inhibiting factors to cooperating more intensively and gaining synergy.

The traditional product unit employed over 100 people. It generated ten times the profit of the other unit. The different operations of the product unit were coordinated by regular meetings, email, and daily ad hoc communication, such as phone calls and face-to-face discussions. The service unit was quite small, employing only 13 people. The service unit did not yet generate much profit as it was recently founded as a development unit. The employees of the service unit worked near each other. Hence, the ways of coordination included mainly informal and ad hoc communication.

The three consecutive action research projects had a common purpose to improve cooperation between the two units in order to serve the common customers better and to find synergies. This was achieved by coordinating the work of selected collaborating operations of the units. In addition, each development project had its own specific goals.

The goal of the first action research project was to build a common frame of cooperation between the business units in order to give the customers a consistent image of the products and services of the units. This challenge required improving communication between the units. The 21 participants of the action research project represented mainly the customer service operations of both units. In addition, 20 common customers participated in the simulation day. In the end of the first action research project, the units realized that the modeling of a common service process could be used as a cooperation process to coordinate the operations of the units.

In consequence, the goal of the second project was to model a common service process that would integrate the marketing and customer service operations of the units. The project also aimed at supporting the formation of common will between the two units. The common will meant that the units could agree on what the units want to be together in the future and how to get there. The project included 27 employees altogether from the customer service and the marketing operations of the units.

In the third action research project, the common service process was extended to cover also the electronic sales and distribution operations of the business units. There was a need to understand the potential advantages of common product and customer relationship management systems. Other goals were improving communication and common understanding between the units as well as finding ways to integrate the electronic sales and distribution operations of the units. Altogether 37 employees participated in the action research project. They came from the electronic sales and distribution operation, customer service operation, and marketing operations.

5 Results

The developed framework for coordinating a collaborative service process is presented in the Fig. 1. The framework contains three zones: the factors supporting cooperation, the prerequisites of common service process, and the operations of common service process and their ways of coordination.

Based on empirical findings, we recognized seven factors that were considered to support cooperation (Outer zone in the Fig. 1): a common profit goal, common understanding about the units' business, common will and values, a common strategic plan, common process simulation projects, internal customership, and communication. Some of the factors were modified from theory (see chapter 2.2).

The results of the case study show that in order to have better service, the separate business units can coordinate their operations by first modeling a common service process together, and then by acting according to it. The common process model defines, e.g., how operations between collaborators' organizations should be coordinated and managed. In addition, the model can define prescheduled planning meetings where coordination challenges related to common business are discussed and solved. Thus, we present that a common service process between organizations can act as a coordination method.

Based on empirical study, the prerequisites of a common service process (Middle zone in the Fig. 1) are a shared service idea, shared knowledge about the products and services, common understanding about the customers' needs, and shared product and customer relationship management systems. These four elements which were modified from theory [1] tie together the operations of two business units.

In the case study, we categorized the operations of the units into three categories: Marketing, Electronic sales and distribution, and Customer service. The researchers recognized ways of coordination (Inner zone in the Fig. 1) that were specifically used inside these operations. Two ways of coordination were used in all of these operations. These were common ways of operating and external integrators. The framework also presents other ways of coordination. (Fig. 1)

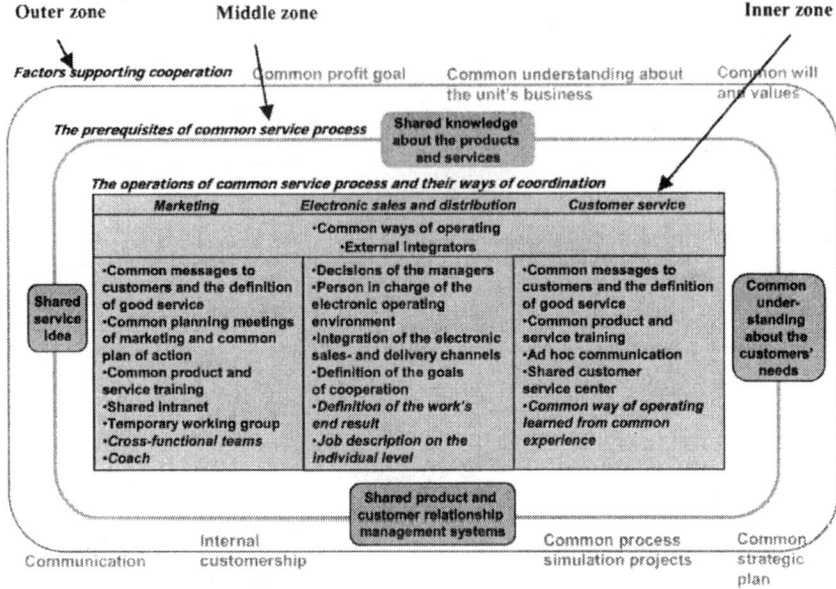

Fig. 1. The framework for coordinating a collaborative service process for better service [3]

The study suggests that a common strategic plan and a common strategic profit goal can be regarded as development and coordination methods on the organization's strategic level. They correspond to Mintzberg's [7] standardization of output.

In addition, the study suggests that the common ways of operating in customer service, common messages to customers, the definition of good service, and ad hoc communication can be regarded as development and coordination methods on the organization's operational level. The common ways of operating in customer service correspond to Mintzberg's [7] standardization of work processes, whereas common messages to customers and the definition of good service correspond to Mintzberg's [7] standardization of skills and norms. Lastly, ad hoc communication can be regarded as Mintzberg's [7] mutual adjustment. (See Fig. 2)

These results support the claim made by Thompson [6], according to which standardization is most efficient on the strategic level of the organization, whereas mutual adjustment is suitable on the operational level.

We state that in order to guarantee the quality of service offering and added value for the end customer, the organizations should organize their work as business processes crossing all the operations. We suggest that a common service process is a way of coordination that unites the strategic level and the operational level of the collaborative organizations (Fig. 2). The common service process also unites the different operations towards a shared business goal. However, the common service process should be managed by having a process owner that takes care of the development of the process [11].

Fig. 2. Development and coordination methods of the collaborative organizations providing services to the common customers [3]

6 Discussion and Evaluation

The study implies that before starting to coordinate the operations of different organizations, cooperation between the organizations should be started and developed. This supports the statement made by Axelsson and Easton [10] according to which integration is needed before coordination can be performed.

The study confirms the known fact that personnel should be involved in the collaborative process early enough to get them committed to changes. It was important for the personnel of the case company that the managers made clear decisions about how to approach cooperation. The personnel needed to understand what kind of advantage could be gained from cooperation.

A message to managers is that successful integration of business units requires that the management is committed to the changes and encourages personnel to start using the ways of coordination agreed together. Our findings suggest that in the early phase of cooperation, the business units should define the common resources required for producing the products and services to common customers. In addition, communication between the business units should flow freely all the time in order to prevent misunderstandings between partners.

Lincoln and Guba [26] present four criteria of soundness of a qualitative study: credibility, transferability, dependability, and confirmability. The credibility of this study was increased by active collaboration between the management of the two business units and the researchers. In addition, triangulation, i.e., the use of different data collection methods, increased the credibility. The transferability of this study was increased by describing the case study in detail. The dependability is related to the objectivity and stability of the research. This criterion was met by describing the context in necessary detail. The confirmability was improved by describing the methods of research, data collection, and analysis in necessary detail.

This study is only a start in understanding how different business units within a company can be coordinated and integrated to serve common customers better. In consequence, the new framework should be developed further in new cases of collaborative service offering. It would be interesting to apply it in a different context, e.g., between business units from different companies.

Acknowledgements

This paper presents the central findings of Södergård's [3] Master's Thesis. The empirical data provides basis for an article in the final report of the Co-Create project that was published in SimLab's Report Series in June 2006.

The research reported in this paper has been conducted in Co-Create and Madeleine research projects at the Enterprise Simulation Laboratory SimLab, Department of Computer Science and Engineering, Helsinki University of Technology. The authors are grateful for the creative research effort of the whole research team which has made this paper possible. The research is financially supported by the following organizations which are gratefully acknowledged: Finnish Funding Agency for Technology and Innovation (Tekes) and partner companies.

References

1. C. Grönroos, Service Management and Marketing, in Finnish (WS Bookwell Oy, Porvoo, 2001).
2. T. W. Malone and K. Crowston, The Interdisciplinary Study of Coordination, ACM Computing Surveys. 26(1), 87-119 (1994).
3. R. Södergård, Developing Collaboration between Two Business Units: Integrating and Coordinating the Service Processes, in Finnish (Master's Thesis, Department of Computer Science and Engineering. Helsinki University of Technology, SimLab Report series 12, Espoo, 2005).
4. P. Lawrence and J. Lorsch, Organization and Environment (Harvard University Graduate School of Business Administration, Boston, 1986).
5. R. Hall, Organizations: Structures, Processes, and Outcomes (Prentice Hall, USA, 2002).
6. J. Thompson, Organizations in Action (McGraw-Hill, USA, 1967).
7. H. Mintzberg, The Structuring of Organizations (Prentice-Hall Englewood Cliffs, NJ, 1979)
8. G. Hoeg, Taken for Granted, Best's Review. 106(1), 101-102 (2005).
9. T. Simatupang, C. Wright and R. Sridharan, The Knowledge of Coordination for Supply Chain Integration, Business Process Management Journal. 8(3), 289-308 (2002).
10. B. Axelsson and G. Easton, Industrial Networks, a New View of Reality (Routledge, London, 1992).
11. J. Hannus, The Keys to Strategic Success. The Effective Strategies, Competences and Business Models, in Finnish (Gummerus, Jyväskylä, 2004).

12. K. Mäkelä, Construction of Common Understanding. Interplay of Organizational Culture, Communication and Knowledge in Inter-company R&D-projects, in Finnish (Master's thesis, Department of Communication, University of Helsinki, 2002).
13. M. Jaatinen and R. Lavikka, Common Understanding as a Basis for Coordination, Journal of Corporate Communications [to be published 2007].
14. H. Kvist, S. Arhomaa, K. Järvelin and J. Räikkönen, Customer processes. How to improve the return by developing processes, in Finnish (Gummerus Kirjapaino Oy, Jyväskylä, 1995).
15. L. Åberg, Communication management, in Finnish (Otavan Kirjapaino Oy, Keuruu, 2000)
16. B. Edvardsson and J. Olsson, Key Concepts for New Service Development, Service Industries Journal. 16(2), 140-164 (1996).
17. B. Danemark, M. Ekström, L. Jakobsen and Jan Ch. Karlsson, Explaining Society, Critical Realism in the Social Sciences (Routledge, London, UK, 1997).
18. M. Grönfors, Qualitative methods for fieldwork, in Finnish (WSOY, Juva, 1985).
19. K. Eisenhardt, Building Theories from Case Study Research, Academy of Management Review. 4(4), 532-551 (1989).
20. R. K. Yin, Case study research: design and methods (SAGE, Newbury Park, CA, 1989).
21. E. Gummesson, Qualitative Methods in Management Research (SAGE Publications, Thousand Oaks, California, 2000).
22. R. Smeds, P. Haho and J. Alvesalo, Bottom-up or top-down? Evolutionary change management in NPD processes, International Journal Technology Management. 26(8), 887-902 (2003).
23. R. Smeds, M. Jaatinen, A. Hirvensalo and A. Kilpiö, SimLab process simulation method as a boundary object for inter-organizational innovation. (The 10th workshop of the IFIP WG 5.7 special interest group on experimental interactive learning in industrial management. Trondheim, Norway, June 11-13, 2006).
24. S. Hirsjärvi and H. Hurme, Theme interview – the theory and practice of theme interview, in Finnish (Yliopistopaino, Helsinki, 2004).
25. C. Marshall and G. Rossman, Designing Qualitative Research (SAGE Publications, Thousand Oaks, CA, 1995).
26. Y. Lincoln and E. Guba, Naturalistic inquiry (SAGE, Beverly Hills, CA, 1985).

Consortium Building in Enterprise Networks to Design Innovative Products

Marcus Seifert, Klaus-Dieter Thoben and Patrick Sitek

Bremen Institute of Industrial Technology and Applied Work Science,
Bremen, Germany
Contact: sf@biba.uni-bremen.de
Tel. +49-(0)421-218 5547

Abstract

The potential of an Enterprise Network is the ability to design and to realize innovative, customized products by selecting and integrating for each order the worldwide leading partners. To exploit this potential, it is too late to configure the network and to search for partners on the basis of an already specified bill of material: Co-operation has already to start with the product-idea, where possible contributions to the planned end- product must be identified while concretizing the bill of material. Only in this way, the network is able to benefit from the expertise of all potential partners and to ensure that the expertise of the planned consortium is also synchronized with the needed capabilities for the requested end-product. In the proposed paper, a method to support the building of consortia within Enterprise Networks on the basis of open product designs will be highlighted. The method starts from an end-product and collects possible contributions from potential collaboration partners.

Keywords

Virtual organizations, Product design, Consortium building, partner selection

1 Introduction

Market success under a worldwide competition depends more and more on the ability to provide customized products. The increasing complexity of these products led to the situation that capital intensive, complex investment goods are almost realized in co-operation between many partners. The single partner focuses on its core competencies while the process diversity is ensured by co-operation. Competition does not happen any more between single companies but between

Please use the following format when citing this chapter:

Seifert, M., Thoben, K.-D., Sitek, P., 2007, in IFIP International Federation for Information Processing, Volume 246, Advances in Production Management Systems, eds. Olhager, J., Persson, F., (Boston: Springer), pp. 121-131.

consortia (Boutellier 1999, S.66). The ability to form excellent co-operations is an important asset for the today's production.

1.1   Collaboration in Enterprise Networks

The today's opportunity to have a worldwide access to resources and capacities enables companies for the first time in history to select for each business opportunity the best suitable partners to fulfill highly customized customer orders. Companies are not dependent any more on existing relationships or regional suppliers-information and communication technologies as well as high performing logistics capabilities encourage the global business which means serving worldwide distributed customers on the one hand and benefiting from global resources and capacities for production on the other hand. The collaboration of legally independent, equal companies in a network to generate customer specific products has been introduced as Enterprise Network. If these networks are short term and dynamic in terms of members, they are called Virtual Organizations (Sydow 2002, p.270).

Within Virtual Organizations, the capability to optimize the existing process chains in a continuous way as required in stable Supply Chains is not any more the main asset. It is rather critical to be able to identify and to select for a certain order the most appropriate partners and to establish in a very efficient way a high performing co-operation (Kemmner 1999, p.33). The today available approaches to identify and to select partners are mainly focusing on the production phase which means that the search and selection criteria are focusing on a companies' capability to provide excellent processes and to integrate themselves into a networked organization. This means that the consortium building process is linked with the realization of an already defined product and founded on an existing bill of material of the desired end-product.

The purpose of the existing approaches can be described as way to set-up high performing production networks. In consequence, many approaches which have been developed in the past provide a wide range of criteria and key performance indicators (KPIs) which are focusing on the performance of companies and their production processes. KPIs are almost the basis for a structured partner search and evaluation. But the approach to select partners according to their process performance does only consider mainly one phase of the product life cycle which is the production phase. This does always lead to optimization tasks to select the best performing partners for a pre-defined set of processes to provide already specified components or sub-systems. The other phases of the product life cycle like the conceptual phase or the after sales phases do have completely different requirements for the partner selection. A service partner for the product support e.g. may be regionally close to the customer. This example illustrates that it is not sufficient to select partners on the basis of their production/process performance.

1.2   The conceptual phase as potential for Enterprise Networks

In Enterprise Networks, the phase with the highest potential is the conceptual phase. Enterprise Networks, esp. Virtual Organizations, do mainly provide complex and customized investment goods (Linde 1997, S.25) which can be described as engineer-to-order products. Considering this aspect, the conceptual phase where the concrete design of the end-product is not defined yet and where the bill of material is rather vague is the phase of the product life cycle with the highest potential for the consortium building: In the conceptual phase, there are still most of the degrees of freedom for the involvement of partners and the design of innovative products.

The chance within Enterprise Networks during the conceptual phase is to be able to enrich and improve the product design by involving the experience and competencies of all potential partners with the challenge to realize excellent products. Free from long-term contracts with suppliers and static processes, Enterprise Networks have the chance to incorporate the worldwide existing expertise into new product designs which enable the network to serve the market with highly customized and reliable products.

This means that it is not the main purpose in the conceptual phase to optimize resources and processes or to identify available capacities but to identify potential beneficial product contributions to the planned end-product and capabilities of possible partnerships. The objective should be to integrate potential partners very early for the concretion of the end-product. Only in this way, possible product innovations can be developed and competencies as well as experiences of all potential partners can be used to concretize the end-product. The chance to get to innovative solutions depends on the ability to gather the knowledge of promising partners already during the conceptual phase: Starting to search for partners when setting up the supply chain means to miss the opportunity to benefit from the knowledge and experience of potential partners for the product design.

Today, it can be recognized that many product designs are not developed from the scratch. Also complex engineer-to-order products are partly composed of available sub-systems and components which have to be adapted and integrated into the new design. Many companies are providing products which are explicitly foreseen to become parts of more complex systems. For example today's mobile phones containing touch-screen, GPS, UMTS and wireless-LAN units on a Windows Mobile platform do only combine and integrate existing sub-systems and components which are available as trading goods on the market and therefore normally well described and known. The main task is to identify beneficial sub-systems to develop innovative products and to integrate them into the planned design.

The ability to make use of those available sub-systems also in complex product designs enables companies in principle to evaluate potential partnerships already during the conceptual phase of the end-product. Against this background, the ability to identify potential contributions to a planned end-product and to initiate very early commercial relationships becomes a crucial asset for Virtual Organizations. This means that beside the evolution of production chains, the fast establishment of temporary trade relations will become more and more important for successful co-operations.

2   State of the Art: Existing approaches for consortium building

The process of consortium building can be divided into two phases which are (1) partner search and (2) partner selection (Mertins, Faisst 1995, p.61ff.). The purpose of the partner search is to identify potential co-operation partners for a specific task within a specific phase of the product life cycle while the objective of the partner selection is to evaluate these potential co-operation partners and to decide for a consortium. Literature and practice provide many methods to support the consortium building. In the following, the most important concepts will be highlighted and structured according the product life cycle shown in Figure 1.

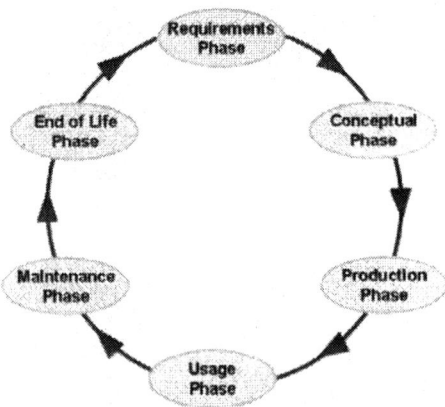

Fig. 1. Product life cycle

2.1   Existing methods for the partner search

Each of the product life cycle phases has different degrees of freedom regarding the partner search: The more indefinite the product design, the larger the pool of potential partnerships. While concretizing the product design and the bill of material, the requirements regarding process capabilities of potential partners and component design become more specific. This leads automatically to a reduction of possible choices for partners. Figure 2 shows the interdependency between the potential partnerships and the product life cycle.

According to Zahn, existing relationships between companies are still the main source for the partner search (Zahn 2001, p.60). This means that companies rely on existing co-operations and try to continue the business with well known suppliers. The decision for co-operation almost bases on private contacts of the company owners (Hoebig 2002, p.43). In consequence, the partner selection process can't be focused on excellent partnerships, but on the maintenance of established bilateral contacts. The advantage of selecting well known partners is the already existing trust between the parties. Disadvantage is on the other hand the reduction on a small group of potential partners which makes it improbable to find the best suitable partner and which impedes the consideration of the available potentials to improve the planned product.

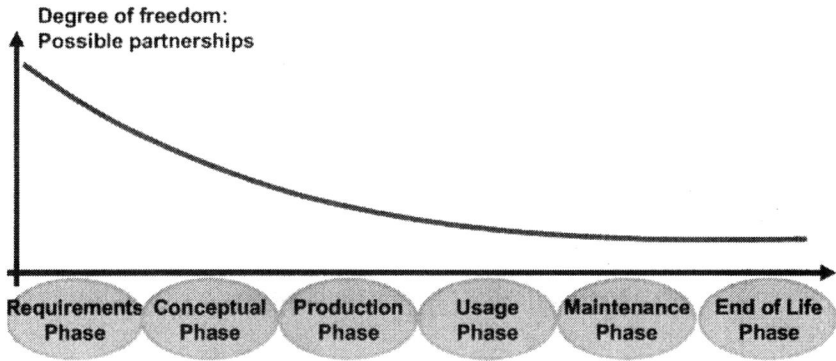

Fig. 2. Potential partnerships during the product life cycle

Alternative concepts coming from literature to search for partners in a more structured way are information brokers and co-operation databases (Kramer 1998, Zahn 2001). Co-operation databases make use of the internet and correspond to virtual public market places for partnerships. Companies willing to co-operate are able to enter their profile into these data bases to offer their capabilities or they can search for partners by defining search criteria. Most of these databases are provided by chambers of commerce. Zahn gives a review on co-operation databases in (Zahn 2001, S.63). To improve the search results, Kramer introduces information broker to identify potential partners by mapping the requested requirements with the offered core competencies.

2.2   Existing methods to select partners

After having identified the potential partnerships, the next task is to select the right partners out of this pool. The selection normally bases on an evaluation process using key performance indicators (KPIs). Today, most of the implemented methods for the partner selection are focusing on finding suppliers – nevertheless, there are also approaches available which are able to evaluate the different phases of the product life cycle. The evaluation can be cost-based, quality-based or process-based (Seidl 2002, p.27). In the following, the most relevant concepts for these three categories to support the partner selection are highlighted. Figure 3 maps these concepts with the product life cycle.

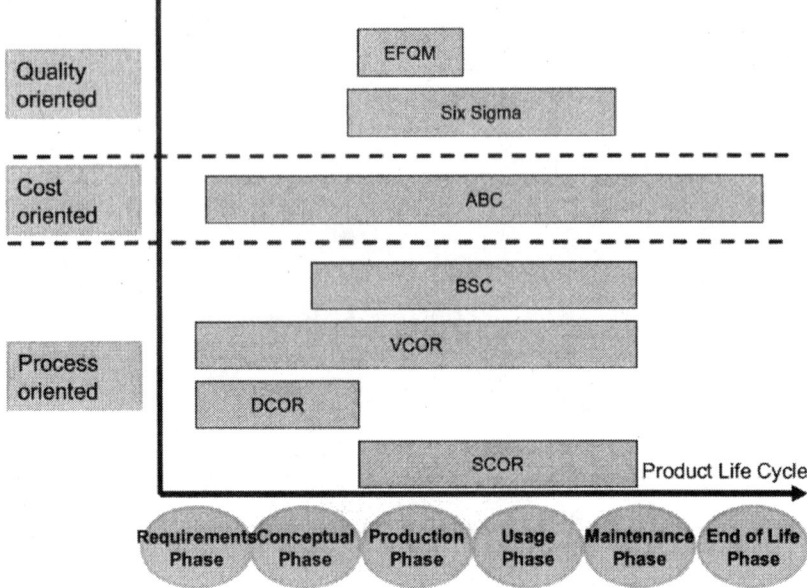

Fig. 3. Concepts for the partner selection in the product life cycle (Seifert 2007, p. 39ff.)

Figure 3 shows that there are specific approaches available to evaluate cost, quality and processes during the different phases of the product life cycle. The most relevant process oriented approaches are the Supply Chain Operations Reference Model (SCOR) for Supply Chain processes, the Design Chain Operations Reference Model (DCOR) for design and conception processes, the Value Chain Operations Reference Model (VCOR) also covering the service development and the Balanced Scorecards (BSC). A well known and often used cost oriented approach is the Activity Based Costing (ABC) able to cover in principle almost all activities related to the product life cycle. Finally, Six Sigma and the European Foundation for Quality Management (EFQM) represent the quality oriented approaches to evaluate industrial processes. Dependent on the phase and the perspective, it is possible to select an appropriate approach for evaluating a potential partners' performance.

3   Research approach

The proposed approach is structured into two parts: Part one (chapter 3.1) is the collection of company profiles of co-operation willing companies to acquire data on their offered sub-systems and components as potential contributions to any kind of planned end-product. These company profiles are the basis for the partner identification. By applying key performance indicators referring to the different phases of the product life cycle to the profiles, it is also possible to take these profiles in a later step as basis for the partner selection in any phase of the life cycle.

Part two (chapter 3.2) uses these profiles to generate in an iterative way possible product structures for a concrete planned end-product. By searching and combining possible components, different product variants can be developed. Collecting and structuring these potential product contributions, the result are the available alternatives in terms of product structure and partner selection. Following this approach, it is possible to contact and involve potential partners for a new customers' order in a very early phase of the product life cycle. A flexible product design is the main basis for collaborative designs and the derivation of innovative products, because innovation requires the involvement of all available knowledge and experience within a potential network and the structured identification and evaluation of potential solutions.

3.1   Company profiles to acquire potential product contributions

To set up company profiles of al co-operation willing companies within the Enterprise Network as starting point to generate flexible product designs, it is important to collect information about the potential partners.

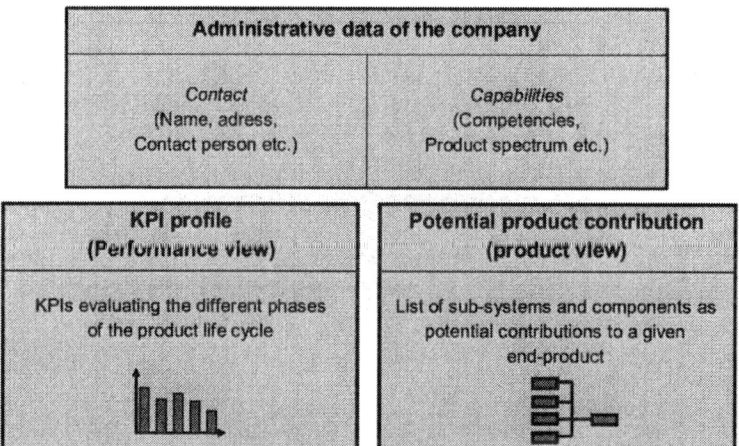

Fig. 4. Components of the company profile

Beside administrative data, information about the offered products have to be collected and stored in a unified, structured way to enable the iterative generation of bill of materials for the desired end-product. The last aspect to be provided by each partner is information about their performance to enable after the partner identification a partner selection on the basis of their KPIs. It is proposed to provide performance data for the different phases of the product life cycle. Figure 4 shows the components of a company profile.

The offered sub-systems and components as potential contributions to an end-product provided by a company have to be described in the way that also the necessary inputs to generate this sub-system/component have to be defined. Figure 5 shows a simple example for the description of a sub-system within a company profile.

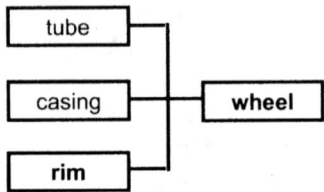

Fig. 5. Components of the company profile

The sub-system provided in this example is a wheel as potential contribution to a bicycle. This sub-system is linked with the necessary input to realise this product which are the rim, casing and tube. Other companies may offer the same sub-system consisting of different inputs which would lead to variants and completely new product designs. Partners providing different variants on the same level enlarge the networks' capability on a horizontal level. Another company may offer the rim as its own potential contribution to a bicycle consisting of the specific input. This kind of vertical contribution is called enrichment of the product scale. Figure 6 shows the difference between horizontal enlargement enabling alternatives and vertical enrichment completing the product structure.

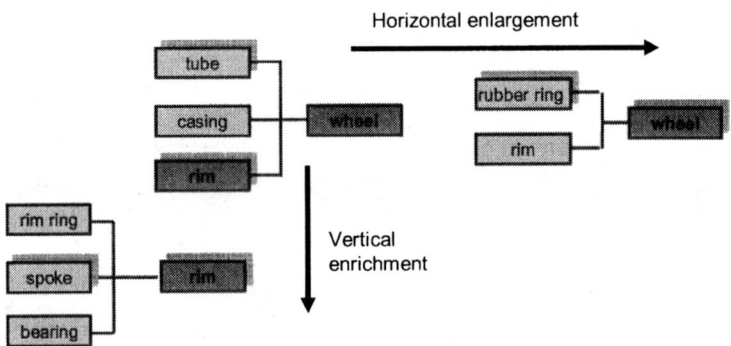

Fig. 6. Horizontal enlargement and vertical enrichment of the product scale

All sub-systems and components of all potential partners have to be provided in this way to be stored in a database. The components are stored within the database independent from their further usage and independent from a specific end-product. In the next chapter, it is described how these sub-systems are used to derive complex product designs

3.2   Iterative generation of possible product structures and derivation of potential partnerships

The generation of possible product structures on the basis of the available company profiles takes its starting point from the desired end-product. By searching in the company profiles, the first vertical level of product contributions is collected and added to the product structure. Different variants of certain product contributions as described in figure 6 may lead to different potential designs. For each product contribution, the necessary inputs are searched within the available company profiles. Step by step, the alternative product designs can be evaluated and completed step by step. Figure 7 shows the mechanism of the iterative completion with a simple example.

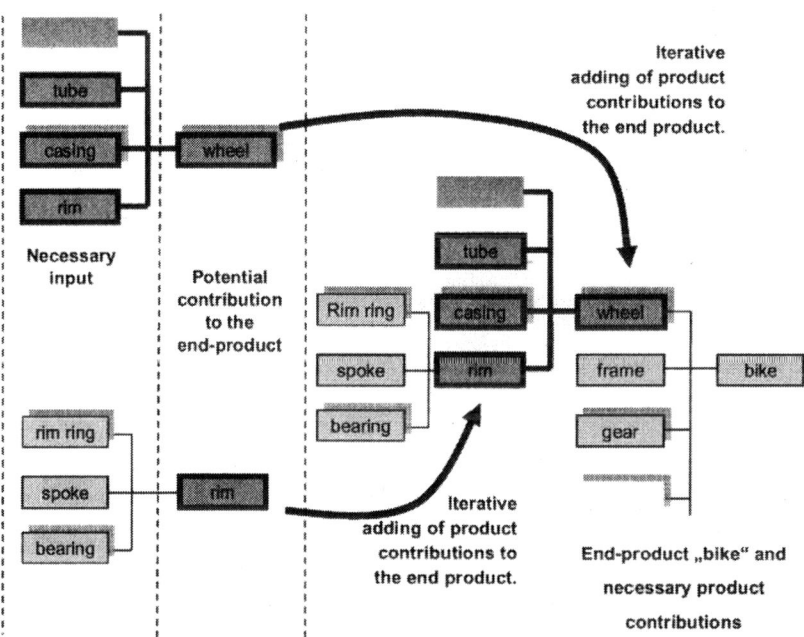

Fig. 7. Iterative generation of a product structure by adding possible product contributions

The described iteration loop is executed as long as there are inputs defined for a product contribution. The result is a multi-dimensional tree structure containing all potential product structures provided by the gathered company profiles. On the basis of the performance indicators, it is now possible to evaluate the potential consortia and to select promising partners to realise the requested end-product.

This tree supports Enterprise Networks not only to identify alternative product designs and to evaluate partnerships. In the case that needed product contributions can't be covered by the capabilities of the network, these gaps are the impact to concretize the strategic competence development of the co-operation willing partners. Missing competencies and resources can be identified already during the early conceptual phase and suitable additional partners can be searched immediately. This improves the reactivity of the consortium and ensures the permanent offering ability.

4   Conclusion

The presented method demonstrated how different product designs can be evaluated in the conceptual phase of the product life cycle and how these alternatives can support the early involvement of promising partners in the product specification. This approach differs from many approaches in practice which require a concrete product design as starting point for the consortium building. The knowledge about potential product variants and available alternatives in the partner selection is a crucial asset to be able to compensate a sudden loss of delivery of a certain partner as fast as possible. The preparedness to generate alternative designs and to involve alterative partners is very important for engineer-to-order products where very late design changes may become relevant in case of changing customer wishes or in case of unpredictable technical problems or malfunctions.

The participation of potential partners in the specification phase and the methodological identification of beneficial partnerships enable the consideration of the available knowledge and experience of these potential partners which is the basis for the development of innovative, competitive products and successful offers. Companies having the chance to participate already in the collaborative product design are in a better position to prepare themselves for the order fulfillment which improves the ability of the Enterprise Network to place offers. Thee ability to provide excellent solutions for customer specific demands and to decline leading offers are crucial for the success of an Enterprise Network - because the production phase and its optimization which is an aspect addressed by many research works, can only take place after a successful and accepted offer after the contract negotiation with the customer.

5   References

Boutellier, Roman: Konkurrenz der Logistikketten, in Logistik Heute, Ausgabe Mai 1999

Höbig, Michael: Modellgestützte Bewertung der Kooperationsfähigkeit produzierender Unternehmen, Hannover 2002

Kemmner, Götz-Andreas, Gillessen, Andreas: Virtuelle Unternehmen – Ein Leitfaden zum Aufbau und zur Organisation einer mittelständischen Unternehmenskooperation, Heidelberg 1999

Kramer, Peter: Die Virtualisierung der Unternehmung: Prozesse, Strukturen und Instrumente eines "grenzenlosen" strategischen Konzeptes, Basel 1998

Linde, Frank: Virtualisierung von Unternehmen - Wettbewerbspolitische Implikationen, Gabler Verlag, Wiesbaden 1997.

Mertins, Peter; Faisst, Wolfgang: Virtuelle Unternehmen – eine Organisationsstruktur für die Zukunft?, in: technologie+management, Nr.2, 1995

Seidl, Jörg: Business Process Performance-Modellbezogene Beurteilung und Ansätze zur Optimierung, in: HMD Praxis der Wirtschaftsinformatik, Heft 227, S.27-35, Oktober 2002

Seifert, Marcus: Unterstützung der Konsortialbildung in Virtuellen Organisationen durch prospektives Performance Measurement, Bremen 2007

Sydow, Jörg: Strategische Netzwerke: Evolution und Organisation, Gabler Verlag, Wiesbaden 2002

Zahn, Erich: Wachstumspotenziale kleiner und mittlerer Dienstleister: Mit Dienstleistungsnetzwerken zu Full-Service Leistungen, Stuttgart 2001

Design Quality: A Key Factor to Improve the Product Quality in International Production Networks

Zhu Yanmei[1], Alard R.[2] and Schoensleben P.[3]

1 Tongji University, Academy of Science & Technology Management,
No.1239 Siping Road, 200092 Shanghai, China,
Tel:+86 21 6598 3690, zhu.yanmei@163.com
WWW home page: http://sem.tongji.edu.cn
2 ETH Zurich, Center for Enterprise Sciences (BWI),
Kreuzplatz 5, 8092 Zurich, Switzerland,
Tel:+41 44 632 0532, ralard@ethz.ch,
WWW home page: http://www.ethz.ch/
3 ETH Zurich, Center for Enterprise Sciences (BWI),
Kreuzplatz 5, 8092 Zurich, Switzerland,
Tel:+41 44 632 0510, pschoensleben@ethz.ch ,
WWW home page: http://www.ethz.ch/

Abstract

As the result of the broken-up value chain of the world, design becomes an independent commodity. Goods often are designed in one company and produced in another company. Although product is shaped in manufacturing companies of supply chain, design quality is the key sticking-point for product quality. Therefore, finished goods quality lies on the quality of Design-Manufacturing Chain. This paper defines Design-Manufacturing Chain Quality Management (D-MCQM), analyses poor quality of design and manufacturing and the controllable factor and noise factor for quality stability. Furthermore, the figure and formula are presented for the leverage relationship among design quality, manufacturing quality and product quality.

Keywords

Design-Manufacturing Chain, Quality Management, Design Quality, Poor Quality, Leverage Relationship

Please use the following format when citing this chapter:

Zhu, Y. M., Alard, R., Schoensleben, P., 2007, in IFIP International Federation for Information Processing, Volume 246, Advances in Production Management Systems, eds. Olhager, J., Persson, F., (Boston: Springer), pp. 133-141.

1   Introduction

With drastic competition from all over the world, the globalization of manufacturing industry , technology complexity increasing, and a strong customer awareness of quality, quality management should be carried out in supply-chain-wide, instead of company-wide. It is indispensable for companies to make the best of resources outside and cooperate with their partners on the supply chain in order to enhance their end-product quality [1]. The cooperation need between the design department and the production department within one company is a well-known topic in practice and research. Nowadays, as the result of the broken-up value chain of the world, technology also becomes an independent commodity [2]. Goods often are designed in one company and produced in another one. Quality management in design and manufacturing phases are not often implemented in a same company, but from the perspective of supply chain. Product quality lies on design and manufacture processes, therefore it is quite significant to study on how to improve product quality by the cooperation between design and manufacture companies from both theoretical and practical sides. This is one of the objectives of the project, Design Chain-Supply Chain-Management (DC-SC-M), focusing the issues between Swiss designers and Chinese suppliers. To be successful in highly competitive global marketplaces where the quality of the designed product is a killer criterion, the designer and manufacturer should improve the end-product quality cooperatively.

2   Design-Manufacturing Chain (D-MC)

From the market demand emerging to finished goods delivery to the end consumer, product goes through several phases from design, manufacturing, transportation, distribution and so on. There into, design phase includes product plan, concept design, detailed design (design specification) and revision, meanwhile manufacturing includes prototyping, test, production plan and full-scale production. Design-Manufacturing Chain (D-MC) [3,4] is a chain or network made up of design and manufacturing companies, in which the final product is designed and produced within different companies. The simplest form of D-MC consists of only two companies: the designer / designing company (e.g. Original Equipment Manufacturer, OEM) and the manufacturer (manufacturing company).

Manufacturing companies possess know-how on manufacture and assembly which designers usually do not know very well. Therefore, product time-to-market reduction and substantial cost saving from higher productivity, lower maintenance and fewer recalls are the results of Manufacturer Earlier Involvement (MEI) in product design and development stages. Essentially, these are based on the philosophy of Concurrent Engineering (CE) [5]. Parts of design and manufacturing process concurrently, rather than sequentially like before. Manufacturing phase can start prototyping and tooling from detailed design phase, not until whole design phase completely finishes (Fig. 1).

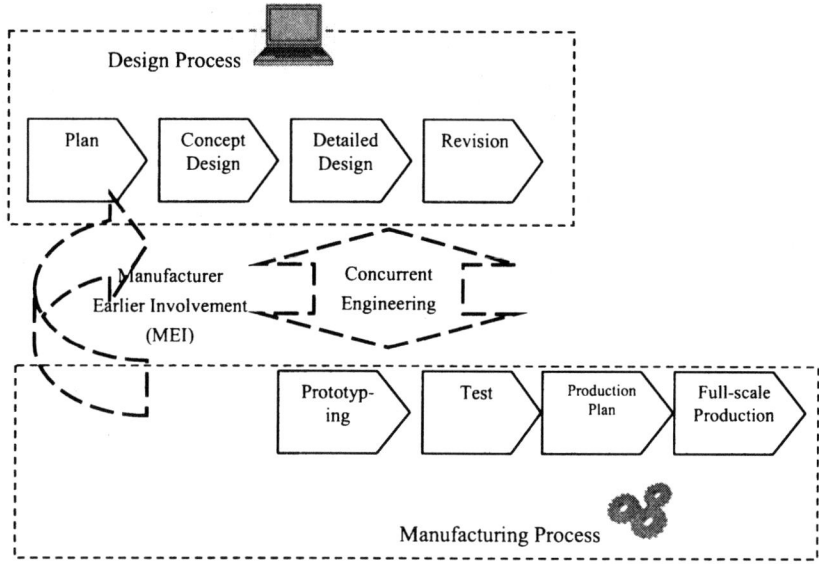

Fig. 1. Design-Manufacturing Chain (D-MC)

Previous research shows that, while product design may account for only 5 per cent of the product cost, it has a key influence on 75 per cent or more of manufacturing costs, and 80 per cent of quality (Huthwaite B.,1988) [6] . Some have asserted that more than 40 per cent of quality problem stems from design aspect of a product (Leonard F.W. et al, 1982) [7]. Moreover, the product design phase drives 70 to 80 per cent of the final production cost, 70 per cent of life cycle cost of product, and 80 per cent of product quality (Dowlatshahi S., 1992) [8].

Thereby, product design, where drives product "innate" quality, is the key to form product quality, and design phase in D-MC is the most important and potential phase to enhance quality and reduce cost (Fig.2).

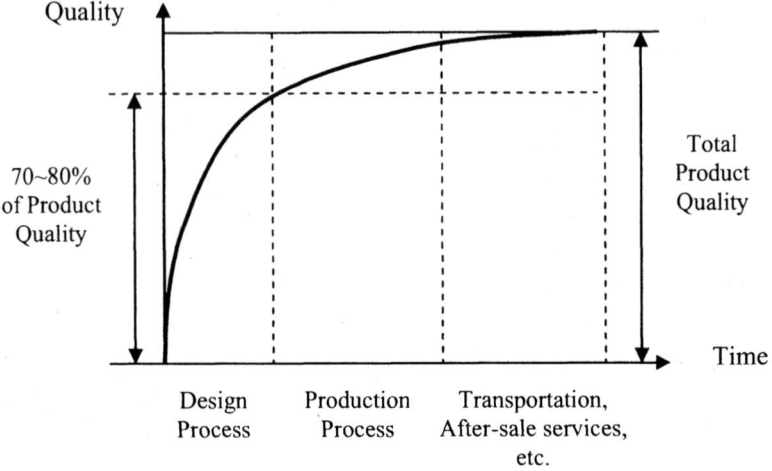

Fig. 2. Incline of Quality

The philosophy of D-MC quality management is controlling product quality from root and focusing on MEI in product design and development, which may accelerate the innovation in manufacturing industry and optimize product quality. D-MC quality management is also based on win-win relationship of supply chain partners. Partners on D-MC think much of quality information/resource sharing about manufacturing, test and so on rather than bargaining.

3   Design-Manufacturing Chain Quality (D-MCQ)

The definition of quality (Crosby, 1986) [9] is conformance to requirement. In design process, quality means that design specification should conform to the requirement of customers, and in manufacturing process, quality means that manufacturing should conform to the design requirement.

Although product quality is related to the level of delivery, after-sale service, maintenance, recyclability, etc, it is mainly shaped in design and manufacturing processes, i.e., product quality depends on D-MC quality.

Based on literature review, desk research and expert interviews involving partners from the design, production and logistics departement in a research project with eight Swiss industrial companies cooperating on an international level, we will demonstrate what the key impact factors for the final product quality are. According to our results these are mainly two factors influencing the final product quality: the design phase of the product (design process) and the manufacturing process itself. Therefore, finished goods quality lies on the quality of Design-Manufacturing Chain. Design-Manufacturing Chain Quality (D-MCQ) includes two parts, design quality and manufacturing quality. Design quality defines that design requirement may reflect the Voice of Customer (VoC) and demand of market. Manufacturing quality means that end-product conforms to the product design requirement and specification, where it is the conformance of quality. If design does not reflect the market requirement, the product can not meet the demand of market even though manufacturing conforms to the design completely, whereas if manufacturing does not conform to the design requirement, the finished product with poor quality can not satisfy customer's needs.

Design-Manufacturing Chain Quality Management (D-MCQM) indicates supervision and controlling the quality of all activities on D-MC. D-MCQM can be depicted by three simple formulae as followed:

Design-Manufacturing Chain (D-MC) = Chain or network made up of design and manufacturing companies and processes ;

Quality(Q) = Conformance to requirement ;

Management(M) = Activities for improving the design and manufacturing quality.

Poor quality of D-MC includes poor quality of design and poor quality of manufacturing. Poor quality of design means that design requirement has not reflected the demand of customer in right way and/or at right cost and/or at right time. Poor quality of manufacturing means that, manufacturing has not completely conformed to the design requirements/specifications so that the finished product can not satisfy market demand at right cost and right time. Design with technological deficiency resembles inferior product, design changed like the product recalled or returned to production line to be worked again. Apparently, all those are of poor quality.

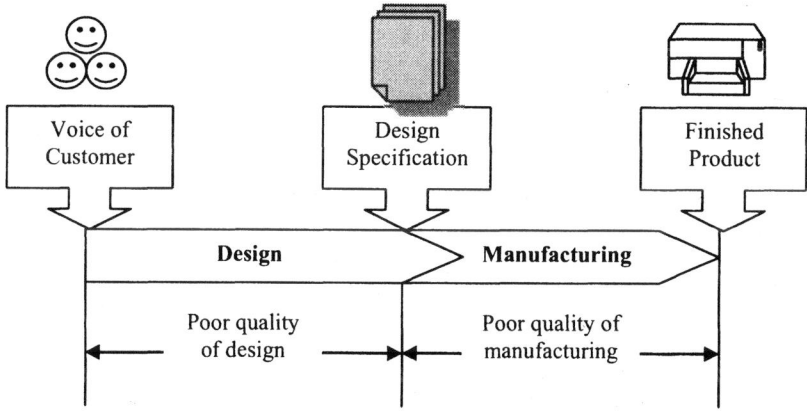

■ **Fig. 3.** Poor Quality of D-MC

Technological deficiency in design, which is "innate" deficiency of product quality, may result in huge quality cost of numerous quality-related maintenance, warranty repair and severe exterior loss [10]. Controlling manufacturing quality usually can not solve the problems which are rooted in design deficiency. Hereby, design quality is decisive to product quality, so in order to gain more customer value it is crucial to manage quality starting from design process instead of focusing on manufacturing process only.

An overwhelming majority of product failure costs and design iterations come from the ignorance of noise factors during the early design stage. The noise factors which crop up one by one in the subsequent product delivery stages cause costly failures. Taguchi Method (TM) presented by Dr. Taguchi [11] may help designer to select appropriate controllable factors so that the deviation from the ideal is minimized at a low cost. Variation reduction is universally recognized as a key to quality reliability and improvement in D-MC.

The number of controllable factors and noise factors for quality reliability are changed along upstream (design process) and downstream (manufacturing process) of D-MC (Fig.4). Noise factors are uncontrollable in natural condition of use. The noise factors can be divided into three groups: production variations, variations in condition of use, and deterioration. Production variations can be exemplified by change of operators or variations in incoming materials. Furthermore, variations in condition of use can be temperature, air pressure, or air humidity and deterioration by wear or aging of material. From upstream to downstream, controllable factors decreases gradually, but noise factors increases. Accordingly, quality control from root is more efficient than downstream.

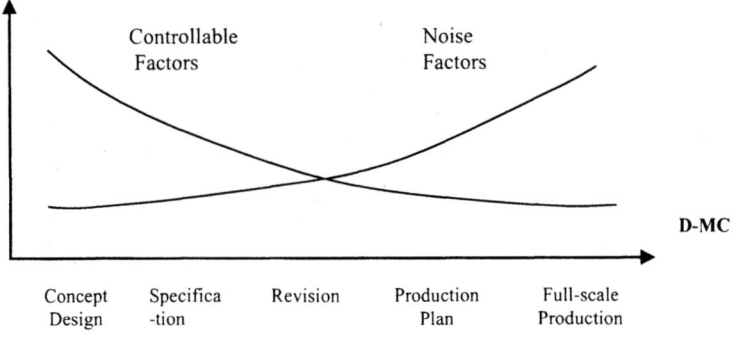

Fig. 4. Controllable Factors and Noise Factors of Quality Stability in D-MC

4   Leverage Relationship among Design Quality, Manufacturing Quality and Product Quality

Therefore, to product quality, influence from design is much bigger than from manufacturing. This relationship can be showed with a figure of leverage (Fig.5).

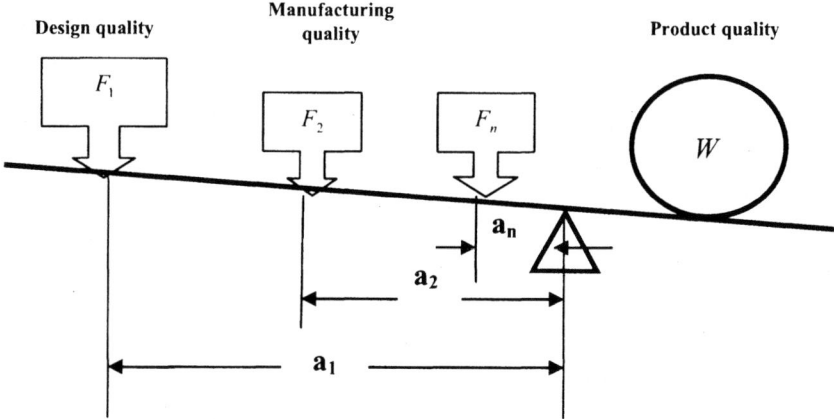

Fig.5. Leverage Relationship among Design Quality, Manufacturing Quality and Product Quality

The followed formula expresses the leverage relationship among design quality, manufacturing quality and final product quality in Fig. 5.

$$W = F1.a1 + F2.a2 + \ldots\ldots + Fn.an$$

Here, W denotes of final product quality. The bigger W , the better final product quality;

F_1 denotes design quality;

F_2 denotes manufacturing quality ;

$F_3 \cdots F_n$ denote respectively the quality of every other process, e.g. delivery, after-sale service, etc.

a_1 denotes the extent of influence on final product quality from design quality. According to Fig.2, the extent of influence from design quality is 70 ~80 per cent, then $0.70 < a_1 < 0.80$;

a_2 denotes the extent of influence on final product quality from manufacturing quality. If the extent of influence from design quality is $0.70 < a_1 < 0.80$, then $0 < a_2 < 0.30$;

Bigger numerical value of a_1 、 a_2 shows stronger the influence of the process to final product quality.

Surely, other processes quality except design and manufacturing, $F_3 \cdots F_n$, also influence product quality. Their influences are $a_3 \cdots a_n$, respectively. If $F_1 \cdots F_n$ are all of the processes influencing product quality, then $a_1 + a_2 + \cdots + a_n = 1$.

$F_1 \cdot a_1$ and $F_2 \cdot a_2$ respectively express the influence from design quality and manufacturing quality to product quality.

Analysis of the formula shows some ways to enhance product quality:

1. Improving design quality (F_1).

① It is imperative that marketing is included as an integral team of design. They are responsible in gather accurate customer requirements for the new product development in order to ensure the delivered product meets the customers' expectations.

② A collection of tools for design engineering (e.g. QFD, DFMA, Tolerance Analysis, Robust Design, DOE, etc.) should be implemented in product development process in order to avoid huge cost of re-design after product has launched.

③ To avoid the expensive re-design processes and improve the manufacturability of the design plan, it is necessary to establish a cross-functional team to facilitate Manufacturer Earlier Involvement (MEI) and/or Supplier Early Involvement (SEI) to initiate the product research and development. The cross-functional team is made up of representatives from such groups as engineering, manufacturing, quality, and supply chain management. Each team member will contribute to specific areas within the product development process.

2. Improving manufacturing quality (F_2).

① To enhance the capability to control quality and strengthen manufacturing process to avoid reworking, manufacturers should adopt Advanced Manufacturing Technologies (AMT) such as robotics and Computer-Aided Manufacturing (CAM) and invest in implementing advanced computer-integrated manufacturing applications such as ERP and MRPII to link with other organizational systems.

□ To offer consistently high quality product and reduce unnecessary cost of rework, two kinds of environmental uncertainty should be paid more attention, one is based on source scarcity and the other based on information complexity.

5 Conclusion

D-MCQ includes two parts, design quality and manufacturing quality. Whereas product is shaped in manufacturing process, product quality stems from design process. In the upstream of D-MC (design process), quality is managed more efficiently due to more controllable factors and less noise factors. The figure of leverage and its formula show clearly the relationship among design duality, manufacturing quality and product quality. Moreover, they direct managers responsible of quality to turn their attention to design process, instead of keeping their eyes on manufacturing all the time. In a whole, to optimize product quality, it is crucial to successfully manage D-MCQ by collaboration of the partner companies on the whole D-MC.

This paper addresses especially industrial practitioners from design and production departements in supply chains. We show the importance of the interaction and organizational cooperation between the design departement and the production departement of different companies in supply chain, especially in international supply chains. The outcome of the paper can be used to sensitise the different actors / practitioners in the industrial companies towards the importance of the coordination between the different involved actors and the relationship between the product quality and the design process (design phase) based on a general and comprehensible model.

Reference

1. Kesheng Wang, George Kovacs, Michael Wozny and Minghun Fang, *Knowledge Enterprise: Intelligent Strategies in Product Design, Manufacturing, and Management*, SV, 2006.

2. Guo, Chongqing., Strategy for improving innovation and marketing capabilities of China, *Proceeding of China Mechanical Engineering Acad.*, 2006(1): 6-12.

3. Zhu Yanmei., Wu Xiaojun, and You Jianxin., Research on the Design - Manufacturing Chain Model Under the Globalization of Manufacturing Industry, *Development and Management*, Shanghai People Press, Shanghai, 2006: 86-93.

4. Zhu, Yanmei, and Alard, R., Shaping the Design-Manufacturing Interface between Swiss Design Department and Chinese Manufacturer, *Journal of Tongji University (Natural Science)*, 2005(9):10-15.

5. Schoensleben, P., *Integral Logistics Management—Planning and Control of Comprehensive Supply Chains (Second Edition)*, St. Lucie Press, 2003:258-259.

6. Huthwaite, B., Designing in Quality, *Quality*, 1988, Vol.27 No. 11, 34-50.

7. Leonard, F.W. and Sasser, W.E., The Incline of Quality, *Harvard Business Review*, 1982(60):163-171.

8. Dowlatshahi, S., Purchasing's role in a concurrent engineering environment, *International Jounal of Purchasing and Material Management*, 1992, Winter, pp21-25.

9. Philip B. Crosby, *Quality Without Tear: The Art of Hassle-Free Management*, McGraw-Hill, New York, 1984:58-86.

10. You Jianxin and Guo Chongqing, *Quality Cost Management*, Pertroleum Industrial Press. 2003.

11. Taguchi, G., Elsayed, E., and Hsiang, T., *Quality Engineering in Production Systems*, McGraw-Hill, New York, NY,1989.

Postponement Based on the Positioning of the Differentiation and Decoupling Points

Joakim Wikner[1] and Hartanto Wong[2]
1 Jönköping University, School of Engineering,
Jönköping, S-551 11, Sweden, joakim.wikner@jth.hj.se
2 Cardiff University, Innovative Manufacturing Research Centre
Colum Drive, Cardiff CF10 3EU, Wales – UK, wongh@Cardiff.ac.uk

Abstract

Structural analysis of supply chains often involves postponement to some degree. The concept has received a lot of attention in the literature but the understanding of how to operationalize the concept still deserves some attention. A framework is introduced that provides an integrative picture of how different aspects of supply chain and operations management interplay in a postponement context. The framework is founded on a process/object perspective where a set of characteristics, properties and concepts are identified for each entity.

1 Introduction

Postponement or delayed product differentiation is an important concept used to accommodate mass customization that has been increasingly receiving attention from researchers and practitioners. It is an effective way to manage the risks associated with proliferating product variety without incurring large operating costs. Although the concept has been around for years, our literature review suggests that there is still a lack of understanding of how the concept is operationalized. Several authors have introduced different conceptual categorization of postponement strategies extending the understanding of where and when postponement is appropriate (see e.g. [1] among others for a comprehensive literature review of postponement).

Despite their differences, most conceptual classifications found in the literature share a commonality in that they all refer to postponement as the delaying of certain operations until customer orders are received. Such a concept suggests that the final

Please use the following format when citing this chapter:

Wikner, J., Wong, H., 2007, in IFIP International Federation for Information Processing, Volume 246, Advances in Production Management Systems, eds. Olhager, J., Persson, F., (Boston: Springer), pp. 143-150.

differentiation process needs to be performed in a make-to-order (MTO) fashion. Ideally, this concept would maximize the profits of postponement as it omits the inventory of the final products. However, it is obvious that in reality it may not always be possible to employ such a postponement strategy especially in the highly responsive environments where the tolerance time that the customer is willing to wait is quite short. In such environments it may be necessary to produce the final products in a make-to-stock (MTS) fashion. From that perspective, we argue that the concept referring to postponement as the delaying of activities until customer orders are received does not always represent the best course of action. This motivated us to undertake a study looking at a more complete set of manufacturing configurations related to the implementation of postponement strategy.

2   Key entities in postponement analysis

Postponement plays an important role in operations and supply chain management as it captures both the issue of where in the supply differentiation takes place and where decoupling points play a strategic role. As a foundation for investigating these different perspectives we have adopted a process/object perspective when developing this framework. In addition to the process we have highlighted three different types of objects of which two are discussed further in the sub-sections below. In summary, the framework focuses on:
- Customer objects, formulating the requirements.
- Product objects, having a level of customization to fulfil customer requirements.
- Resource objects, providing a level of flexibility to enable customization.
- Processes, embracing the level of uncertainty by exploiting resource flexibility to fulfil customer requirements.

2.1   Product object modelling

It can be argued that the increasing pressure to become more and more customer-centric has forced manufacturing firms to formulate their marketing and operations strategies around mass customization [2]. Mass customization has been defined as the technologies and systems to deliver goods that meet individual customers' needs with near mass production efficiency. A quintessential feature of mass customization is the proliferation of product variety. The prevailing view is that offering larger variety allows firms to increase both demand and market share [3]. However, it has also been recognized that a large number of product variety is associated with diseconomies of scale and increases in production and distribution costs. *Postponement*, also termed as *delayed product differentiation* or *late customization* represents a way to implement mass customization without incurring large operating costs associated with managing proliferating product variety. This is done by properly designing the product structure and the manufacturing and supply chain process so that one can delay the point in which the final customization of the product is to be configured. Fig. 2 and Fig. 3 contrast two different configurations that are different from each other in terms of the position of the differentiation point.

The first configuration (Fig. 1) is the single-stage system where end product configurations (PCs) are processed and customized through a single-stage production originating in raw materials (RM). The second system employing postponement consists of two stages. Stage 1 produces the generic components and Stage 2 differentiates the final product configurations.

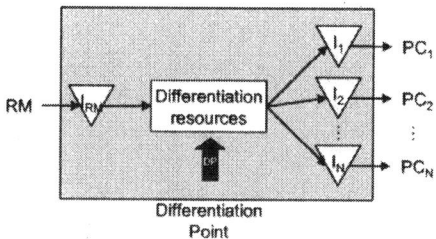

Fig. 1. Single-stage early differentiation

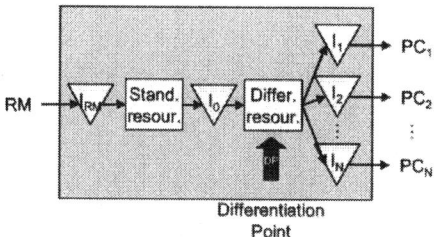

Fig. 2. Two-stage delayed differentiation

2.2   Resource object modelling

Our focus of the resource object modelling is on the flexibility, which is an important resource characteristic when dealing with the proliferation of product variety. Flexible manufacturing systems (FMS) allow a firm to cope better with demand uncertainty when manufacturing several products in the same facility. This also relates to the well-known result in the queuing systems literature specifying that employing a flexible resource (Fig. 4) instead of several dedicated resources (Fig. 3) may offer so called pooling benefits resulting in improved system's performances such as reduced expected waiting times in the queue. These benefits, however, could be distorted by the presence of non-negligible changeover times. Diseconomies of scope associated with the increase of product variety are particularly true for manufacturing environments with significant changeover times. In the presence of significant changeover times products are usually manufactured in batches to achieve economies of scale, which would in turn lengthen the manufacturing lead times and consequently increase the inventory and backorder levels. This unfavorable effect will be more pronounced whenever the system produces more product configurations.

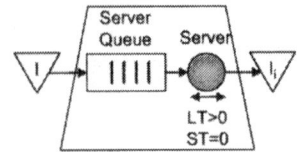

Fig. 3. Dedicated resource

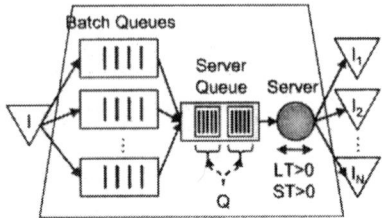

Fig. 4. Flexible resource

Four different configurations, as shown in Fig. 5, emanate from combining product and resource objects.

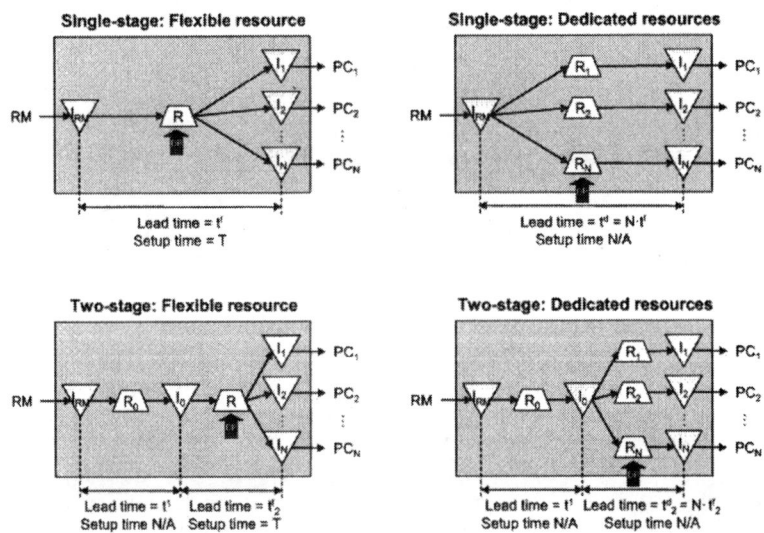

Fig. 5. Product-resource combinations

It is worth noting that from a system perspective, common material (generic component) used in the configurations employing postponement (the two-stage configurations) can also be seen as an analogy to a flexible resource with no changeover time being required when switching production from one product to another. The main difference is that the resources stays in that state during production of a batch and then is set up for a new assignment whereas the common material enters a product and then usually stays in that state during the rest of its life cycle. But when deciding on how to use the object the time from decision to operation does not require any delay for set up as the resource/material is flexible.

2.3   Process modeling

Fisher [4] proposed a model for the matching of product characteristics and supply chain design. A supply chain has two generic functions. First, it has a physical function of transforming physical goods promoting an efficient flow of products with high resource utilisation. Second, a supply chain has a market mediating function of conveying information from the point of sales to upstream of the supply chain, promoting quick response to variations in customer demand and a high level of flexibility through the supply chain. The focus of the supply chain in terms of physical efficiency and market mediation will affect the way the supply chain is designed. In relation to this, we highlight an important concept that relates to the position of the buffer usually referred to as a decoupling point.

Two decoupling points are of particular interest [6]. The first is the physical supply decoupling point (PSDP) also called the customer order decoupling point (CODP) or order penetration point (OPP) [5]. At a more detailed level the process can be characterized as supporting different product delivery strategies based on the position of the CODP. The second decoupling point is called the demand mediation decoupling point (DMDP), which is not related to a buffer in the conventional physical sense but to a decoupling point from an information perspective. The DMDP simply indicates where demand information is available but provides no direction for how the information is supposed to be used in the physical supply of products [6]. Combining the two decoupling points result in a decoupling point framework illustrated in Fig. 6. In Fig. 7 we show the demand mediation process for the flexible and dedicated resources.

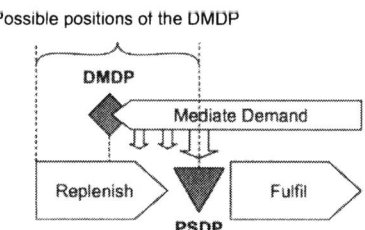

Fig. 6. Process configuration framework [6].

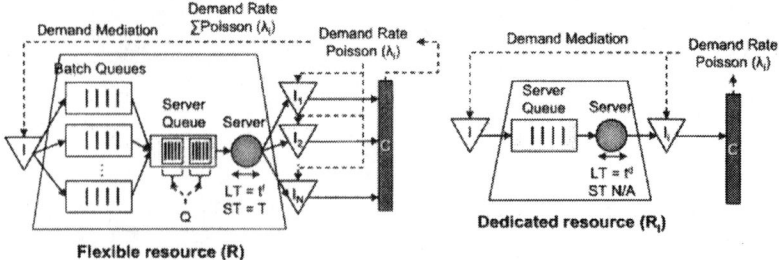

Fig. 7. Demand mediation and resources.

A more complex approach would allow the possibility to produce some of the product configurations to stock and others to order resulting in a wide range of possible system configurations. Some examples of possible configurations are shown in Fig. 8 for the case of two-stage flow with flexible resource at stage 2. Since there are three possible positions of the CODP/OPP decoupling buffers (at OP in Fig. 8) there are three possible configurations. In total there are 10 possible configurations considering the differentiation point, the type of resource configuration and the positioning of the CODP/OPP (see Table 1).

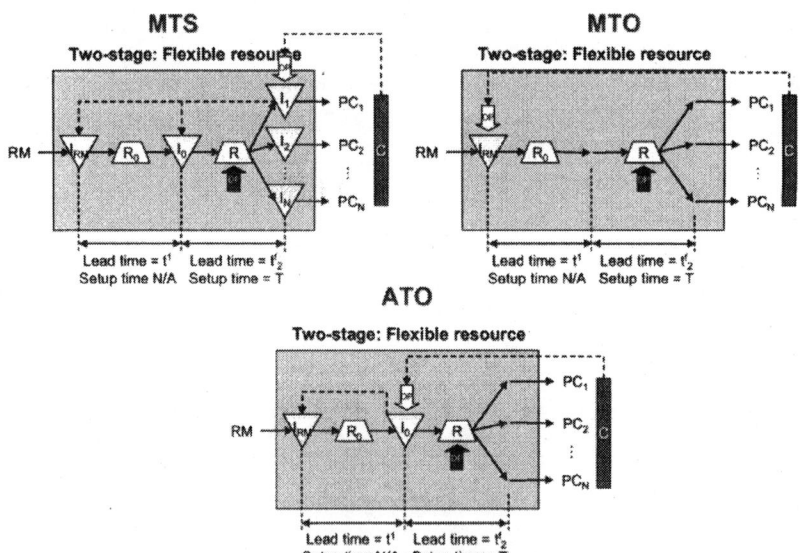

Fig. 8. Example of system configurations with different OPP/CODP

Table 1. Number of possible positions of the CODP/OPP

	Single-stage flow	Two-stage flow
Flexible resource	2	3
Dedicated resource	2	3

3   Some findings

In [7-8] we developed analytical models and carried out numerical experiments for making the evaluation of different system configurations. Some important findings are highlighted in the following.

- Delaying product differentiation in postponement does not necessarily suggest that the differentiation is only processed after customer orders are received. Different system parameters have an impact on the choice of optimal system configuration.
- Our numerical results confirm that the diseconomies of scope associated with the increase of product variety occur even though the total volume of demand does not change. The total costs increase very rapidly as the changeovers take longer. It becomes obvious that the marketing strategy, which is usually in favour of offering more product variants to achieve a higher degree of customization must be aligned with the manufacturing strategy in order to minimize costs or maximize profits.
- Sharing a common material as allowed by postponing product differentiation i.e. moving from a single-stage system to a two-stage system, is always beneficial while sharing a common production resource may be beneficial but may also be disadvantageous depending on the changeover time incurred. It is however important to note that there is usually a premium cost associated with redesigning the product to accommodate material commonality that needs to be balanced against the benefits of postponement.
- The presence of significant changeover times associated with the use of a flexible resource reduces the possibility to employ the most desirable system configuration where the CODP/OPP is positioned far enough upstream allowing final products are made with certain demand. This observation suggests that continuous efforts in reducing changeover times are essential for maximizing the benefits of postponement.

4   Conclusion

Postponement is an important concept in operations and supply chain management as it highlights the strategic importance of the positioning of differentiation and decoupling points. These aspects are however not usually discussed from a modelling perspective highlighting the key constructs used in process modelling. In this paper we have introduced a postponement framework based on products, resources and processes. The framework highlights and positions a set of characteristics, properties and concepts in response to customer requirements for different levels of customization, as summarized below.

- Product Objects
  - Key product characteristic: Level of customization
  - Key customization property: Number of variants
  - Important concepts: Differentiation point and commonality
- Resource Objects
  - Key resource characteristic: Level of flexibility
  - Key flexibility property: Volume and mix flexibility
  - Important concepts: Changeover time and capacity
- Process
  - Key process characteristic: Level of uncertainty
  - Key uncertainty property: Demand uncertainty in process driver
  - Important concepts: Physical supply decoupling point (CODP/OPP) and Demand mediation decoupling point

In summary, these three entities and their respective attributes in terms of characteristics, properties and concepts provides a framework that supports the operationalization of postponement by enabling an integrative perspective of postponement from an object/process perspective.

References

1. R.I. Van Hoek, The rediscovery of postponement: a literature review and directions for research, *Journal of Operations Management* 19, 161-184 (2001).
2. M. Agrawal, T.V. Kumaresh, and G.A. Mercer, The false promise of mass customization, *The McKinsey Quarterly* 3, 62-71 (2001).
3. P. Kotler, *Marketing Management* (Prentice Hall, 2002).
4. M. Fisher, What is the right supply chain for your product?, *Harvard Business Review* 75(2), 105-116 (1997).
5. G. Sharman, The rediscovery of logistics, *Harvard Business Review* 62(5), 71-80 (1984).
6. J. Olhager, E. Seldin, and J. Wikner, Decoupling the value chain, *International Journal of Value Chain Management* 1(1), 19-32 (2006).
7. H. Wong, J. Wikner, and M. Naim, Analysis of form postponement based on optimal positioning of the differentiation point and stocking decisions, Working Paper, Cardiff Business School, 2006 (unpublished).
8. H. Wong, J. Wikner, and M. Naim, Evaluation of postponement in manufacturing environments with non-negligible changeover times, Working Paper, Cardiff Business School, 2007 (unpublished).

PART III

IS/IT Applications in the Value Chain

Closed-loop PLM of Household Appliances: An Industrial Approach

Jacopo Cassina[1], Maurizio Tomasella[1], Andrea Matta[1], Marco Taisch[1]
and Giovanni Felicetti[2]
1 Politecnico di Milano
Piazza L. da Vinci,
32 I-20133, Milano
[jacopo.cassina, maurizio.tomasella, andrea.matta, marco.taisch]@polimi.it
WWW home page: http://www.polimi.it
2 Indesit, Via Lamberto Corsi 55 60044 Fabriano (AN) ITALY
Giovanni.Felicetti@indesitcompany.com
WWW home page: http://www.indesitcompany.com

Abstract

This paper will present how the technologies and specifically the PLM data management semantic object model developed within the PROMISE project have been successfully applied to a concrete business scenario in the household sector, allowing the development of new services and functionalities to the customers, creating a real life demonstrator of closed-loop PLM.

1 Introduction

In the context of the globally scaled scenario, the product and its related management has unavoidably become a key-aspect. The customers are becoming more and more demanding, asking for better products, "extended" with related services.

Even if the services can be outside the core business of manufacturers, and can be provided by different enterprises, the product's value perceived by the customer is a sum of both the physical product itself and its correlated services. Among the main existing approaches *PLM (Product Lifecycle Management)* and *Product Extensions* are considered the most promising. The former has in particular emerged as an enterprise solution and implies that all software systems/methods/tools, such as CAD (Computer-Aided Design), PDM (Product Data Management), CRM (Customer Relationship Management), ERP (Enterprise Resource Planning), etc., used by the various departments throughout the product lifecycle, have to be integrated, in a way that the information managed by these systems can be promptly and correctly shared among different people and application packages. Nevertheless, PLM is not

Please use the following format when citing this chapter:

Cassina, J., Tomasella, M., Matta, A., Taisch, M., Felicetti, G., 2007, in IFIP International Federation for Information Processing, Volume 246, Advances in Production Management Systems, eds. Olhager, J., Persson, F., (Boston: Springer), pp. 153-160.

primarily an IT (Information Technology) problem but, at first, it represents a strategic business orientation of the whole enterprise [Garetti 2004].

At the same time, the explosion of information technologies has created a new kind of concept, defined as Extended Product, where the product is more then a simple artefact, but it is a complex result of tangible and intangible components. The extension is usually related to the functionality or a new business process around the product. According to Jansson [Jansson 2003] and Hirsch [Hirsch 2001] tangible extended product can be intelligent, highly customized, and user-friendly; an intangible product is mostly the business process itself.

The paper will present how these possibilities have been addressed within the PROMISE project, creating a closed loop PLM that enables the creation of product extensions.

2   The PROMISE approach to closed-loop PLM

The PROMISE project's approach to closed-loop PLM aims at developing a new-generation product information tracking and data management system during all lifecycle, seen as divided into BoL (Beginning of Life), MoL (Middle of Life) and EoL (End of Life) . The project is developing appropriate technologies, including product lifecycle models, Product Embedded Information Devices (e.g. RFID systems and bar-code systems) with associated firmware and software components, and tools for decision making based on data gathered during a product's life. The aim is to enable and exploit the seamless flow, tracing and updating of information about a product, after its delivery to the customer and up to its final destiny (e.g. deregistration, decommissioning), and then back to the designer and producer.

The prototypical PROMISE closed-loop PLM system is being applied to ten application scenarios, covering the whole set of product lifecycle phases in the automotive, railway, heavy-load vehicles, EEE (Electronic and Electrical Equipment), instrumental and white goods sectors (in particular the one discussed in these pages). The PROMISE closed-loop PLM system is composed of many software and hardware systems and related infrastructures, widely explained in previous papers of the same authors [Cassina 2006]; some of them are explained in the following.

The PROMISE PDKM (Product Data and Knowledge Management) system is devoted to the management both of product data collected from the field, via smart product-embedded devices, and of the knowledge created and updated starting from this data, in order to enhance e.g. the design of new products in the future. The conceptual (semantic) data model behind the PDKM system (and behind its technical data schema as well) was develop using as a basis, an existing approach, based on the Holon concept [Morel 2003], and developed within Politecnico di Milano, with the support of the CRAN research lab [Terzi 2004].

The present paper highlights how this conceptual model was applied to a real industrial case in the white goods sector and how it is being tested to support predictive maintenance of refrigerators at INDESIT Group.

The PROMISE DSS (Decision Support System), which is part of the PDKM system, is devoted to the support of lifecycle decision making activities, thus providing the analytical basis to the PROMISE closed-loop PLM system.

A set of PEIDs (Product Embedded Information Devices), i.e. RFID (Radio Frequency IDentification) active and passive tags, bar-codes, sensors and on-board computers, with the related embedded and backend software systems, are finally the means by which field data is collected in the different product lifecycle phases.

3   The PROMISE approach becomes reality: predictive maintenance at INDESIT Group

The main focus of Indesit in PROMISE is to develop a test application in the Middle of Life (MOL) phase of the product lifecycle. It specifically deals with Predictive Maintenance services related to a domestic refrigerator, capable to generate data through a PEID and thorugh a so-called SA (Smart Adapter), and then to send such data to a remote Maintenance Centre. More precisely, the kinds of data involved in the application are generic statistical data and generic diagnostic data considering the PEID, and specific statistical data on energy consumption and on data related to the electric loads for the SA.

To implement the system solution enabling its new closed-loop PLM approach, INDESIT Group is currently developing, in the context of the PROMISE project, together with Politecnico di Milano and Helsinky University of Technology, demonstration software/hardware tools covering the different levels/components of the PROMISE system architecture (Figure 1).

The physical elements of the demonstration case are a refrigerator, indicated as DA (Digital Appliance), an interface device, indicated as SA (Smart Adapter), which is placed between the power cable of the household appliance and its electric plug (Outlet) and finally a wireless communication link (RF comm. system) between the SA device and the remote monitoring centre, where the Decision Support System (DSS) runs predictive maintenance algorithms, allowing both a long term diagnostic of the product, the management of maintenance missions and of spare parts.

The main goals of these tools are:

- To demonstrate the improvements brought by the project activities into the in-line testing process (BOL phase) of the Refrigerator, where the product is tested in order to check its proper functioning. During this phase, some of the measured parameters (electric load characteristic parameters) are stored into the memory device of the PEID, where they will be later read by the predictive maintenance algorithms as one of their fundamental inputs.
- To demonstrate the improvements brought by the project activities into the product installation process, where the Refrigerator and the Smart Adapter (SA) are installed by a technician in a domestic environment and automatically recognized by a home network controller (emulated by a local PC in the demonstrator) using UPnP (Universal Plug-and-Play) technology.

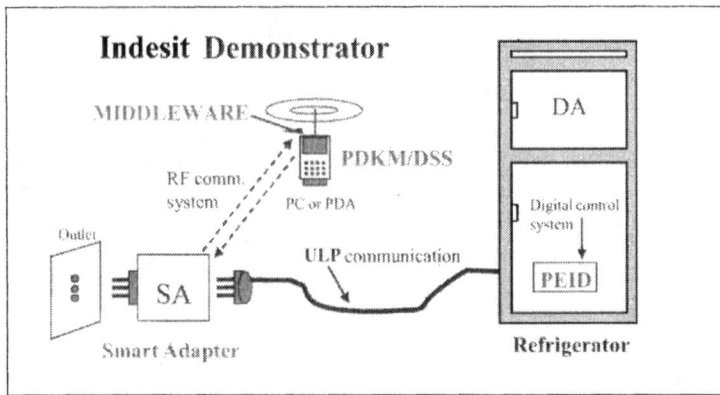

Fig. 1: Structure of the Indesit demonstrator in PROMISE

- To demonstrate the breakthrough innovation represented by the Predictive Maintenance operations, where field data, sent by means of the SA, are received through the PROMISE MIDDLEWARE (installed in a local PC in the demonstrator), stored in the PROMISE PDKM database, and processed by the PROMISE DSS (also both installed in a local PC in the demonstrator).

One of the most important results for Indesit is related to the development of a very powerful PEID included in the digital control system of the refrigerator, which adds the following new important features to a domestic Refrigerator:

- An auto diagnosis system, properly connected to sensors and actuators
- Very low cost communication capabilities using Indesit's ULP (Ultra-Low cost Power) technology
- Predictive maintenance capabilities at a very low cost

Another breakthrough innovation concerns the communication between the SA and the DA, provided by means of Indesit's ULP technology, whose key points are:

- the DA can exchange data with the SA at zero added-cost, since there is non need of using a conventional (expensive) communication node;
- the SA can virtually be interfaced to any communication node (using any communication protocol and any physical media), independently of the characteristics of the DA.
- Indesit can extend the use of its ULP technology to the whole set of product families, without affecting the industrial cost of its products.
- the SA would become a tool for maintenance operators, able to communicate with a local predictive maintenance system (portable PC or PDA), in order to perform additional on-site analyses.
- the SA would become a new product itself, marketed as an optional device for connecting to a home network any of the Indesit products (with ULP inside) already sold

Field data (statistical and diagnostic data) flow can be schematically described as follows. Field data, coming from sensors and actuators, are stored by the control system of the refrigerator into its non-volatile memory. These data are first sent to the SA (that provides data related to appliance energy consumption) and later to the PROMISE back-end, where they are stored into the database of the PROMISE PDKM, and then analyzed by the PROMISE DSS, in order to eventually find out malfunctioning problems on one or more of the refrigerator components. If an incipient failure is detected, an e-mail is sent to the Service Company which is so enabled to perform predictive maintenance actions on the refrigerator.

3.1 Management of the refrigerator's MOL structure

To enable IDESIT Group's approach to PLM, the focus must be shifted from information on product types to information on product items, virtually each product item of any given product type. The new approach requires the identification and tracing of each physical product entity, the access to all of the data available on it, in particular data collected from the field while the product is being operated/used, and finally on the use of this data by the decision support systems to be adopted to support decision makers in the value creation process. This is possible only if product items at the different levels of the product structure can be identified, and the related information can be properly collected and managed. It must be also possible to manage information on product structures related to both products "as-designed" (those typically managed by currently available PLM/PDM systems) and to physical products. This last type of information must carry within itself an always-up-to-date description of the identities of each component/assembly/subassembly presently part of the product. The problem of correctly identifying and tracing each item during its life must be also properly tackled.

Figure 2 shows the PROMISE solution to INDESIT Group's problems from this data management perspective. The figure presents an UML 2.0 object diagram showing an exemplificative instantiation of the classes of the PROMISE PDKM conceptual data model. The bold objects of the PHYSICAL_PRODUCT class represent the different components of a typical refrigerator which are involved in the implementation of a real-world predictive maintenance application, namely the fridge itself (as a whole), and its different components:

- the electronic control board, with the related sensors, i.e. the freezing temperature probe and the ambient temperature probe
- the smart adapter, fundamental enabler for communication with the predictive maintenance central platform
- the compressor electric motor
- the fan electric motor
- the set of resistors, whose status is fundamental for running the predictive maintenance algorithms, itself divided into the Drip resistor and the defrosting resistor
- the lamp

The objects of the ID_INFO class are there to keep record of the identifiers of the product and or the components, where necessary. Moreover, to keep record of the design data related to the product and its different components, permanent links to proper objects of the AS_DESIGNED class (not represented explicitly in the diagram)are present by means of attributes of the objects of the PHYSICAL_PRODUCT class.

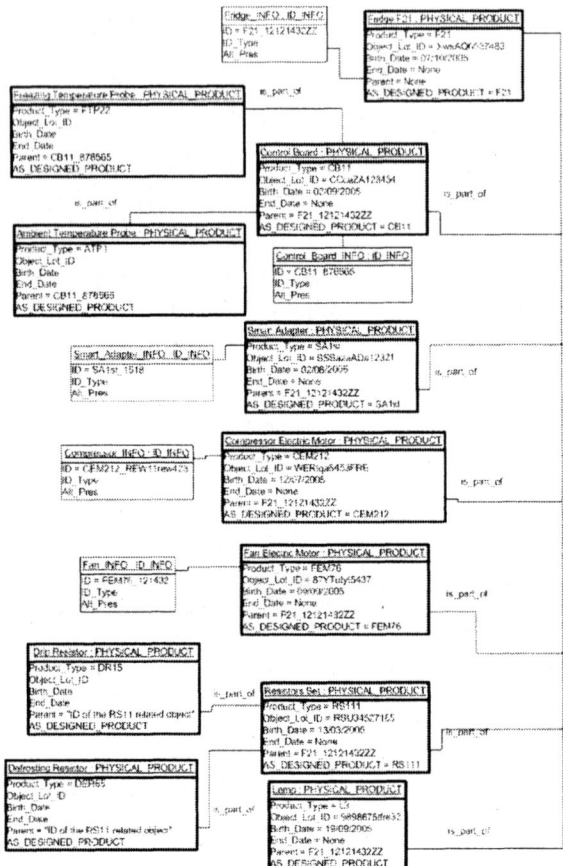

Fig. 2: MOL structure of a typical refrigerator

3.2. Closed-loop maintenance management of a refrigerator

Figure 3 represents an instantiation of objects related to a realistic case of maintenance of a typical refrigerator. Two components of the refrigerator are indicated, namely its compressor and its control board.

The upper portion of the figure indicates that the considered compressor (see the related PRODUCT_BOL_SUPPLY object) was reworked while being produced, and that the results of the final in-line tests are available in a specific file of the file system. More precisely, after the rework the compressor was certified to be correctly functioning and (see the LIFE_CYCLE_PHASE object attached to this PRODUCT_BOL_SUPPLY object) then its "Age" attribute was appropriately set to '0'. Though not explicitly shown in the figure, the same setting was done for all of the rest of the refrigerator's components, after passing the final in-line inspection, by changing the attributes of the related objects.

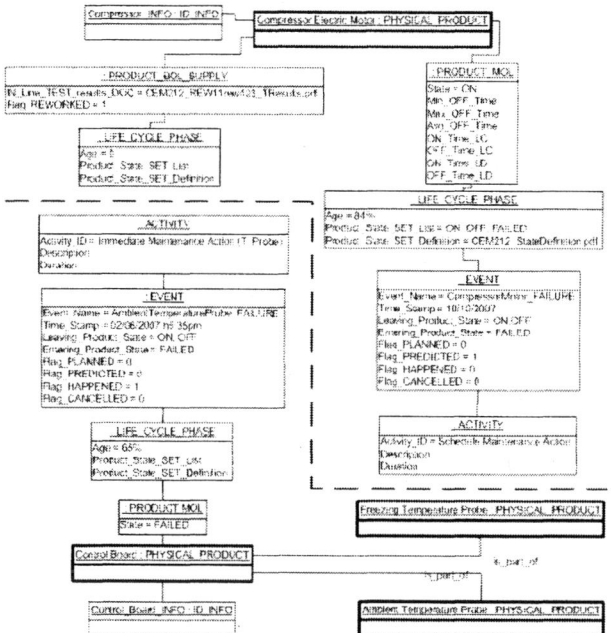

Fig.3: Supporting predictive maintenance

The rest of the objects regarding the compressor say that a failure was predicted to happen (EVENT object) around late October 2007, and that for this purpose a proper maintenance action must be scheduled (ACTIVITY object). The current state of the compressor is 'ON' (PRODUCT_MOL object), i.e. it is correctly functioning, and its age is about 84% (LIFE_CYCLE_PHASE object). The failure prediction as well as the estimation of the aging of the compressor were performed in the past by the decision support system (DSS), connected to the PDKM, and the related objects were created/modified as a consequence.

The lower portion of the figure shows that a failure of one of the temperature probes attached to the control board happened on June 2nd, 2007. In particular, the failure is related to the Ambient Temperature Probe. Thus, the control board of the refrigerator results to be (PRODUCT_MOL object) 'FAILED'. The LIFE_CYLCE_PHASE object also shows the additional information that, at the moment of the failure, the age of the probe was about 65%. The failure happened to the refrigerator requires an immediate action.

As a consequence, the central management division of the company is notified about this need and, already aware of the forthcoming potential failure of the compressor (and thus of the related thermodynamic circuit), can appropriately schedule a maintenance action with the related visit to the customer's site, where both problems can be solved together, saving time, cost and avoiding mishaps to the customer.

4 Conclusions and further research

In this paper has been explained how the PLM data and knowledge management object model developed within PROMISE has been applied to the concrete business scenario of Indesit creating a test application of closed loop PLM. This allows the creation of new services and functionalities, enriching the product and creating an added value to the customer.

The PLM data object model has been also applied to other nine business application cases, ranging from electronic devices to cars and trucks. Due to these successes this model has been proposed as a possible standard for PLM data management of products with a low and medium complexity, such as the consumer goods.

5 Acknowledgements

This work has been partly funded by the European Commission through the FP6-IST Project entitled PROMISE: PROduct lifecycle Management and Information tracking using Smart Embedded systems (No. IST-2004-507100).

6 References

Cassina J., Taisch M., Tomasella M., Matta A., Metin A., Marquard M., "Development of the semantic object model for a PDKM system", Proceedings of the International Conference on Concurrent Engineering, ICE2006.

Garetti M. Terzi S., Product Lifecycle Management: definition, trends and open issues, Proceedings at III International Conference On Advances In Production Engineering, 17 - 19 June 2004, Warsaw, Poland

Jansen-Vullers J., A. van Dorp, B. Beulens, 2003, Managing traceability information in manufacture, 2003, International Journal Of Information Management 23 : 395-413

Hirsch B. E., K.D. Thoben and J. Eschenbaecher, 2001, Using e-business to provide Extended Products, Automation, Automation days, Helsinki

Morel G., H. Panetto, A. Zaremba, G. Mayer, 2004, Manufacturing Enterprise Control and Management System Engineering: paradigms and open issues, IFAC Annual Reviews in Control, 27 : 199-209

Terzi S., Cassina J., Chiari G., Panetto H. 2004, "Traçabilite Des Produits : Une Approche Holonique", Modélisation et simulation pour l'analyse et l'optimisation des systèmes industriels MOSIM'04 – du 1er au 3 septembre 2004 - Nantes (France)

Global Supply Chain Control
A Conceptual Framework for the Global Control Centre (GCC)

Heidi C. Dreyer [1], Ottar Bakås [2,] Erlend Alfnes [1], Ola Strandhagen [1] and
Maria Kollberg [1]

[1] Department of Production and Quality Engineering, Norwegian University
of Science and Technology, 7491 Trondheim, Norway,
phone (+47) 73 59 38 00, fax (+47) 73 59 71 17.

[2] Department of Operations Management, SINTEF Technology and
Society, 4765 Trondheim, Norway,
Phone (+47) 73 59 03 00, fax (+47) 73 59 25 70.
{ Heidi.C.Dreyer; Ottar.Bakas, Erlend.Alfnes; Jan.Strandhagen;
Maria.Kollberg } @sintef.no

Abstract.
The operation of global manufacturing network is challenging due to the complexity in product and information flow, diversity in sites, localization and processes and the information processing needed for control. Thus information technology has to be developed to cope with this complexity and to develop decision support for controlling the network. In this paper the concept of the Global Control Centre for manufacturing activity is developed based on research on a Norwegian supplier of fish hooks. The main elements of the GCC is found to be the global control model, performance measurement system, ICT solutions and the organization and the physical environment. In order to realize the GCC the main challenges are ICT investments and standardization, and the management of change and organizational resistance. The findings presented in this paper are not yet collectively implemented and tested and must therefore only be viewed as conceptual proposals.

Key words:
Operations management, case study, supply chain control, global operations

1. Introduction

Supply chain management (SCM) is the concept of how to orchestrate and operate the broad set of activities, resources and companies in the supply chain (SC) in order to assure competitiveness and efficiency. The complexity most companies faces due to outsourcing, globalization and just-in-time deliveries makes the SC an interwoven

Please use the following format when citing this chapter:

Dreyer, H. C., Bakås, O., Alfnes, E., Strandhagen, O., Kollberg, M., 2007, in IFIP International Federation for Information Processing, Volume 246, Advances in Production Management Systems, eds. Olhager, J., Persson, F., (Boston: Springer), pp. 161-170.

network, often with a geographical decentralized structure. However, even if there exist knowledge of how to organize the SCM activities, many companies finds it challenging to coordinate the SC. The consequence of poor coordination is a less responsive, cost efficient and service oriented SC (Frost and Sullivan, 2006).

Access to real-time information and development of information and communication technology (ICT) is expected to revolutionize SCM and to build a new fundament for improving productivity and competitive strength (Hansen and Nohria, 2005; Beardsley et al., 2006; European Commission, 2006). The next generation ICT based SC will consist of RFID and track and trace technology, real-time monitoring and visualisation, and control systems which will lead to more intelligent logistical solutions (European Commission op cit.). However this will challenge our traditional SC models. New operation and control models, work processes and integration and collaboration models have to be developed (Beardsley et al., 2006). Successful operation requires that decision making at all levels can be performed in a setting where relevant information is transparent and can be accessed from any place in the SC in real-time (Strandhagen, Alfnes and Dreyer 2006). The challenge is to establish ICT-based concepts where electronic information and signals replace the traditional manual and physical control processes, and to utilise the best available real-time technology in integrated teams independent of localisation.

The aim of this paper is to develop a concept of the Global Control Centre (GCC) for manufacturing. The research question we address is how to control and make decisions in a geographical spread and decentralized manufacturing network? The authors are involved in development of such centres in Norwegian companies with globally distributed SC. Through modern collaborative technology, these centres will enable a holistic insight of global SC control and global, real-time communication.

The paper is a conceptual paper. Based on state-of-the-art theory and empirical data from a Norwegian case company, we discuss and suggest a concept for a GCC. Data is collected and analyzed through traditional methods as observations, interviews and extracts from the company's databases. The theoretical fundament which forms the GCC is SCM and operations management, production planning and control and organisational theory.

2.   Theoretical Framework

The trend in manufacturing has moved towards the optimisation of the total SC instead of the single company. Operations of manufacturing networks cause significant complexity both related to the number of actors, processes and information involved, and to the geographical distance (Cooper and Gardner, 2003; Chopra and Meindl, 2007). It is claimed that such networks often suffer from insufficient co-operation and integration (Sanders and Premus, 2005). Inefficient information and communication processes combined with historical and static information often causes limited performance knowledge and reducing the ability to control the network activities.

In global and geographical distributed SC, information sharing and access to real-time information becomes essential. Decision making and control has to be performed in a setting where relevant information is accessible and updated, and can be accessed from any place in the SC in true time (Holweg et al., 2005). Today a vast majority of the information and numerical demand analysis is based on forecasts, historical information and assumptions rather than *real-time information* and dynamic facts. Real-time information leads to higher predictability and insight into the demand situation which will prevent the "bullwhip" and artificial demand amplification. Several studies have identified the challenges caused by a lack of information and to what extent competitive advantages can be gained from a seamless SC (Forrester, 1961; Lee et al., 1997; Chen et al., 2000). Improved information quality will among other factors lead to a reduction in inventory level and stock-out situation which will increase the service level.

Real-time information is considered to be point-of-sale (POS) data, updated stock level information, bare code systems and scanners. However, RFID (Radio Frequency Identification) technology will provide much more accurate and real-time information due to the number of data registration points, data capturing frequency, reading quality, etc (Hedgepeth, 2007). Even though such solutions will create formidable amount of information there is still is a challenge related to the absorption, utilisation and grasping of the information. Thus in order to grasp large and complex amount of information there is a need for processing instruments and ways to represent and visualise information (Liff and Posey 2004).

The proliferation of ICT, e-business and internet technology has enabled new ways of working and sharing information, and changed the way organisations communicate. This have significant implications for operation of the SC leading to reduced inventory and cost levels, improved transactions efficiency and planning processes and better performance (Busi & Dreyer, 2004; Silberberger, 2003; Thoben et al., 2002).

A recent strategy for utilising the potential of ICT is Integrated Operations (IO). IO aims to improve integration and interactivity between the different off-shore and on-shore activities in the oil and gas industry. New data transmitting technologies, intelligent sensors, advanced monitoring and visualising technology, and internet based solutions for communication and information sharing, is integrated in advanced control centres that enable efficient collaboration and remote control of off-shore operations (OLF – The Norwegian Oil Industry Association, 2003).

A similar strategy is emerging in the manufacturing industry. Based on recent technology and software, Supply Chain Studios (Jonsson and Lindau, 2002; Strandhagen et al., 2006) are developed that enables close collaboration and effective operations in geographically dispersed SC. The authors are currently involved in several research projects that aim to realise this strategy. One of the projects is performed in collaboration with the company Mustad, which aims to develop a GCC to synchronise the operations in their global manufacturing network.

3.   The case company – a global supplier of fish hooks

Mustad is the world leading supplier of fish hooks and fishing tackle. Mustad has facilities for manufacturing, assembly, packing and distribution in eight counties worldwide, including Norway, China, Singapore, Philippines, USA, Dominican Republic, Brazil and Portugal. Mustad is the no. 1 selling hook brand world wide, and products are exported to more than 160 countries. Customer requirements are differing within different geographical regions, application type (recreational, sport, industry, sea) and customer type (wholesalers, retailers, OEMs).

Traditionally, the supply chain has been characterised by fully decentralised control. Decisions regarding inventory levels and product programs are made independently on each site. Production and market forecasts are shared only to a small degree, and there is an overall lack of coordination across the different SC sites. Mustad's fish hooks have a Y-shaped product variant structure, with very few raw materials (mainly steel wires) and a large number of sizes, shapes, surface treatments and packaging, totalling up to about 12.000 finished product variants. Mustad is also offering trading products of complementary fishing equipment, in order meet requirements from retail chains that demand supply of complete range of fishing products. Further, new hooks are introduced to the market frequently, but exclusion of products from product programmes are challenges as customers require full product series of a hook type, even though some variants are seldom sold. Together, this has lead to a situation where Mustad now has about 20.000 stock keeping units. This is adding further complexity Mustad's supply chain management.

The company has been facing major logistics challenges. The total stock turn is low, at about 1.5 years, meaning that a product is kept in stock for about 35 weeks in average. The lead times in their supply chain are large, with an average manufacturing lead time of 8-12 weeks, and transportation lead times between 1 – 7 weeks (depending on whether air or sea transport is used). The material flow in their supply chain is complex. Certain products can be produced only at certain sites, and there is a large degree of internal transportation. A mapping of material flows in their supply chain showed a true "spaghetti" structure (Figure 1).

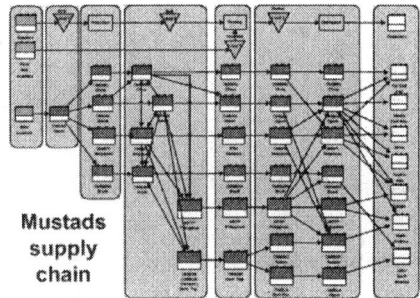

Figure 1: Mustad's material flow

The company therefore wants to build a supply chain studio that utilises modern ICT to integrate and control their global operations. They term this studio the Global Control Centre (GCC). The GCC will become the central information hub in the Mustad supply chain, and is based on a combination of various computer-based systems and information technology tools to capture, process, present, store, and distribute the information needed. The control centre that is being developed is further described in the next chapter.

4.   Conceptual framework for the Global Control Centre

GCC is a concept for integrated operations and synchronized SC control. The main aim is to enable holistic and integrated control in the decentralized and global network of the entire Mustad supply network. In this case The GCC constitutes a central information and communication node that enables integrated operations and remote control in the network (see Figure 2).

The GCC enables decisions makers at each node to collaborate virtually and make decisions based on updated information. The GCC also constitutes a single-point of control where a team takes overall decisions regarding Mustad's global operations. Most decisions regarding operations are still performed locally at each node. However, some overall control tasks are allocated to a GCC team that are co-located and collaborate virtually with other decisions makers in the network.

The GCC consists of four core elements, the global control model and decision support, ICT, organization and the physical environment. In addition, potential benefits of the GCC are described.

Figure 2: GCC position in the SC

4.1 Global control model and decisions support

The primary role of the GCC is to provide global control and decision support for a set of defined areas at Mustad. These areas are typically related to the operational activities for control of materials and information flows in the SC, including forecasting, planning and replenishment, inventory management, product management, sales and operations planning and transportation management. Furthermore, more tactical and strategic decisions regarding for instance SC structure, design and internal pricing will be supported by the GCC.

Decision support will be provided to support the following tasks: centralized forecasting, planning and replenishment; global stock program coordination; global product program management; sales and operations planning; global transport tracking and costing; and performance management.

4.2 Information and communication technology

The technology that will be used are business intelligence applications, ERP-systems, production and inventory management systems, CRM systems, forecasting and planning tools, etc. Additional equipment and tools includes PCs, large projected displays, interactive whiteboards, wireless technologies, portable devices, video conference systems, chat, SMS, etc. In addition, a performance dashboard will be developed to represent all the collected information in an intuitive and visual manner. The performance dashboard is a key element in the decision support involving the performance parameters for control. This computer-based solution ensures that control decisions are based on indicators related to performance objectives of the SC. Combined, these technologies facilitate the flow of information across different control situations. The team will be able to communicate and share information with managers at the other Mustad sites through virtual meeting rooms.

4.3 Organization

The organization of the GCC activities will primarily be team oriented. A team oriented structure ensures an efficient decision process and flexibility in allocation of tasks and responsibilities and limits vulnerability and dependence on key personnel Kristensen et al., 2005; Teasley et al., 2000). This structure will further help to support the SC in meeting defined customer service requirements such as improved responsiveness and delivery precision. The GCC will not only serve as a single-point-of-contact for internal communication but also as an important interface towards customers and suppliers of Mustad. The following are examples of tasks that will be performed in the centre: SC design and analysis; coordination of production and packing activities; coordination of a global stock program; establishment of plans and forecasts, adjustments of control parameters; central coordination of purchasing from suppliers; and coordination of deliveries. The GCC will provide high level of service to all parts of Mustad's supply chain.

4.4 The physical environment

The GCC will be located at Mustad Groups headquarters in Gjøvik, and will be a management centre with physical, virtual and visual elements. The Mustad SC control team is physically located at the centre, in an open office environment ensuring efficient communication among team members. The team is in direct contact with people at the various SC sites through virtual workspaces/interfaces. The decision support solution ensures the visual display of all information regarding the supply chain that is necessary for the team to make right decisions. If needed, the team can also use wireless and portable tools for remote access to the centre.

4.5 Realisation - potential benefits

On a general level, the realization of the GCC is expected to contribute to the overall aims of SC orientation. Some central areas for performance improvements are specified in Table 1.

Table 1: Potential benefits of the GCC (excerpts)

Supply chain measure	Reduced costs/ increased efficiency	Improved service/quality
Supply chain coordination and integration between sites	- Increased utilization of production and planning resources (economies of scale) - Increased efficiency in delivery and shorter lead times - Improved efficiency in inventories and transportation	- Opportunities for developing additional services such as transportation tracking, track-& trace - Opportunities to centralize common functions e.g. sourcing of products - Redesign of SC structure - reduced complexity
Forming a GCC team – combination of competences	- Reduced need for moving tasks between multiple functions - Increased efficiency in decision-making processes - Increased logistics support to	- Higher level of responsiveness, better positioned for dealing with deviations - Single-point-of contact: One interface towards external partners - Combination of competences give

	internal sites	better/ more informed decisions
Focus on market needs - integration of functions	- Shorter time for decision-making - Reduced administrative/ transaction costs	- Increased focus on customer needs - Quicker response to market demand and improved customer service levels - Timely orders and delivery
A common platform for decision support	- Increased efficiency in decision-making processes - Increased coordination of IT, information and communication structures	- Information gathered at one single place - Improved data quality - Improved coherence - common understanding

5.   Discussion – major challenges

The realization of the GCC requires that a number of essential elements are in place. In addition, the most critical conditions for success should be targeted. The most critical obstacles to the realization and success of the GCC are:

- ICT investments and standardization
- Change management and organizational resistance

These critical areas for success are discussed below.

5.1   ICT investments and standardization

As the GCC decision support primarily depends on information, issues regarding information and data management need to be dealt with. Data quality, trust and willingness to share information, confidentiality, data security, availability of data and the increase in data volume are all issues that need to be considered in order to be a reliable decision support system. The selection and implementation of technology is critical. The IT platform must have enough capacity to deal with the extensive amounts of data gathered from various sources. The system must permit easy integration with other systems and interfaces and there must be an overall coherence in information and IT support throughout the SC. Existing systems may need to be substituted entirely or can be integrated with new applications.

The standardization element is considered to be critical for SC integration. It is not until Mustad has reached a common understanding of processes, systems, roles, KPIs etc. on a global scale that the major achievements from the SC orientation will be achieved. However, the GCC can serve as a central support for this standardization process.

A major task is to develop a performance measurement system for the entire SC with uniform indicators that will constitute an essential input to the decision support solution at the GCC. Indicators must be valid and unambiguous for the SC to permit aggregation and facilitate comparison of measures across the SC. Indicators must be linked to vision, strategies and performance objectives defined for the entire system.

5.2   Change management and organizational resistance

For Mustad, it is important to establish an organizational structure that truly supports the process towards becoming a SC oriented company. The organisational model will have to clearly define the relationships between the individual sites, the GCC team, the headquarters, etc. It is important that the sites are organised to support the work towards optimisation of the performance of the whole system and do not seek to optimise their own individual performance on the expense of the SC. The organisational structure will therefore be based upon the global vision, strategies and objectives for all parts of the SC. All employees at each site must be aware of their specific role in the system.

A major challenge may be the potential reorganisation of individual sites from business/production units into supply units. When management responsibility is transferred to the GCC team and the frame of operations is defined centrally, the individual sites may be turned into demand-oriented supply units. This change needs to be done in a setting where the individual site gets incentives for carrying out their core business which will be the daily operations.

An assessment of the current SC structure will detect potential inefficiencies and propose necessary measures for redesign of the structure. This work also involves reconsideration of site location, links between sites and specific details regarding material and information flows.

6.   Conclusion

In today's global economy, manufacturing enterprises must be viewed in the context of their contribution to the total SC. Collaboration in the SC has a wide range of forms with one common goal: to gain information and to create a transparent, visible demand pattern that paces the entire SC. Even though formidable amount of information is created there is still is a challenge related to the absorption, utilisation and grasping of the information in such a complex network.

As a contribution to close this gap a concept for a GCC that supports the monitoring, analysis and management of the SC performance has been developed. It supports control and decision making by visually displaying in true time leading and lagging indicators in a SC process perspective. The GCC will help to decrease the level of complexity and improve control of operating environment and the main benefits are:

- The access to true-time monitoring facilities at a high level
- A true SC perspective (different from a single actor perspective)
- Speeding up recognition and decision making
- Integrated decision making (for instance purchasing and production control)

The findings presented in this paper are not yet collectively implemented and tested in a SC context. Therefore, the findings can only be viewed as conceptual proposals. Thus, further research is needed within the following areas to realise this strategy:

- Development of Supply Chain Studios in other companies and industries.

- Development of ICT infrastructures supporting integrated global operations
- Development of a performance measurement system for the entire SC with uniform indicators. Indicators must be valid and unambiguous.
- Development of an organisational model that defines the relationships between, the GCC team, individual sites and HQ. The organisational structure should be based upon common vision and for all parts.

References

Beardsley, S.C., Johnson, B.C & Manyika, J.M, (2006), "Competitive advantage from better interactions", *The McKinsey Quarterly*, No 2, 2006

Busi, M, Dreyer, H.C. (2004), "Collaboration or business as usual? Suggested matrix to categorize and develop collaboration". *Proceedings of the 16th Annual Conference for Nordic Researchers in Logistics*, NOFOMA 2004, Linköping, Sweden

Chen, F. Drezner, Z., Ryan, J., Simichi-Levi, D. (2000), "Quantifying the Bullwhip Effect in a Simple Supply Chain: The Impact of Forecasting, Lead Times, and Information", *Management Science*, Vol. 46, No. 3, pp. 436-443

Chopra, S. & Meindl, P. (2007), *Supply Chain Management – Strategy, Planning, & Operation*, 3rd Ed., Pearson Prentice Hall, New Jersey

Cooper, M. C, Gardner, J. T. (2003), "Strategic Supply Chain Mapping Approaches", *Journal of Business Logistics*, Vol. 24, No. 2, pp. 37-64

European Commission (2006), "Collaboration@Work. The 2006 report on new working environments and practices", *IST and European Commission publication*, October 2006

Forrester, J. W. (1961), *Industrial Dynamics*, MIT Press, Cambridge, MA

Frost & Sullivan (2006), "Meetings Around the World: The impact of Collaboration on Business Performance", *Frost & Sullivan white paper*, sponsored by Verizon Business & Microsoft, www.verizonbusiness.com/us/resources/conferencing/impactcollab.pdf

Hansen, M.T & Nohria, N. (2006), "How to Build Collaborative Advantage", *Sloan Management Review*, special monograph for World Economic Forum, Davos, Switzerland, January 2006

Hedgepeth, W O. (2007), *RFID Metrics – Decision Making Tools for Today's Supply Chains*, CRC Press Taylor & Francis Group, Boca Raton, Florida

Holweg, M, Disney, S, Holmström, J, Småros, J. (2005), "Supply Chain Collaboration: Making Sense of the Strategy Continuum", *European Management Journal*, Vol. 23, No. 2, p. 170

Jonsson, P., Lindau, R. (2002), "The supply chain planning studio – utilising the synergetic power of teams and information visibility". *Proceedings of the Annual Conference for Nordic Researchers in Logistics*, NOFOMA 2002, Trondheim, Norway.

Kristensen, K., Røyrvik, J, Sivertsen, O.I. (2005), "Applications of the Physual Designing Network in Extended Teams", *Proceedings of the 11th International Conference on Concurrent Enterprising*, Munich, Germany

Lee, H. L., Padmanabhan, V., Whang, S. (1997), "Information Distortion in a Supply Chain: The Bullwhip Effect", *Management Science*, Vol. 43, No. 4, pp. 546-558

Liff, S. & Posey, P. (2004), *Seeing is believing – how the new art of visual management can boost performance throughout your organisation*, Amacom, New York

Niven, P. (2002), *Balanced scorecard step by step: maximising performance and maintaining results*. John Wiley & Sons, p.116

OLF: Oljeindustriens Landsforening – The Norwegian Oil Industry Association (2003), "e-Drift på norsk sokkel" – e-Operations on the Norwegian continental shelf, thematic report from OLF, available at: www.olf.no/hms/aktuelt/?19031

Silberberger, H. (2003), *cBusiness. Erfolgreiche Internetstrategien durch Collaborative Business*. Spinger, Berlin 2003

Strandhagen, O., Alfnes, E., and Dreyer, H.C. (2006), "Supply Chain Control Dashboards", *Conference proceedings Production and Operations Management Society (POMS)*, Boston, US.

Sanders, N. R., Premus, R. (2005), "Modelling the Relationship between Firm IT Capability, Collaboration, and Performance", *Journal of Business Logistics*, Vol. 26, No. 1, pp. 1- 23

Teasley, S, Covi, L, Krishnan, M.S., Olson, J.S. (2000), "How Does Radical Collocation help a Team Succeed?", Computer Supported Cooperative Work. *Proceedings of the 2000 ACM conference on Computer supported cooperative work*, Philadelphia, Pennsylvania, US

Thoben, K.-D., Jagdev, H., Eschenbaecher, J. (2003), "Emerging concepts in E-business and Extended Products". In: Gasos, J., Thoben, K.-D. (Eds.): *E-Business Applications - Technologies for Tomorrow's Solutions;* Advanced Information Processing Series, Springer

Integrated Approach for Self-Balancing Production Line with Multiple Parts

Daisuke Hirotani[1], Katsumi Morikawa[1] and Katsuhiko Takahashi[1]

1   Aritificial Complex Systems Engineering, Graduate School of Engineering, Hiroshima University,
1-4-1, Kagamiyama, Higashi-Hiroshima, Japan
Tel: +81-82-424-7703 Fax: +81-82-422-7024
{dhiro,mkatsumi,takahasi}@hiroshima-u.ac.jp
WWW home page: http://www.pel.sys.hiroshima-u.ac.jp

Abstract

In a "Self-Balancing Production Line", each worker is assigned work dynamically, thus they can keep the balanced production under satisfying the specific conditions. For structure of line, in-tree assembly network line has been analyzed in previous paper. In that paper, line are virtually integrated to one and slowest to fastest sequence can be balanced under the integrated line. However, if an item consists of multiple parts and parallel work is possible, a new approach is applicable under the condition, and performance measure increase comparing to integrated line. In this paper, new integrated approach for both previous self-balancing line and buffer is proposed, and we compare the line that had been proposed in the previous paper.

Keywords
Self-balancing, production, blocking, multiple parts

1   Introduction

In the traditional assembly line, each worker is usually assigned to a fixed work, and each worker iterates the assigned work continuously as assembly line balancing. For this line, assigning workers to the balanced work is studied in the previous research, (for example, [1]). When imbalance of speed of workers exists in this kind of line, the slowest worker will delay the overall work. As a result, the production rate of the production line will also decrease. For solving this problem, "Self-Balancing Production Line" was introduced. The utilization of the mentioned method is reported in at least two commercial environments: apparel manufacturing and distribution warehousing [2]. In this type of production line, each worker is assigned

Please use the following format when citing this chapter:

Hirotani, D., Morikawa, K., Takahashi, K., 2007, in IFIP International Federation for Information Processing, Volume 246, Advances in Production Management Systems, eds. Olhager, J., Persson, F., (Boston: Springer), pp. 171-178.

to work dynamically, and when the last worker completes an item, he/she walks back and takes over the next item from his/her predecessor. Then, the predecessor walks back, takes over the next item from his/her predecessor, and so on until the first worker walks back and starts a new item. Since faster workers are assigned more work in processing an item, and slower workers are less, they can keep the balance. For this line, it has been found that the maximum production rate can be achieved if the workers are sequenced from slowest to fastest [3]. Also, the other conditions for three workers have been found numerically by simulation [1], and the performance of production line with n workers have been analyzed mathematically [4].

Only one paper related to this line with multiple parts has been published [5]. In that paper, if all the line are virtually integrated to one line and workers are sequenced to slowest to fastest, then they can keep the balance and the maximum production rate can be achieved. However, if an item consists of multiple parts and parallel work is possible, a new approach is applicable under the condition, and performance measure increase comparing to integrated line. In this paper, new integrated approach for both previous self-balancing line and buffer is proposed, and we compare the line that had been proposed in the previous paper. This targeted line is a special case for self-balancing production line. However, using this approach, if there are many lines (ex., to make upper and lower parts of an item for each line) in that factory, it can be applied easily, and can be achieved higher production rate with lower flow time.

This paper is organized as follows: In section 2, assumptions, characteristics of this production line, behavior and formulation of this model are derived. In section 3, new rules for integrated approach are shown. In section 4, we compare and analyze the rules for integrated approach. Finally, concluding remarks is described in section 5.

2 The Production Line

In this section, assumptions and workers' behavior are explained, and characteristics of this line are also shown.

2.1 Assumptions

In this research, the production line with the following assumptions as shown in Fig. 1 is considered.

1. There are multiple lines, and there is a buffer in middle of the whole line to stock materials/parts. Capacity for each buffer is infinity.
2. Each worker processes only one identical item sequentially.
3. Workers are sequenced from one to n on production line, and each worker never passes over the upstream and downstream workers.
4. Worker processes his/her work while moves along the line, and worker i processes at constant velocity v_i in the production line. In this paper, a continuous line is considered. This is different from the previous papers [2, 3, 5].

5. When the last worker finishes processing an item, worker n walks back to worker n-1 and takes over the next item from worker n-1. Then, worker n-1 walks back to worker n-2 and takes over the next item from worker n-2. Similarly, all workers walk back to their preceding worker and take over the next item from the preceding worker, and worker 1 introduces a new item into the system. The time required to walk back and take over is ignored.
6. The total length of line is 1. Under this condition, the position of worker i when he/she starts to process is given by x_i. Then, the position at iteration t is defined as $x_i^{(t)}$. Note that $x_1^{(t)}=0$ for any iteration t. This is because the first worker starts to process an item.
7. There is one over-lapping zone. Starting and terminal position for upstream side is defined as A_0 and A_1, for downstream side, B_0 and B_1, respectively. Also, $A_1=B_0$ for convenience, and moving time to another line is ignored.

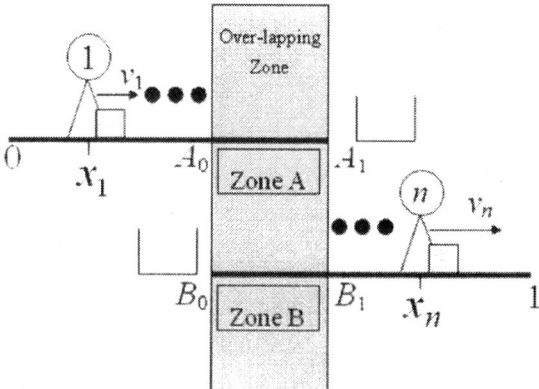

Fig. 1. Production line and position of n workers.

2.2 Self-Balancing and Convergence

It has been proved that the production line can maintain balance when workers are sequenced from slowest to fastest [3]. Subsequently, the position of workers will converge to a unique fixed point defines as x_i^* for worker i as follows.

$$x_i^* = \frac{\sum_{k=1}^{i-1} v_k}{\sum_{k=1}^{n} v_k} \tag{1}$$

Under this condition, the production rate can be calculated as the sum of each worker's velocity v_i (i=1, 2, ..., n) of each worker. Hirotani *et al.* [4] find the convergence condition for n workers as follows:

$$\frac{\sum_{k=1}^{i-1} (-1)^{i+k-1}}{v_n} < 1 \tag{2}$$

If this condition is satisfied for all workers, worker's starting point converges and the line can be balanced.

2.3 Imbalance

When workers are not sequenced that the line can maintain balance, a slower worker prevents the preceding faster worker to continue processing. This condition is called blocking. After blocking occurs, the faster worker moves at the same velocity as the slower worker until the last worker finishes his/her item. In this condition, the position of workers will not converge to a fixed point, and thus, the production rate decreases. Two kinds of blockings exist, one is blocking caused by worker's initial position, and another is blocking caused by worker's velocity.

2.4 Behavior and Formulation of the Model

Fig. 2 shows the time chart for three workers. In Fig. 2, horizontal axis represents worker's position and vertical axis represents time. Zero at the horizontal axis means the head of line and one shows the end of line. Since each worker works with changing his/her position down-stream, diagonal lines, according to their speeds, can represent the worker that is working. Also, since the time to walk back and take over is ignored, a horizontal line can represent the worker that is walking back and taking over. When worker 3 finishes an item, i.e. the position is one, he/she walks back to worker 2 and takes over the next item from worker 2. At the same instant, worker 2 walks back to worker 1 and takes over the next item from worker 1. Then, worker 1 walks back and the position reaches zero, where he/she starts to process a new item.

The position of worker i at iteration t is defined as $x_i^{(t)}$, and the worker i's velocity is defined as v_i. Using these notations, when no blocking occurs, position of the worker at one iteration changes, as follows:

$$
\begin{cases}
x_1^{(t+1)} = 0 \\
x_i^{(t+1)} = x_{i-1}^{(t)} + v_{i-1}\left(\dfrac{1-x_n^{(t)}}{v_n}\right) \quad (i=2,3,\cdots,n)
\end{cases}
\tag{3}
$$

Iteration time is derived by calculating the time spent in one iteration. Iteration time of each worker $a_i^{(t)}$ is shown, as follows:

$$
a_i^{(t)} = \frac{x_{i+1}^{(t)} - x_i^{(t)}}{v_i} \quad (i=1,2,\cdots,n)
\tag{4}
$$

Using this, the production rate can be calculated as the reciprocal of the iteration time, as follows:

$$
\min_i\{1/a_i^{(t)}\} \quad (i=1,2,\dots,n)
\tag{5}
$$

It should be note that when no blocking occurs, the iteration time of each worker i is the same.

In Fig. 2, at the square area, blocking only caused by workers 1 and 2. After the blocking, worker 1 processes at the same velocity as slower worker 1, until worker 3 finishes an item. Therefore, iteration time increases a lot because of slower worker 2, and the production rate decreases. If blocking does not occur, self-balancing and convergence of positions of each worker is obtained. Thus, the production achieves the maximum production rate.

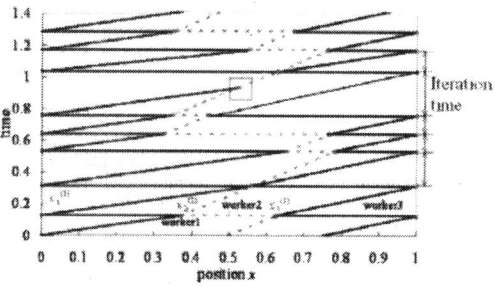

Fig. 2. Time chart (☐ :blocking).

3   New Worker's Behavior Rule

Considering integrated approach, rule for worker's behavior should be defined. Therefore, we propose two new worker's behavior rules in this paper. For these rules, performance measures are better than that of derived by previous papers [2-5]. Worker's behavior changes in the over-lapping zone as follows.

- Behavior of worker when processing
  - ➢ When any worker i finishes processing until A_1, he/she goes to B_0, and starts to process to remaining work unless worker $i+1$ walks back.
  - ➢ When worker finishes processing in zone A, he/she puts an item in buffer and walks back to preceding worker.
  - ➢ When worker who finishes processing in zone B, he/she takes an item in buffer and continues to process. If there is no item in buffer, he/she must wait until an item is put in buffer.
- Behavior of worker when walking back
  - ➢ When worker i takes over an item to worker $i+1$, worker i walks back to worker $i-1$ and takes over an item from worker $i-1$ and starts to process the item if worker $i-1$ is in the downstream from B_0.
  - ➢ When worker i takes over an item to worker $i+1$, worker i walks back to B_0 and starts to process a new item if worker $i-1$ is between A_0 and A_1.
  - ➢ When worker i takes over an item to worker $i+1$, worker i follows one of two rules if worker i is in the upstream from A_0. These rules are shown in following subsections.

3.1 Rule 1 (Zone Rule: ZR)

This rule is that when worker i takes over an item to worker $i+1$, worker i walks back to worker $i-1$, and takes over an item from worker $i-1$ if worker i is in the upstream from A_0. In applying this rule, advantage is that not only time for completing an item is smaller than that of the previous researches [2-5] but also the maximum production rate maintains the same as previous researches by processing at the same time in the over-lapping zone.

3.2 Rule 2 (Fixed Rule: FR)

This rule is that when worker i takes over an item to worker $i+1$, worker i walks back to B_0 if worker i is in the upstream from A_0. In applying this rule, advantage is that not only it is easy to balance by separating the working area since there is a fewer worker in one line but also time for completing an item is smaller than that of the previous researches [2-5].

4   Comparison and Analysis

In this section, we compare two rules for integrated approach. In previous researches [2-5], only production rate is considered as performance measure. However, in this paper, flow time should be considered since an item consists of multiple parts, and thus an item can be made earlier than integrated line. Therefore, we use two measurements: production rate and flow time. Flow time means that time for completing an item. If flow time is low, an item is completed earlier. We use two figures to explain. Fig. 3(a) and (b) are examples of time chart for three workers ($v_1=1$, $v_2=3$, $v_3=5$). For these Figures, $A_0=0.4$, $A_1=B_0=0.5$, $B_1=0.6$. Fig. 3(a) is for ZR and (b) is for FR.

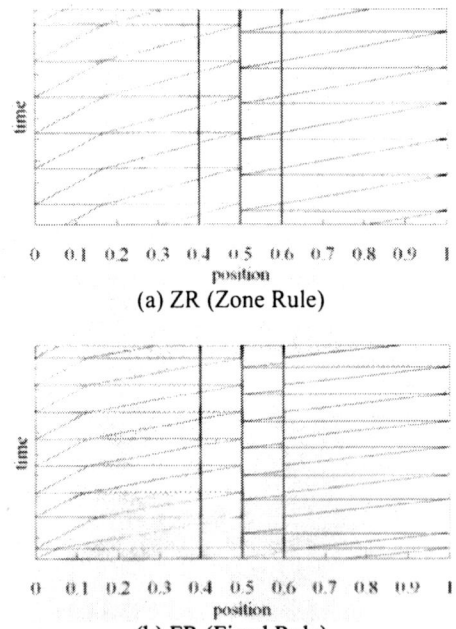

Fig.3. Example of time chart of steady-state for three workers with multiple parts

4.1 Production Rate (PR)

First, we analyze production rate. In Fig. 3(a), two workers process in the overlapping zone alternatively. In previous rule, if the line is balanced, all the workers process the same area in steady-state. However, in Fig. 3(a), slower worker do not processes the same area at any iteration. In stead, down-stream faster worker processes. Under this phenomenon, the production rate is the same as maximum because a faster worker can make up for a slower worker. On the other hand, there is a possibility that production rate for ZR is less than that of FR. In Fig. 3(b), the fastest worker cannot process more work because of rule for FR. This is why production rate is not maximum (sum of the velocity of all workers) unless the line can be balanced. This is shown defined as PR_{ZR} using equation (4) and (5) as follows:

$$PR_{ZR} = \min_{i}\left\{\frac{x^*_{i+1} - x^*_i}{v_i}\right\} \tag{6}$$

where,

$$x^*_i = \begin{cases} A_1 \cdot \dfrac{\displaystyle\sum_{k=1}^{i} v_k}{\displaystyle\sum_{k=1}^{l} v_k} & (\text{if } i \leq l) \\[2em] (1 - A_1) \cdot \dfrac{\displaystyle\sum_{k=1}^{l} v_k}{\displaystyle\sum_{k=l+1}^{n} v_k} & (\text{if } i > l) \end{cases} \tag{7}$$

Equation (7) is different with equation (1). Equation (7) indicates each worker works according to their working velocity in each line while each worker works according to their working velocity in one previous line.

Note that if faster worker in the downstream from B_0 exists, there is a high possibility that item is empty in buffer. Therefore, the worker must wait until an item is put in buffer, and thus production rate decreases because of waiting time. On the other hand, if slower worker in the downstream from B_0 exists, this problem does not occur.

4.2 Flow Time (FT)

Next, we compare and analyze flow time. In fig.3(a), at any iterations, worker 2 finishes to process in zone A, at the same time, worker 3 finishes to process in zone B. This means that in flow time, processing in zone B can be neglected, and thus flow time is smaller than previous research. On the other hand, in fig.3(b), worker 3 must wait because of no items in buffer. However, flow time is shorter than that of derived in previous research because worker 3 can start to process in zone B while worker 2 processes in zone A. Comparing these results, in flow time, FR is better than ZR. Flow time for ZR and FR defined as FT_{ZR} and FT_{FR} can be calculated as follows:

$$FT_{ZR} = n \left/ \left(\frac{n}{\sum_{i=1}^{n} v_1} - \frac{B_1 - B_0}{v_i} \right) \right. \tag{8}$$

$$FT_{FR} = n \left/ \left(\sum_{i=1}^{n} \frac{v_i}{x_i^*} - \frac{B_1 - B_0}{v_i} \right) \right. \tag{9}$$

where, x_i^* is shown in equation (7). Above equations mean that time for over-lapping zone can be neglected. Therefore, flow time is smaller than that of previous research.

5   Concluding Remarks

In this paper, we propose integrated approach that combine buffer and traditional self-balancing line approach for self-balancing production line with multiple parts. Considering integrated approach, two rules (Zone rule and Fixed rule) are proposed. As a result, maximum production rate is the same as previous research, and if worker have to wait, production rate decreases according to waiting time. On the other hand, flow time is smaller than that of previous research. This result indicates that an item can complete earlier.

In our research, we assume there is one over-lapping zone. Considering multiple over-lapping zones is future research works.

References

[1]   A. Scholl, *Balancing and Sequencing of Assembly Lines*, (Physica-Verlag, New York, 1995).

[2]   J. J. Bartholdi , and L. A. Bunimovich, Dynamics of two- and three-worker "Bucket Brigade" production lines. *Oper. Res.*, **47**(3), 488-491 (1999).

[3]   J. J. Bartholdi, and D. D. Eisenstein, A production line that balances itself. *Oper. Res.*, **44**(1), 21-34 (1996).

[4]   D. Hirotani, Myreshka, K. Morikawa, and K. Takahashi, Analysis and design of the self-balancing production line, *Computers and Industrial Engineering*, **50**(4), 488-502 (2006).

[5]   J. J. Bartholdi, and D. D. Eisenstein, Bucket Brigades on In-tree Assembly Networks, *European Journal of Operational Research*, **168**, 870-879 (2006).

Changeability of Production Management Systems

M.S. Hoogenraad and J.C. Wortmann
Faculty of Management and Organization, University of Groningen
Landleven 5, P.O. Box 800, 9700 AV Groningen, the Netherlands
Tel. +31 50-363 3864
j.c.wortmann@rug.nl
Marco.Hoogenraad@infor.com

Abstract.
Modern production management systems consist of transaction processing
systems and decision enhancement systems. A clear example of two such
components are an ERP systems and APS systems. These systems are often
standard software systems, and therefore suitable for many different
situations. This paper analyses the combination of ERP and APS from the
perspective of change. The paper builds on a classification of ERP-APS
integrations and an analysis of changeability of ERP systems. The paper
elaborates on issues encountered when changing ERP-APS integrations.
The main conclusion is that such changes are technically complex because
of different technologies and different data models used. Therefore, end
user education and training is emphasized Moreover, they are
organizationally complex because many stakeholders are involved.

1 Introduction

Production management systems (PMS) have been around for several decades now.
In the last 20 years, most implemented information systems for production
management were ERP systems (Klaus et al. 2000). These systems are standard
software packages aiming at integrated transaction processing across a company
(Wortmann, 1998). Because of their focus on transaction processing, ERP systems
can be called *data-oriented*. Because of their standard software nature, ERP systems
are highly parameterised.

Despite their focus on transaction processing, many ERP systems have their roots
in production planning and control (PPC). The abbreviation *Enterprise Resource
Planning* is derived from *Manufacturing Resource Planning* (MRPII), which in turn
is derived from MRP I – *Material Requirements Planning*. These planning systems
and frameworks were well established before ERP became mature. Production

Please use the following format when citing this chapter:

Hoogenraad, M. S., Wortmann, J. C., 2007, in IFIP International Federation for Information Processing, Volume 246,
Advances in Production Management Systems, eds. Olhager, J., Persson, F., (Boston: Springer), pp. 179-187.

planning and control applications are delivered together with transaction processing as integral part of an ERP package suitable for the market of manufacturing industries.

However, these PPC applications as parts of an ERP suite of software are not always satisfying the needs of production management. ERP planning software is essentially not interactive software but requires batch runs, which has to occur over night or in the weekend. In addition, ERP planning software does not use advanced user interface (UI) capabilities. Moreover, ERP planning software cannot deploy advanced iterative algorithms. Finally, it is restricted to the transaction processing as modelled in the ERP package, and, it is therefore limited to the application areas chosen in a particular ERP-package. Consequently, it is not easy to use ERP planning functionality in areas such supply chain management e.g. for what-if analysis or real-time control of large data volumes.

Because of the drawbacks of ERP software planning functionality, other PPC software has come to the PMS market. Under the name of *Advanced Planning Systems (APS)*, standard software has been offered which is interactive (Stadtler and Kilger., 2002). APS offers the user a high-resolution UI, fast iterative algorithmic calculations, and provides options to go beyond the domain covered by ERP transactions. APS software modules operate on a fast (InRam) database, often called an *object store*, which is extracted from the transaction database.

Consequently, many production management systems have now advanced planning software modules integrated with ERP planning and transaction processing (Tarn et al., 2002). This paper analyses such production management systems consisting of ERP and APS from the perspective of *change*. The relatively vast amount of literature on ERP implementations has shown that enterprise information systems are never "ready": systems continue to be changed (Botta-Genoulaza et al, 2005).

For production management systems, there are several reasons for change. First of all, the fact that systems are being used, leads to requirements for more advanced or sophisticated functionality. Secondly, the nature of planning problems changes when companies invest in (or withdraw from) their product portfolio, market approach, governance and collaboration structure or manufacturing resources. Thirdly, change may induced by developments in ICT, such as upgrades in one of the employed standard software packages. These reasons constitute a background for the change perspective taken in this paper.

2   Problem statement and approach

This paper will investigate the changeability of production management systems consisting of a transaction processing backbone and one or more planning modules. The transaction processing backbone is implemented by an ERP package. The planning functionality is *partly* implemented by APS software. However, as will be discussed later, there is usually a mixture of ERP planning software, APS planning software and other means of planning (Liu et al., 2002; Wiers, 2002). Against this background, the problem statement can be phrased as follows. *Given an implemented production management system consisting of ERP and APS, which issues have to be addressed when there is a need for change?*

The above problem statement can be decomposed into several smaller questions:

- What types of interfaces exist between ERP transaction processing and APS?
- What are the changeability issues of implemented ERP packages for transaction processing?
- What are the changeability issues of implemented planning modules? Is there a difference between APS and ERP planning modules?

The first question, on ERP-APS integration, was investigated by the authors, and the results have been laid down in an earlier paper. This paper is summarized below, in Section 3. The second question, on changeability of ERP for transaction processing, was also researched by the authors (and others). The result is presented in Section 4. The third question is the main contribution of this paper, and will be answered in Section 5.. Section 6 presents conclusions.

3   The nature of APS-ERP integration

The first question is concerned with the types of interfaces encountered when APS planning software is interfaced with ERP transaction processing. This investigation was also done by the authors and laid down in an earlier paper (Hoogenraad and Wortmann 2007). The main results of this earlier paper are first outlined here.

When combining APS and ERP modules, the resulting architecture takes a form as depicted in Figure 1. This figure illustrates a number of points. At the right hand side, software modules of ERP transaction processing are depicted. The figure suggests that many users work in parallel, interacting with a single database. In the middle part, several ERP batch planning software modules are shown, which may or may not be implemented in a particular organisation. These modules run on the same ERP transaction database. At the left hand side, one or more APS modules are in use.

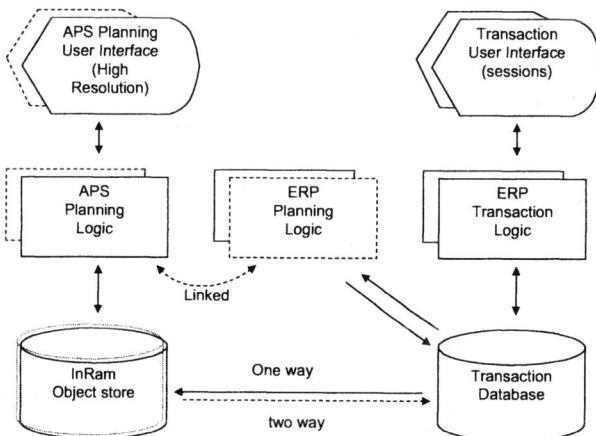

Figure 1. Collaborating software modules of ERP transaction processing, planning and APS

Each APS module has its own (InRam) object store. The relation between an APS planning module and the ERP transaction database may be *one-way* or *two-way*. A one-way relation means that the object store is created by taking a periodic snapshot of the ERP database and by transforming this snapshot into the requires set of data objects needed for APS. A two-way relation means that the results of planning decisions made are automatically transformed inversely into meaningful ERP data and placed in the ERP database[1]. The arrow between APS Planning Logic and ERP planning logic with the label '*linked*' represents the fact that APS planning software may either replace ERP planning software or enhance it, for each relevant planning function. In other words, most planning functions can be supported by ERP planning software, or by APS planning software or by both (or even by none of these).

When classifying interfaces between APS and ERP, there are two functional issues and two technical issues involved. As for the *functional issues*, the first question that should be answered is concerned with the *one-way* versus *two-way* integration (cf. Figure 1, arrows at the bottom). When the integration is one way, the APS planning software is able to perform calculations based on uploaded data, but the software cannot download any results back to ERP. Therefore, planning results have to be entered manually by the planner into ERP. In a two-way integration, the planning modules will provide ERP automatically with planning decisions. Obviously, this type requires more alignment between ERP and APS software. In particular a data model transformation during upload has to be followed by an inverse data model transformation during download. This is a complex requirement.

The second functional question pertains to the *link* between APS planning software and ERP planning software for a particular planning function (cf. Figure 1, arrow labelled "link"). If APS planning software replaces ERP planning software partially for a particular planning function, then the software requirements on APS can be somewhat relieved – after all, ERP planning functions will take care of many details and data modelling peculiarities

The first technical question is concerned with the complexity of the data model mapping during upload and download. As demonstrated in Hoogenraad and Wortmann 2007, these mappings can become quite complex, because there may be considerable difference in data models between ERP transaction processing and APS planning software. The second issue pertains to the timeliness constraints of upload and download. If these constraints are very tight, all kinds of provisions are needed in the interface software to guarantee data integrity

---

[1] An example of such data may be an agreed Master Production Schedule, a set of firm-planned orders, or a sequence of operations of different orders on a bottleneck machine.

Table 1 below summarized these four issues.

	Type	Question (for each planning role where APS planning software is applied):
1	functional	Is it a one-way or two-way interface?
2	functional	Does APS replace or enhance ERP planning software?
3	technical	How different are the data models? How complex is the mapping therefore?
4	technical	What timeliness constraints have to be taken into account?

If the integration is one-way, if APS replaces ERP planning software, if the data models are identical, and if there are no timeliness constraints, then the integration between ERP and APS is relatively straight forward. In all other cases, integrations become complex, due to the different technologies in use, and due to the amount of specific software engineering work needed.

4   Changeability of ERP for transaction processing

The first question on ERP changeability, was investigated on substantial literature review and case studies. This research work was documented in Wortmann et al, 2007. In this paper, changeability of ERP is treated both from a vendor perspective (denoted as ERP *flexibility*) and from a customer perspective (denoted as ERP *adaptability*). In this paper, only the customer perspective will be discussed.

When considering ERP adaptability, a key question is whether the ERP package is used as designed by the vendor, or whether it has been modified beyond the vendor's responsibility. If the package is not modified, it is *configured* by setting parameters. Therefore, the package may be changed by assigning new values to these parameters (Soffer et al., 2003). If the customer cannot obtain the desired functionality by setting parameters (i.e. by configuration) the package has to be modified. This happens in many cases, which implies that it cannot be avoided (Holland and Light, 1999). A modified package cannot be changed by merely changing parameters: also the modifications have to be assessed and redone.

In addition to configuration and modification, an important property of ERP systems is that they are data-oriented. Change of the ERP package may therefore lead to changes in persistent data. Because organisations need to continue the use of their data from the past, these past and current data have to be *migrated* to new data structures. These data migration problems are typical for transaction processing systems in business. However, when migrating ERP systems, (new) parameters with their values may have to change and logic may have changed by the vendor, which adds to the complexity of migration. When an installed ERP package is upgraded to a new version, the vendor will normally provide software tools for data migration. Despite of these tools, data migration can be a considerable effort. When ERP systems have been modified heavily, this effort can even become prohibitive. The above discussion is illustrated by Figure 2, which is taken from Wortmann et al, 2007. Section 5 will identify more factors.

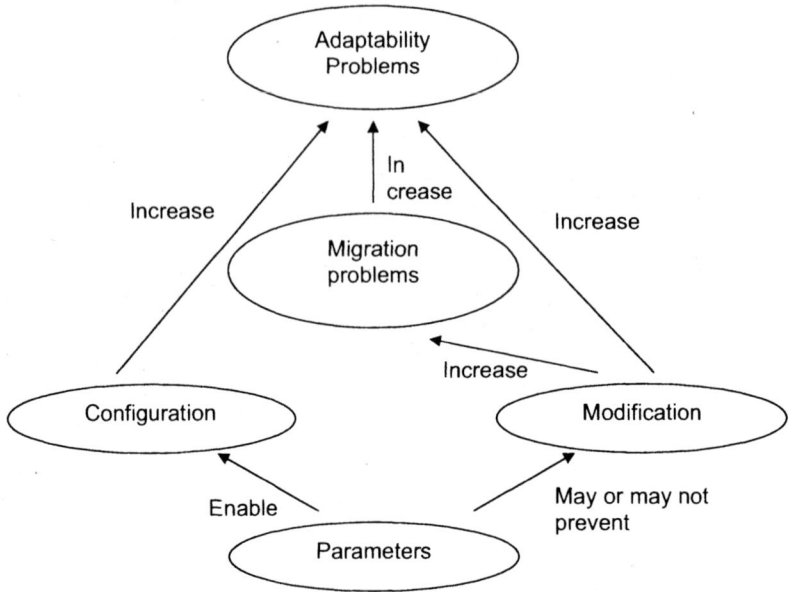

Figure 2. Factors influencing ERP adaptability

5   Analysis of changeability of planning modules

5.1. Introduction

Now the question can be addressed, whether planning modules have the same changeability issues as transaction processing systems. Looking back to Section 4, it is clear that planning modules own very limited persistent data. Therefore, the data migration problems discussed there will probably not occur with planning software modules. Therefore, the migrations issues are depicted by a dotted ellipse in Figure 3 below.

However, *parameters* play an important role in planning modules of standard software packages. Actually, planning software has an abundant number of parameters, ranging from different ways to handle lead times and calendars to weighting factors when prioritising. Business changes in planning will lead to changing answers on each of the questions of Table 1. These changing answers will lead to changing parameter values in the applicable planning modules – from APS, from ERP, or both. When using both APS and ERP planning software, the parameters have to be mutually aligned, which is in itself a mind-boggling job. Therefore, the changeability issues following from parameters shown in the lower part of Figure 2 may be expected to be present in the case of planning. These changes are driven by business developments.

Moreover, the literature studied when investigating the changeability of ERP transaction processing revealed two more factors than shown in Figure 2. These factors are related to *multiple stakeholders and their roles* and to *education and training*. These factors are arguably even more important for planning than for transaction processing. Planning is an organizational function where many stakeholders play a role. Therefore, subsection 5.2. analyses the effect of multiple stakeholders on changing a production management system as depicted in Figure 1. Finally, planning is a function requiring multiple skills. The role education and training is discussed in subsection 5.3. The resulting conceptual framework for change of PMS is depicted in Figure 3.

5.2. The role of stakeholders when changing production management systems

When implementing production management systems, the interpretations and meaning that different stakeholders attach to systems play an important role. Boonstra (2006) analyses ERP systems as redistributors of power, by interpreting implementations from a stakeholder perspective. He provides an overview of studies focusing on the meaning that people attach toward a particular technology. Boonstra concludes that: "different stakeholders can interpret ERP systems in different ways, given their own histories, interest, self-images, prospects and views. Some groups perceive the system as a means to realize certain new company objectives, while others see the system as a way to regain lost power or as a threat to legitimate local interest" (pp. 23-24). Moreover, politics and power relationships among stakeholders are found to be a key factor when implementing an ERP system. Stakeholders have an impact on the politics, which increases ERP upgrading problems because the various stakeholders may have conflicting interests. This also applies for the planning domain, especially when different stakeholders have more interest in the supply chain planning domain or in the local planning domain.

5.3. Education and training

Education and training of managers and users is one of the mostly recognized critical success factor, according to Macris (2004). Therefore, Macris proposes a virtual lab implementation for these two target groups. As mentioned by Macris, the full benefits of a system can be experienced only if managers and users are using it properly. Macris remarks that costs of education and training are often underestimated: If 10-15% of the ERP implementation budget is reserved for training, this will give an organization a high chance of implementation success. Following Macris, two levels of training are required: ERP parameterization training (for the implementation professionals, providing knowledge on parameterized options), and end-user training. Since the planning modules are complex, especially when optimization engines of APS are used, it is crucial to have the involved employees trained for an implementation success. Therefore, it seems that these conclusions can safely be generalized to planning software modules, and therefore to PMS.

5.3. Education and training

Education and training of managers and users is one of the mostly recognized critical success factor, according to Macris (2004). Therefore, Macris proposes a virtual lab implementation for these two target groups. As mentioned by Macris, the full benefits of a system can be experienced only if managers and users are using it properly. Macris remarks that costs of education and training are often underestimated: If 10-15% of the ERP implementation budget is reserved for training, this will give an organization a high chance of implementation success. Following Macris, two levels of training are required: ERP parameterization training (for the implementation professionals, providing knowledge on parameterized options), and end-user training. Since the planning modules are complex, especially when optimization engines of APS are used, it is crucial to have the involved employees trained for an implementation success. Therefore, it seems that these conclusions can safely be generalized to planning software modules, and therefore to PMS.

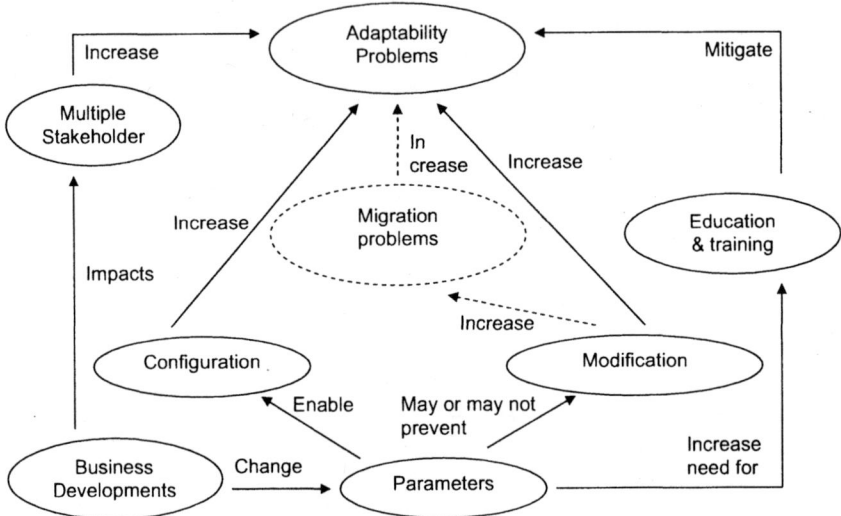

Figure 3. Factors influencing changeability of PMS systems

6   Conclusion

PMS systems are important for many stakeholders in an organization and its environment. Therefore, changing such systems is often a highly political process, which has to be managed properly. Moreover, is difficult to understand planning software, because both APS and ERP planning software is based on numerous parameters. Therefore, education and training of planners and other stakeholders cannot be neglected. Because of the different technologies used, changing an APS integration with ERP transaction processing is never really easy. However, the change becomes more difficult when:

- the integration between APS and ERP transaction processing is two-way
- both APS and ERP planning software are supporting the same planning function
- the data models differ substantially, such that a complex mapping is realized
- strict constraints on timeliness have to be adhered to.

All factors above, should be taken into account when considering the changeability of production management systems.

References

1. Boonstra, A.: "Interpreting an ERP-implementation from a stakeholder perspective", *International Journal of Project Management* 24(1), 2006, pp. 38-52.
2. Botta-Genoulaza, V., P.-A. Milleta, and Grabot, B.: "A survey on the recent research literature on ERP systems", *Computers in Industry*, Volume 56, Issue 6 , August 2005, Pages 510-522
3. Grabot, B., and Botta-Genoulaza, V., "Editorial – special issue on Enteprise Resources Planning", *Computers in Industry,* 56 (2005), pp. 507-509
4. Holland, C. and Light, B.: "A critical success factors model to ERP implementation". *IEEE Software,* 1999, p. 60-66.
5. Hoogenraad, M.S., and Wortmann, J.C.: "APS applications integrated with ERP: problems and solutions". Submitted to *International Journal of Production Planning and Control,*2007.
6. Klaus, H. Rosemann, M. and Gable, G.G.: "What is ERP?". *Information Systems Frontiers,* 2(2), 2000, pp. 141-162.
7. Liu, W., Chua T.J., Lam, J. Wang, F.Y., Cai, T.X., Yin, X.F., "APS,ERP and MES systems integration for Semiconductor Backend Assembly". In Seventh International Conference on Control, Automation, Robotics and Vision (ICARCV'02), Dec 2002, Singapore.
8. Macris, A.: "Enterprise Resource Planning (ERP): A virtual lab implmentation for managers and users training". *SPOUDAI ,* October-December (2004), pp. 13-38.
9. Stadtler, H. and Kilger, C. (eds.), *Supply chain Management and Advanced Planning,* 2nd ed., 2002 (Springer, Berlin).
10. Soffer, P., Golany, B.,and Dori, D.: "ERP Modelling: a comprehensive approach", *Information Systems* 28 (2003), pp.673-690.
11. Tarn, J. M., Yen, D. C. & Beaumont, M., "Exploring the rationales for ERP and SCM integration". In *Industrial Management and Data Systems,* **102** (1/2), 26-34, 2002.
12. Wiers, V.C.S., A case study on integration of APS and ERP in a steel processing plant. *International Journal of Production Planning and Control* **13** (no 6), 552-560, 2002 (Taylor & Francis, London).
13. Wortmann, J. C. :"Evolution of ERP systems". In U. S. Bititchi & A. S. Carrie (Eds.), *Strategic management of the manufacturing value chain,* APMS 1998, (Kluwer Academic Publishers, Amsterdam).
14. Wortmann, J.C., Hoogenraad, M.S. and Maruster, L.: "Changeability of ERP systems: a paradox". Paper submitted to *Journal of Information Technology,* 2007.

OEE Monitoring for Production Processes Based on SCADA/HMI Platform

Lenka Landryová [1] and Iveta Zolotová [2]

1 Department of Control Systems and Instrumentation, Faculty of
Mechanical Engineering, VŠB – Technical University Ostrava, 17.listopadu
15, 708 33 Ostrava-Poruba, Czech Republic, tel. ++420 597 234 113, E-
mail: Lenka.Landryova@vsb.cz

2 Department of Cybernetics and Artificial Intelligence, Faculty of
Electrotechnics and Informatics, Technical University Kosice, Letná 9, 042
00 Kosice, Slovak Republic, tel. +421 556 022 570, E-mail:
Iveta.Zolotova@tuke.sk

Abstract

This article aims to show the problem of Overall Equipment Effectiveness
(OEE) monitoring and performance measurement on a case study done in a
small company. Losses occur during the production process and the ways of
detecting them must be analyzed. The methodology for calculating total
Overall Equipment Effectiveness is followed while applying monitoring tools.
Some existing methods used for production control, the flow of material and
production facilities control and monitoring, and the methods used for
maintenance control are demonstrated as examples regarding data acquisition,
collecting idle time or occurred accidents. The tools and methods implemented
for maintenance control can vary depending on, among other areas, the
operational conditions, but when visualization application can be designed for
monitored technological process on the SCADA/HMI platform, the
implementation can be described as shown in this article.

Keywords

Monitoring, performance, equipment effectiveness.

1 Introduction

Although each production is specific, all of them deal with downtimes, losses, which
unable us to reach the maximum theoretical production performance. What can be
considered as a loss can be further explained and divided into these four basic
categories based on the reasons for their occurrence:

Please use the following format when citing this chapter:

Landryová, L., Zolotová, I., 2007, in IFIP International Federation for Information Processing, Volume 246, Advances
in Production Management Systems, eds. Olhager, J., Persson, F., (Boston: Springer), pp. 189-196.

1. Planned downtime includes days of work off, state holidays, time for overhaul, time for cleaning the workplace, but even development and testing, in some cases, can be considered as a loss;
2. Operational loss includes time for setting up machines, time for changes in production, waiting time for material delivery, and time wasted with bottlenecks or by operator's mistakes, as well as breakdowns;
3. Performance loss occurs due to wrong machine settings, slowing down the production cycle or causing failures;
4. Quality problems with material defects and product rejects, production imperfections, reworks, and repairs decrease efficiency and production effectiveness.

As demonstrated on the above mentioned categories, some causes produce losses qualified in more than just one way. They can touch availability, performance and /or quality issues at the same time since downtime losses, for example, can cause losses in speed and quality at the same time.

2 Methodology for OEE Calculation

Wauters and Mathot [4] describe two types of OEE, the total OEE, which has to do with the theoretical production time, and the OEE dealing with available production time. In other words, the total OEE is the sum of available production time (OEE value) and external losses.

Available production time consists of valuable operating time and technical losses. The OEE coefficient calculated is then a quotient of the valuable operating time against the available production time and speaks about technical losses only, when it decreases and equals less than 100%.

External losses are planned and unplanned:

- Planned losses consist of revisions done, lack of demand for the products, authorized breaks or free weekends for employees;
- Unplanned losses include insufficient material delivery, lack of personnel (sick leaves and epidemics), additional administration or ecological agreements.

The valuable operating time against the theoretical production time results in the total OEE as an indication of how effectively machines have been used compared to how their use could theoretically be maximized. This tells us now about not only the technical issues, but also organizational and social consequences within the production.

3 Monitoring of OEE

When talking about production optimization, standards should be implemented, so that it would be possible to measure and express problems in numbers and based on knowledge. Correct, complete and up-to-date data about production events are necessary.

3. 1 A manual method

Monitored values are written by foremen, workers setting the machines and equipment, etc. into working sheets. Calculations of OEE are performed most commonly in MS Excel. The first occasion to make mistakes occur when transferring data from the workbooks into the PC. Furthermore, daily/weekly values are rounded up and in terms of months and years the final report may show much differentiated data from the reality.

Manual records about downtimes put an extra load on operators from production lines. Therefore, companies introduce easier ways for reporting, such as establishing average times for a concrete type of breakdown, neglecting shorter stoppages, etc. These result in incorrect calculations of overall equipment effectiveness.

For example, a company produces its products on two production lines, but for the optimization reasons, only a line producing more results will be analyzed. A certain time period is analyzed; during April till December of 2007 **the theoretical production time** is 274 days. This corresponds to 274 X 24 = 6576 hours. The available production time is calculated as 274 days minus days, when work was not planned for weekends (79), state holidays (8) and annual leave (15) of employees, resulting in 172 days. For two shifts of 8 hours per day in total there are 172 X 2 X 8 = 2752 hours of work during shifts. If work stoppages and pauses last about 1 hour every day for both production lines, then it is 2752-172 = 2580 hours. The lack of personnel or raw materials causing the rest of the time losses cannot be directly obtained; an estimate has to be made. The head of production decides upon 5%, so **the available production time** 2752 X 0.95 = 2614.4 hours. Then **the valuable operating time** was calculated from the throughput of production lines and the amount produced resulting in 1205 hours.

Table 1. Manually calculated coefficients

		OEE	Total OEE
Valuable operating time	1205 hours	1205/2614=0.46	
Available production time	2614 hours	46%	1205/6576=0.18
Theoretical production time	6576 hours		18%

This simple way of calculation is not sufficient for many companies, which are, for many other various reasons, introducing Enterprise Resources Planning (ERP) type of information systems. With the ERP systems the process is gradually semi- or completely automated.

3. 2 Semi- automated method

At the present time, the small company (where the case study was run) still keeps operation books as the primary sources of data. Reports into the BRAIN ERP system introduced not long ago are sent by certain workers usually at the end of each shift. The reports consist of:
- Amount of good products produced

- Produced amount of rejects
- Time needed for production of good products
- Time spent on producing rejects
- Codes indicating the way how a reject was made
- Stoppages, idle time and extra work with their codes (see Fig. 1).

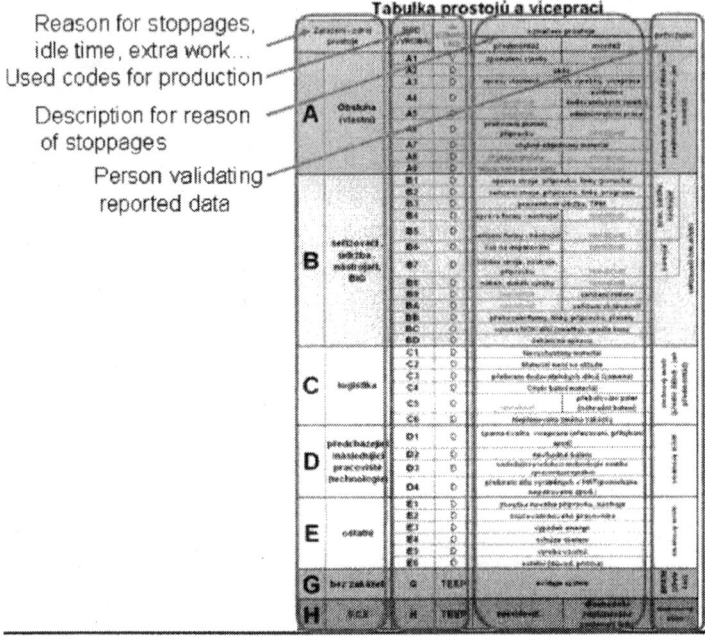

Fig. 1. Table of reported data

The database then provides the fundamental source of production data needed for indicator calculations, such as:

- Fulfillment performance standards
- Calculation of product prize
- Capacity burden
- Production efficiency, overall equipment efficiency
- Production rejects
- Production planning
- KPI (Key Performance Indicator) reports and statements.

Data is read into the Merit controlling system, which processes it resulting in graphical outputs (see Fig. 2) and stoppages and extra work breakups. This data is also provided further as an information source for other departments and units within a company.

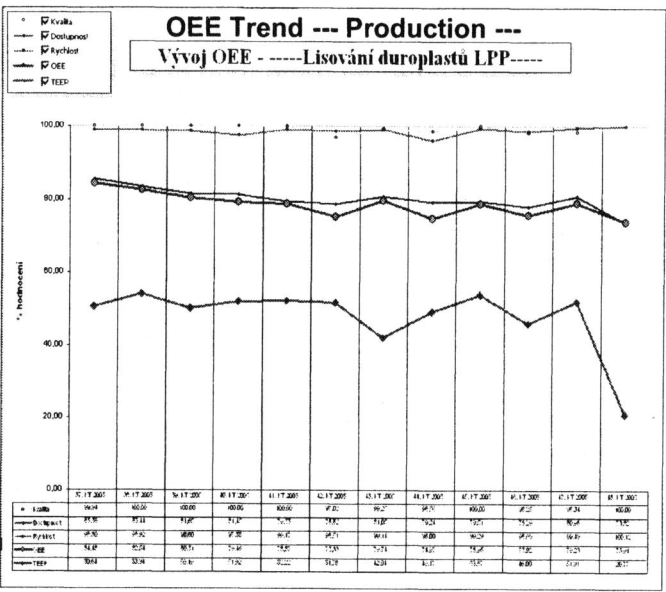

Fig. 2. Tools modeling, executing, and tracking information associated with material and control flow across the plant.

By defining views and requests, the system processes data and filters it according to the time period requested, where the weekly report is the minimum time period, regarding each workplace or a group of them, technology, and a company.

3. 3 Integrating control systems with equipment and maintenance monitoring

Real time monitoring enables recording causes of stoppages of a different kind, which can occur even within a small time period in different production technologies and areas of a shop floor; the real causes can be identified then. These monitoring systems help find:

- a hidden production capacity for even better utilization of production equipment
- reason for investments into new machines,
- more accurate maintenance intervals and OEE measurement setting.

Using standard IT technologies makes integration of a control system easier with the company's higher level information system (MES, ERP aj.)

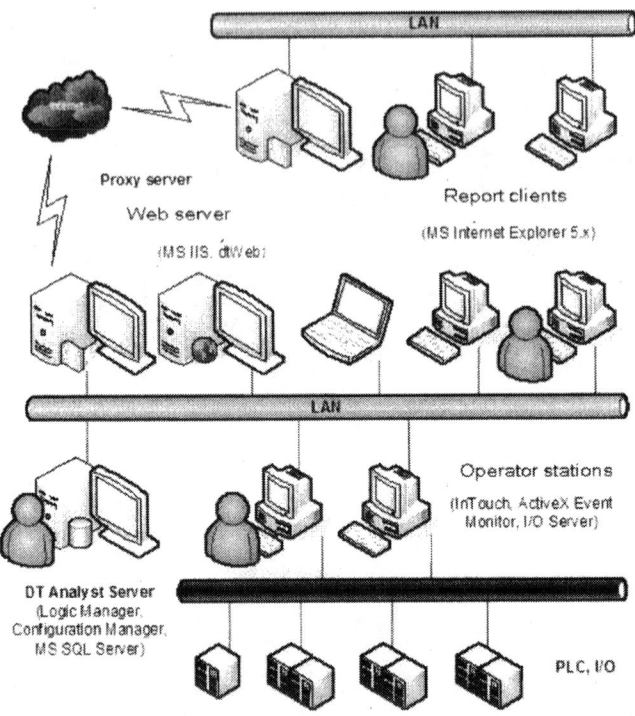

Fig. 3. An architecture of a control system configured with downtime analytical and reporting tools. [3]

The simulated application was designed in InTouch visualization environment to demonstrate the process. [1, 2]The visualized process is simulated by running several scripts: the application script for animating process, conditional scripts linked to simulated breakdowns, and other scripts. For monitoring downtimes and total OEE calculations in the environment of InTouch supervisory control system, the DT Analyst 2.2 software [5] was configured. The Configuration Manager of DT Analyst enabled the definition of stoppages and their causes and the configuration of the entire system. The graphical user interface (GUI) enabled defining new causes and new causalities and combinations of downtimes to tune the system up. The ActiveX Event Monitor (see Fig. 4) is running on an operator station and allows a system operator to execute and manually enter additional information about downtime events. An operator can run scheduled downtimes, which did not result from PLC indicated alarms, such as production changes, cleaning machines, and other scheduled maintenance. The user application also defines entirely automated execution based on occurred downtimes from the controlled process.

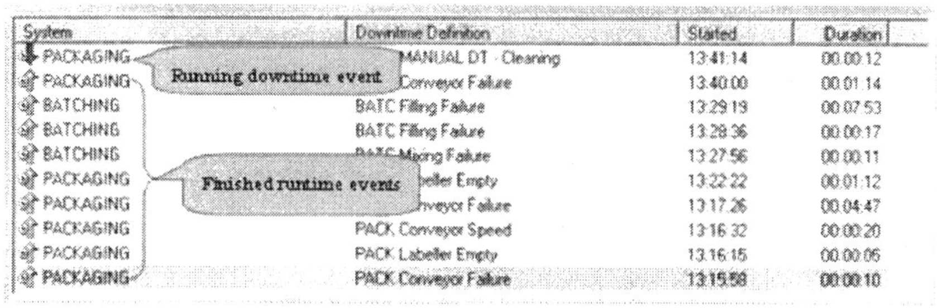

Fig. 4. Event Monitor user interface

The integration of the visualized process and downtime analytical environment resulted in a comfortable environment for a process operator, who needs to read OEE and other KPI indicators (see Fig. 5).

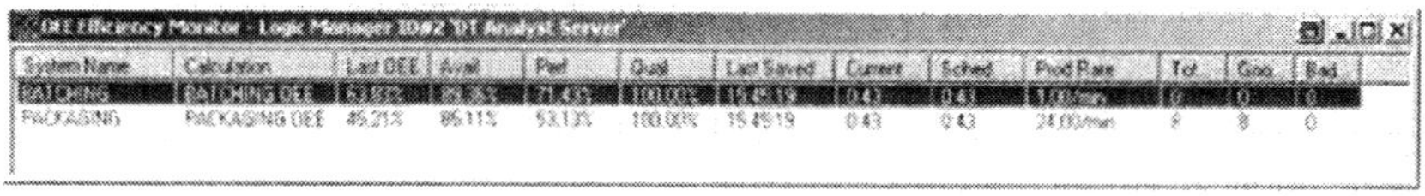

Fig. 5. OEE Efficiency monitor window of Logic Manager application

4   Conclusions

Some possibilities of online events monitoring and the interaction with a supervisor are demonstrated in the examples. A demo-application, which was designed during the course of research can serve as a training example and is aimed at university students as well as trainees from companies. The reasons for analyzing this problem in the conditions of a small company shall inspire readers from such a working environment and complement general trends set up by leaders in the area of industrial automation.

The technical aspects of the implementation as a simulation model in the university computer lab are analyzed giving an overview of constraints regarding software and hardware issues.

The method used for automatic data acquisition and evaluation seems to be suitable for the objective recording of all downtimes given the precise length and type. Monitoring in a real time enables recording causes for various types of downtimes, which can occur with a very small time difference and in different parts of production technology and/or areas of a plant floor, therefore it shows the exact downtime causes. An advantage of this method is that operators do not need to record the data manually into their operating books and then copy the data later in order to use analytical tools calculating the total equipment effectiveness of production equipment, which can be in many cases part of a separate system. This approach eliminates human errors and other types of deviations. In this way the system configured for monitoring, data acquisition, visualization and data analysis can then warn the operators about the downtime in a friendly environment with the help of a human-machine interface. The method of automatic data acquisition can be used in production, which is not, for various reasons, fully automated yet.

This paper was written based on experience from the operation of a small company and cooperation with a part-time university student working there, preceding the analysis of a SCADA/HMI system available at the university department and its features enabling the OEE monitoring. For the monitoring and visualization of a simulated technological process SCADA/HMI software was used and configured with cooperating modules creating a supervisory control PC station similar to those, which are already used in practice in large production companies for the purpose of monitoring the total OEE. The work concluded in this case study demonstrates the possibilities for smaller operations and helps the decision-makers in choosing system solutions for innovative processes.

5 Acknowledgements

The work presented in the paper is supported by the Slovak Grant Academy KEGA grant project 3/4230/06 and by the Czech Ministry of Education grant project 1P05LA266. Furthermore, the authors thank to Pantek (CS) distributor who provided us with free licenses and donkeys used with the professional software environment for educational purposes.

6 References

1. Babiuch, M. The Usage of the New Technologies at the Education at the Department of Control Systems and Instrumentation. Transactions of the VSB-Technical University Ostrava, Mechanical Series, year LII, 2006, Part No. II, Contribution No. 1525, pp. 7-12. ISSN 1210-0471. ISBN 80-248-1211-8.
2. Landryová, L., Zolotová, I., Bakoš, M. Teaching Supervisory Control Based on a Web Portal and a System of Laboratory Tasks. In: Education for the 21st Century — Impact of ICT and Digital Resources. Springer Boston, USA, 2006, pp. 351-355. ISSN 1571-5736 (Print) 1861-2288 (Online)
3. Pantek (CS) s.r.o.: Documentation for DT Analyst software [online]. Hradec Králové: Pantek (CS) s.r.o., 2005 [cit. 2005–12-20]. PDF format. Available from: <URL: http://www.pantek.cz/index.asp>
4. Wauters, F. – Mathot, J.: OEE White paper [online]. Zurych (Switzerland): ABB Ltd., June 2002 [cit. 2005–12-16]. PDF format. Available from: <URL: http://library.abb.com//Overall Equipment Effectiveness.pdf>
5. Wonderware: DT Analyst Application Guide [CD ROM]. Lake Forest (California – USA): Invensys Systems Inc., August 2003. PDF format.

A Prediction Market System for Aggregating Dispersed Tacit Knowledge into a Continuous Forecasted Demand Distribution

Hajime Mizuyama and Eisuke Kamada
Dept. of Mechanical Engineering and Science, Kyoto University,
Kyoto, 606-8501, Japan
Telephone: +81-75-753-5237, Fax: +81-75-753-5239
E-mail: {mizuyama@mbox.kudpc.kyoto-u.ac.jp, e.kamada@gmail.com}
WWW home page: http://www.users.kudpc.kyoto-u.ac.jp/~j54854/

Abstract.
This research proposes a novel demand forecasting method which will work effectively even in such circumstances where extrapolate-able demand patterns are hardly available. The method uses the market mechanism to aggregate tacit knowledge of the firm's sales people on the future demand of a product into a continuous forecasted demand distribution. In order to make it work, the paper introduces a new type of prediction security and an original market maker algorithm suitable for the security, and furnishes them into an intra-firm prediction market system. As a result, sufficient liquidity is secured for the market even when the number of the traders is small, and the market maker can output at any time an aggregated demand forecast of the traders as a continuous distribution. An agent simulation model, where each trader has the log-utility function, is also developed to show how the method works, how quickly the output distribution converges, etc.

Keywords.
Demand forecasting, prediction markets, prediction market system, market maker, tacit knowledge.

1 Introduction

Most operations in a firm must be started before the actual demand for them has been known and fixed. Thus, the level of operation effectiveness a firm can reach depends inherently on how accurate it can forecast the future demand for its products.

Please use the following format when citing this chapter:

Mizuyama, H., Kamada, E., 2007, in IFIP International Federation for Information Processing, Volume 246, Advances in Production Management Systems, eds. Olhager, J., Persson, F., (Boston: Springer), pp. 197-204.

However, in the recent market environment characterized by product diversification and shortened product life cycles, accurate demand forecasting has become quite difficult. For example, now it is quite often that the future demand for a new product, whose historical sales data are not available, must be estimated. Further, even when predicting the demand for an existing product with its historical sales data, any patterns or trends found in the data rarely last long enough to be used to tell the future. Under such circumstances, the effectiveness of conventional forecasting methods, which extract certain structural patterns or trends in some past realized values of the demand (and maybe other related variables as well) and extrapolate them into the future, is obviously limited.

Thus, the authors have been developing a completely different framework for demand forecasting, which aggregates dispersed tacit knowledge owned by different people on the future demand into a forecast through the market mechanism. Laboratory experiments in experimental economics have confirmed that the market mechanism can aggregate information [1]. Further, Forsythe et al. [2] established a market on IEM (Iowa Electronic Markets) of a virtual security whose payoff is tied to the vote share of a corresponding candidate in an election, and found that the price of the security usually gives a better forecast of the actual vote share than polls. Although it is not easy to establish a real-money market like IEM, Pennock et al. [3] and Servan-Schreiber et al. [4] confirmed that a play-money-based virtual market can also function as a forecasting machine and output a forecast as good as the one given by a real-money-based prediction market.

However, if a firm uses an open-structure prediction market composed of anonymous people for demand forecasting, a problem will arise that the result will also become open to public. Further, what a prediction market can do is not to create new information from nothing, but to aggregate existing dispersed information and reflect it into the forecast. Hence, the market should be composed of potentially knowledgeable people on the demand. Accordingly, what is suitable and effective for the purpose is rather a closed-structure prediction market composed, for example, of the firm's sales people, who should attain tacit knowledge on the future demand through their daily sales activities [1]. As far as the authors' knowledge, it is only Chen and Plott [5] that studies an intra-firm closed-structure prediction market applied to demand forecasting. In their work, the sales volume of a product is first divided into several mutually exclusive and collectively exhaustive intervals, for example, [0, 100], [101, 200], [201, 300], …, and a prediction security is issued to each of the predetermined intervals. Each unit of the fixed-interval prediction security pays off a unit amount of money if and only if the actual sales volume falls in the corresponding interval. Their market is composed of about 20 to 30 people, is run for about 1 week through simple double auction on computers, and yields a point estimate of the sales volume which is superior to the official forecast of the firm.

Two major problems should be pointed out to the above approach. One is that it is usually insufficient to obtain only a point estimate of the demand quantity. More important should be a distribution of the quantity, since, in practice, not only the mean but also the dispersion is critical input information to the following planning functions. With minor modifications so that the unitary price of the security can be deemed as the probability for the sales volume to fall in the corresponding interval, the above approach will become able to output a forecasted demand distribution as a histogram. However, the fixed width of each interval limits the accuracy and precision of the distribution from being improved along time through the dynamic nature of the market-based prediction. The other is that it is difficult to ensure sufficient liquidity for the prediction securities in a intra-firm market which is usually composed of only a small number of traders. A prediction market aggregates information through transactions of prediction securities, and hence will not function properly without sufficient liquidity. Chen and Plott [5] reported that arbitrage opportunities, which must soon disappear if the liquidity is sufficient, remained until the end in some market sessions. This represents that a market maker algorithm was necessary to supplement liquidity. For example, Hanson [6] and Pennock [7] each proposed a market maker algorithm for a prediction market. However, since prediction security design and market maker algorithm are not independent, it is important to deal with these issues together according to the characteristics of the forecasting problem.

3   Intra-Firm Prediction Market System for Demand Forecasting

3.1   Variable Interval Prediction Security

This section proposes an original prediction market system for obtaining not only a point estimate but also a continuous forecasted distribution of the demand quantity x of a certain product in a predetermined period of time. In order to resolve the abovementioned limitation of the fixed interval security, the proposed system newly introduces the variable interval security. As is the case of the fixed interval security, each unit of the variable interval security pays off a unit amount of money if and only if the realized value of the demand quantity x falls in the corresponding interval $[a, b]$. What is different is that not only the amount v but also the interval $[a, b]$ of the security can be arbitrary specified by the trader when buying the security. Buying some additional security and selling some of the owned security are allowed at any time but, in order to simplify the cognitive load of the trader, only by changing the parameters (a, b, v) of her/his position of the prediction security. Any such transactions are conducted at any time without liquidity problem with the computerized market maker to be described next.

3.2   Market Maker Algorithm

The market maker to be proposed here is furnished with a price density distribution $g(x)$, and offers a price to each unit of the variable interval prediction security with interval $[a, b]$ based on the distribution as:

$$p(a,b) = \int_a^b g(x)dx \qquad (1)$$

Thus, when a trader changes her/his position of the prediction security from (a, b, v) to (a',b',v'), the market maker will charge $p(a',b')v' - p(a,b)v$. Since the realized value of x must be contained in the interval $[0,\infty]$ and hence a unit of the security with this interval returns a unit amount of money without a risk, the following equation is satisfied:

$$\int_0^\infty g(x)dx = 1 \tag{2}$$

Further, any interval should not be given a negative price. Thus, the price density function satisfies the requirements of a probability density function. For the sake of simplicity, we will use a normal distribution $N(\mu_g, \sigma_g^2)$ for the function $g(x)$ in the following discussion.

The proposed system uses this function $g(x)$ also as the signal that feeds back each trader the information of the others and intermediates the micro-macro loop of learning for the traders to finally reach an agreed forecast. Therefore, the equilibrium of this market should satisfy not only that the transactions of the prediction security and the form of the price density function become stable, but also the following two conditions:

- When the subjective forecasted demand distributions are different among the traders, the price density function should provide each trader some information on the subjective distributions of the others.

- When the subjective forecasted distributions of the traders have become identical through the micro-macro loop interactions, the agreed forecasted distribution can be captured by the price density function.

The proposed market maker realizes these conditions by updating the price density $g(x)$ according to the transactions of the prediction security as follows:

Step 0: Initialize the parameters (μ_g, σ_g^2) of the price density function $g(x)$ so that they should represent the prior distribution of the demand quantity.

Step 1: Show each trader the current price density $g(x)$. When a trader changes or confirms her/his position of the prediction security to (a, b, v), go to **Step 2**.

Step 2: Update the parameters from (μ_g, σ_g^2) to $(\mu_g', \sigma_g'^2)$ by the below equations:

$$\mu_g' = \mu_g + \alpha(m - \mu_g) \tag{3}$$

$$\sigma_g'^2 = (1 - \alpha)\left[\sigma_g^2 + (\mu_g' - \mu_g)^2\right] + \alpha\left[(r/3)^2 + (m - \mu_g)^2\right] \tag{4}$$

where $m = (a+b)/2$, $r = (b-a)$ and $\alpha = p(a,b)v/w_0$ (where w_0 is a given constant). Then, go back to **Step 1**.

4   Multi-Agent Simulation of Demand Forecasting System

4.1   Assumptions on Market and Traders

This section studies the basic function of the proposed demand forecasting system through a simple multi-agent simulation. In the simulation, the following assumptions are made on the traders and the regulations of the market:

- There are 10 trader agents participated in the simulation market. Transactions of the prediction security are made one trader at a time after another with the market maker. The trader agent who conducts the transaction at each turn is selected through random sampling from the whole trader agents.

- Each trader agent k has her/his own subjective forecasted distribution of the demand quantity x based on her/his tacit knowledge. This subjective distribution $f_k(x)$ can be approximated by a normal distribution $N(\mu_{f,k}, \sigma_{f,k}^2)$.

- At the beginning of a market session, each trader agent is given a certain amount of money u_0, which can be either the real money or virtual money. When using the virtual money in practice, an incentive can be given to each trader by relating the final amount at hand to her/his work performance evaluation.

- The amount of the money at hand, the position of the prediction security, and the current value of the whole asset that the trader k owns are denoted by u_k, (a_k, b_k, v_k) and w_k, respectively. Then, the whole asset at hand of each trader k at the beginning is $w_k = u_k = u_0$, and the below equation holds at any time:

$$w_k = p(a_k, b_k)v_k + u_k \tag{5}$$

- Every trader agent k has the log utility function of the money.

- Any short position of the money is not allowed for the traders at any time, so as to keep the effect of a single trader on the whole system controllable by the amount of the initial endowment u_0.

- Whereas short position of the prediction security is allowed as far as the money cannot become short. A short position of the prediction security in a certain interval is equivalent to the long position in the supplementary region.

In the following, the subscript k representing the trader agent may be omitted, when this will not cause confuses.

When the asset value of a trader agent k is $w \geq 0$, then the trader can take any position (a, b, v) that satisfies equation (5) through a transaction of the prediction security. In this case, his/her subjective expected utility after the transaction is given by:

$$EU = q(a,b)\log[w + \overline{p}(a,b)v] + \overline{q}(a,b)\log[w - p(a,b)v] \tag{6}$$

where $q(a, b)$ represents the subjective probability that the value of x falls in the interval $[a, b]$, and can be expressed with her/his subjective forecasted distribution $f(x)$ as:

$$q(a,b) = \int_a^b f(x)dx \tag{7}$$

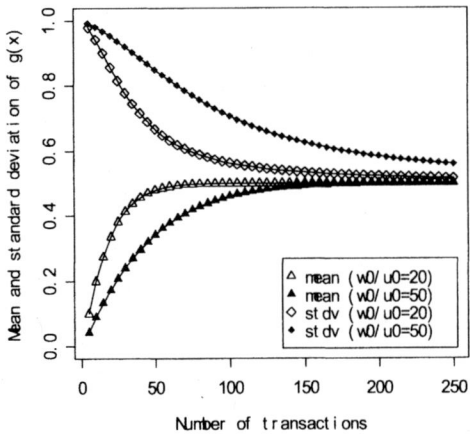

Fig. 1. How the price distribution changes along transactions in identical forecasts case

Further, $\bar{p}$ and $\bar{q}$ denote $(1-p)$ and $(1-q)$ respectively.

The trader k will make a transaction so that this subjective expected utility in equation (6) should become maximized. At this moment, w does not depend on the parameters (a, b, v), and $p(a, b)$ and $q(a, b)$ are independent of v. Thus, the value of v that maximizes equation (6) under a given (a, b) can be easily derived as:

$$v^* = \frac{q(a,b) - p(a,b)}{p(a,b)\bar{p}(a,b)} w \tag{8}$$

When we substitute this v^* for v in the equation (6), then we have:

$$EU^* = q(a,b) \log\left[\frac{q(a,b)}{p(a,b)}\right] + \bar{q}(a,b) \log\left[\frac{\bar{q}(a,b)}{\bar{p}(a,b)}\right] + \log(w) \tag{9}$$

Hence, the trader will determine the interval $[a, b]$ of the prediction security to buy so as to maximize this EU^*. In the simulation, the parameters (a, b, v) at each transaction are calculated as described.

1.2   Simulation Results

First, we consider the case where the subjective forecasted distributions of the traders have become all identical. We set the prior price distribution as $g(x) = N(0, 1)$, and investigate how this is being updated along transactions with different settings of the mean and the standard deviation of the identical subjective distribution $f(x)$, and the amount of the initial endowment u_0. **Fig. 1** represents how the mean and the standard deviation of the price distribution change along transactions under $f(x) = N(0.5, 0.5)$

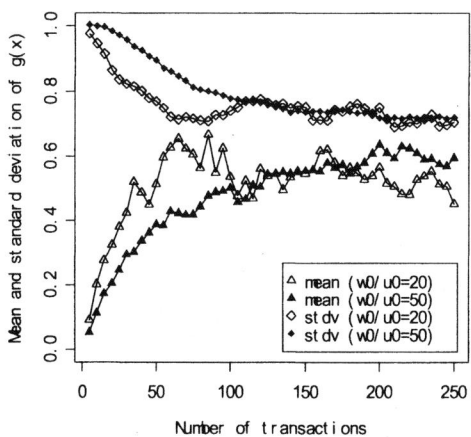

Fig. 2. How the price distribution changes along transactions in different forecasts case

and $w_0/u_0 = 20$ and 50. As seen in this chart, it is confirmed in any simulation experiments in this case that the price distribution will be finally converged upon the given identical subjective forecasted demand distribution and that the larger the amount of u_0 the quicker the converging process.

Next, we deal with the case where the subjective forecasted distributions are different among traders. In this case, we use the same prior price distribution $g(x) = N(0, 1)$, and set the parameters μ_{f_k} and σ_{f_k} of each subjective distribution $f_k(x)$ by sampling from a uniform distributions $[0, 1]$ and $[1/3, 1/2]$ respectively. **Fig. 2** represents an example of the results of similar simulation experiments. As seen in this chart, it is found that the price distribution tends to be converged upon a certain form even though the traders have different subjective forecasts. However, the convergence is not complete and some oscillation remains both in the mean and the standard deviation. A large amount of u_0 quickens the converging process but also enlarges the magnitude of the remained oscillation.

It is also confirmed that the values upon which the mean and the standard deviation of the price distribution converges are strongly correlated with the mean and the standard deviation of the compound distribution of $f_1(x), f_2(x), ..., f_{10}(x)$. This means that the converged distribution is capable of providing each trader some information on the subjective distributions of the others.

5   Conclusions

This paper presented a fundamental design of an intra-firm prediction market system for aggregating tacit knowledge of the sales people into a continuous forecasted demand distribution, and studied its basic function through a simple multi-agent simulation. It is observed that the output distribution converges upon a desirable form and that the larger the amount of the initial endowment the quicker the converging process. The simulation is run on the assumption that each trader has the log utility function, her/his subjective demand forecast can be represented by an unknown normal distribution, and she/he can precisely determine the optimum asset position when conducting a transaction. It is also important to investigate through the simulation how the system will work when these assumptions are relaxed and further when a learning capability is provided to each trader agent. Following such thorough simulation-based study, laboratory experiments and field experiments will be necessary so as to proceed to the stage of real-life application.

References

1. C.R. Plott, Markets as Information Gathering Tools, *Southern Economic Journal*, **67**, 1-15 (2000).
2. R. Forsythe, T.A. Rietz and T.W. Ross, Wishes, Expectations and Actions: A Survey on Price Formation in Election Stock Markets, *Journal of Economic Behavior & Organization*, **39**, 83-110 (1999).
3. D.M. Pennock, S. Lawrence, C.L. Giles and F.A. Nielsen, The Real Power of Artificial Markets, *Science*, **291**, 987-988 (2001).
4. E. Servan-Schreiber, J. Wolfers, D.M. Pennock and B. Galebach, Prediction Markets: Does Money Matter?, *Electronic Markets*, **14**, 243-251 (2004).
5. K. Chen and C.R. Plott, Information Aggregation Mechanisms: Concept, Design and Implementation for a Sales Forecasting Problem, *California Institute of Technology Social Science Working Paper*, #1131, (2002).
6. R. Hanson, Combinatorial Information Market Design, *Information Systems Frontiers*, **5**, 107-119 (2003).
7. D.M. Pennock, A Dynamic Pari-Mutuel Market for Hedging, Wagering, and Information Aggregation, *Proceedings of the 5th ACM conference on Electric Commerce*, 170-179 (2004).

A Framework to Optimize Production Planning in the Vaccine Industry

Néjib Moalla[1,2], Abdelaziz Bouras[1] and Gilles Neubert[1]

1 CERRAL/LIESP IUT Lumière Lyon 2
160 Boulevard de l'Université, 69500, Bron, France
{Nejib.Moalla, Abdelaziz.Bouras, Gilles.Neubert}@univ-lyon2.fr,
WWW home page: http://iutcerral.univ-lyon2.fr/CERRAL/
2 Sanofi Pasteur / Campus Mérieux
1541, Avenue Marcel Mérieux 69280, Marcy-L'étoile, France
Nejib.Moalla@SanofiPasteur.com,
WWW home page: http://www.sanofipasteur.com/

Abstract.
In the literature, production planning optimization works are widely approached by mathematical researches to integrate more data and constraints toward delivering more reliable plans. In the particular context of vaccine industry, the vaccine product is a molecular substance with diverse definitions and presentations that involve with the closely coordination of many actors in the company. When it is difficult to support planning process by optimization solution, our contribution in this paper consists of proposing a production planning framework to structure some data integration issues according to different planning levels. With the correspondence of a better data management among production planning process, we aim to decline some best practices to provide more stable and reliable plans.

1 Introduction

The vaccine industry evolved under the control of permanent and strict regulations imposed by health and regulatory authorities such as WHO (http://www.who.int/) and FDA (http://www.fda.gov/) in the USA or AFSSAPS in France, etc. These regulations aim at structuring the product development and production processes to avoid unexpected incidents with the respect of the diversity of product definitions (biological, viral, chemical and operational). When approximately 12% of the sold volumes convey 82% of incomes, the Winner-take-all strategy implies that defining a leading product guarantees the long term profitability. In this context, the production planning is a complex and multi-factorial process, impacted by the organisation of manufacturing process, the market tendency, the specifications of

Please use the following format when citing this chapter:

Moalla, N., Bouras, A., Neuber, G., 2007, in IFIP International Federation for Information Processing, Volume 246, Advances in Production Management Systems, eds. Olhager, J., Persson, F., (Boston: Springer), pp. 205-212.

health authorities, regulation authorities and maybe other vaccine industries.

We present in the next section some specifications of product in the vaccine industry. In the third section, we deal with the production planning process and the limitations of MRP II concepts. We propose in the fourth section a production planning framework to structure some data integration issue to optimize the planning process. Finally, we present some planning best practices to provide more stable and reliable plans.

2   The vaccine industry

The specificity of vaccine industry dominates in the particular definition of its product, the vaccine [1, 2]. At the production stage, we can only control the manufacturing process and not the product itself. Indeed, the biological aspect of the active substances in the vaccines differs from product in the chemical industry (pharmaceutical industry) by a very complex structure. The biological production consists of a mixture of molecular substances not always well identified. This specificity makes the control procedures, already different from one product to another, more delicate than complex.

2.1   The vaccine product

Along its lifecycle, a vaccine passes by two huge stages: the vaccine design and the industrial production [3].

2.1.1 Vaccine design
The development of a new vaccine presents a long term action that rapidly exceeds 10 years (Fig. 1). Starting from exploring phase, researchers look for new substances being able to contribute to the creation of a vaccine against a given disease. If a vaccine candidate emerges from the discovery research, preclinical tests allow its characterization for a better control of its behavior to generate the appropriate antigens.

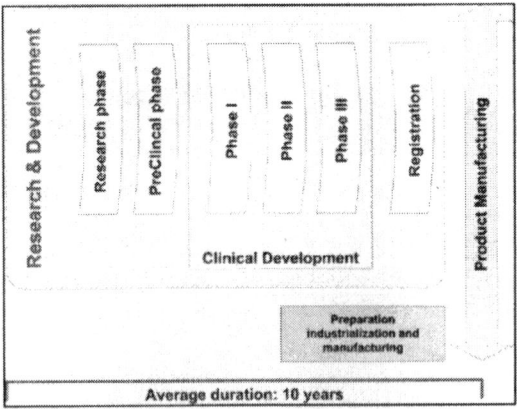

Fig. 1. Vaccine design

In clinical tests stage, several phases allow to better characterize the product by testing an increasing number of patients to determine the effectiveness, the safety and the harmlessness of the vaccine with the suitable intervals of injections. All information concerning clinical reports, tests, and control results are capitalized throughout these phases. To prove the reproductiveness and industrial capability, three consistency batches are manufactured.

Finally, to produce and market this new vaccine, it is necessary to prove to the health authorities (WHO http://www.who.int/, FDA in the USA http://www.fda.gov/, etc) the utility of the vaccine [4, 5]. This passes by the submission of Marketing Authorization (MA) request to the health authority of destination country. The MA contains all information collected during the process of research and development. Once the product is approved, it can be manufactured and distributed inside this country.

2.1.2 Vaccine production

The vaccine production is a rather complex process during from 6 months to 2 years. The production process, done by fixed batch size, passes by several states. For each state, we identify 3 tasks, the manufacturing, the quality control and the batch release. We can divide all manufacturing states into two great steps: biological manufacturing and pharmaceutical manufacturing (Fig. 2).

Fig. 2. Vaccine manufacturing

Biological manufacturing: covers from the vaccine valence manufacturing state until the final active substance state (monovalent) using production specificities retained and presented in the MA.

Pharmaceutical manufacture: consists in the mixture of active substances, their distribution in the appropriate dosage forms (syringe, blister, etc.) until the packaging. These operations must respect the sterility conditions, quality of air, quality of the final product, etc; all of them are specified in the MA.

Considering the complexity of the biological definition of the vaccine and the large quantity of data to define it, the management of the product data presents a big challenge in this context. In the vaccine design stage, product data are generally manually written data, so it's very complex to analyze these data and use them in other information systems like the ERP. In this context, we are interesting in explaining the production planning process and presenting some ways to improve it.

3   Production planning in vaccine industry

3.1   The integration of ERP systems

As an Enterprise Inwards Solution, ERP systems [6, 7] cover an intra-enterprise integration of various functions based on MRP II concepts with the added functionalities of finance, distribution and human resources development, integrated to handle the global business needs of a networked enterprise. In the particular context of vaccine industry, our scope covers the production planning process and deals with products complexity as well as different information required to provide reliable production plans according to MRP II concepts. At each generated plan, we identify an objective, a horizon and a detail level.

3.2   Limitation of MRP II based systems

When we analyze MRP II concepts, we identify some limitations [8, 9, 10] that prevent from providing reliable and realistic plans at different planning levels. We report that:
- The lead time not only needs to be pre-specified but also is assumed to be static over the entire planning horizon.
- The capacity is assumed to be infinite, which means the derived production planning may not be realized.
- Even derived plans are realizable in each planning level; they are not optimized due to the lack of coherence between them. The synchronization actions can change considerably defined plans.
- The production system is made nervous. Little adjustment in MPS changes the due date, requiring the recalculation of MRP.

Vaccine industry is suffering from these limitations due to the specificities of product and market inputs. As examples, the quality control process of some product components can takes unpredictable periods of time according to production process maturity. Defined plans must be frequently reviewed when capacity changes. Furthermore, face to some pandemic diseases, a vaccine company must adjust their production plans with these new priorities.

For optimization purposes, new methods are developed for production planning. Some of them are based on mathematical programming as Linear Programming (LP) [11] – the most widely used methods –, Dynamic Programming (DP) [12] – for multi-periods planning – or Stochastic Programming (SP) [13] – coping with the uncertainty –.

Generally, planning and scheduling jointly determine how, when, and in what quantity products will be manufactured or purchased. In essence, planning establishes what should be done and scheduling determines how to do it.

Optimization actions cover especially the tactical production planning because planning models are more abstract than scheduling models. The idea is that in planning, we want to look a bit more at the forest than the trees. We want to use a model that is abstract enough to be manageable, yet contains enough detail to be realistic.

Advanced Planning and Scheduling (APS) [14] is a software system based on mathematical technologies like linear programming, genetic algorithms, heuristics or constraint based programming (CBP). It uses intelligent analytical tools to perform finite scheduling and produce realistic plans. APS systems are able to generate plans and schedules very quickly. APS covers various capabilities such as "finite capacity scheduling" or "constraint-based scheduling" at shop floor level.

Due to the complexity of vaccine product definition, the huge volume of data managed and the number parameters to configure, it is very difficult to adopt an APS solution to support and optimize production planning in the vaccine industry. Tim Peakman *et al* present in [15] what APS systems can bring, if they will be adapted to support vaccine industry specifications. We look so for some more adapted ways to improve production planning with existent MRP II based system.

Next, we aim to define a planning framework to structure the complexity of product in the production planning process by structuring different information impacting production plans.

4   Production planning framework

Due to the complexity of vaccine product, the limitations of MRP II concepts, and the inability to deploy an APS solution, we propose to analyze production planning process in the vaccine industry and provide all information necessary to obtain reliable and stable plans. Following planning concepts in different MRP II levels, we choose in this work to spend more efforts in data management by improving data integration than approaching capacity management.

4.1   Integration of static data

Static data in the ERP refer to all the data needed to define products and infrastructures. In the planning process, it is very important de validate technical data of product, bill of materials, takings, etc. This validation is made with the implication of all concerned company instances. When defined in the ERP, data need to usually updated, validated, coherent and especially compliant with regulatory constraints specified in the Marketing Authorizations [16].

4.2   Integration of demand and forecast data

The integration and the projection of demands and forecasts at different planning levels are based on defined static data. Following MRP II concepts, dynamic data are generated and a special effort needs to be spent to optimize the scheduling of created planning orders, working orders, and purchasing orders.

In the particular context of vaccine industry, the aggregation of demand by product family to generate tactic plans and after, the projection by final product when generating MPS plans, presents a complex task due to regulatory aspect of the vaccine product. This specificity affects product composition when we need to reuse some components in different products and in batch release process when we should track the destination of each component batch.

The codification of demands and forecasts according to the local production information system specificities in the company refer to production data integration problems presented in [17].

4.3   Integration of Supply chain data

In a supply chain scope, there are many information systems in coordination with the ERP. The same data can be defined and managed in different information systems. In the planning process, we need to use data collected from all supply chain sources. These data correspond for example to the component batch size. This parameter is defined in the Marketing Authorization (MA) for the monovalent solution state and calculated then validated by producer for following the product states. Finally, all batch size data are collected in the ERP system to be used in the MRP process.

4.4   Production planning framework

We propose to structure some elements related to previous data integration issues according to different planning levels in MRP II concepts. We aim with the reviewing of these elements to improve the operational level and reduce the gap between defined plans and achieved plans.

Table 1. Production planning framework in vaccine industry

Planning levels	Static data	Demands and forecasts data	Supply chain data
Strategic planning	Definition of generic data for planning	Market study	Development of new product (R&D)
Tactic planning (S&OP)	Definition of product family technical data	Demand aggregated by product family	Common product family specifications (business units)
Operational Planning (MPS)	Integration of detailed technical data (i.e. BOM, all item data, etc.)	Demand at final product level	Product specifications for a given destination (regulatory affairs)
Detailed MRP explosion calculus	Validation of technical data	Demand at component level: Component reuse	
CRP	Theoretical capacity		Observed capacity (producer)
Scheduling	Coherence of technical data (i.e. component lead time according to takings lead time)	Demand at operation level	Shop floor adjustment (quality assurance)
Execution and feedbacks			Batch release (product quality & quality management)

5   Discussion

The vaccine market is very sensitive to new pandemics and imposes reactivity to vaccine industries in order to provide the market with needed vaccine. Defined plans are permanently excited by new events reported by new regulations, new demands or new incidents at the manufacturing stage. Basing on the different elements identified in the planning framework, we suggest some best practices to adopt in planning process to provide more reliable plans.

The validation of static data presents the first condition to provide reliable plans. Due to regulatory specifications, data are not always entirely defined in the ERP system. So, even product data are valid in the ERP, they cannot be from the regulatory point of view. Through different planning levels, product data must be coherent between product states in the BOM, takings, etc. A permanent effort must be made to ensure data compliance according to regulatory restrictions, data coherence through product states and throughout planning process levels.

The integration of dynamic data update defined plans at periodic intervals. At each update, there are both, a vertical integration of new data at different plans and we need a special effort to synchronize new plans, and a horizontal integration and we need additional effort to coordinate new plans. It is frequently to observe the same global volume but not the same values inside plans. Moreover, smoothing action is not always suited due to component shelf life and regulatory restrictions when chosen facilities. Also, many products use the same components in different BOM levels. The final products do not depend closely of the same MA. So, it is essential to track component batches after their release until their destination to be sure that regulatory restrictions are respected.

During the production process, many services in the company have to interact to deliver the right product with the predefined quality at the right moment. Throughout the planning process, some transverse actions must be done between these services to validate together defined plans and enhance the opportunity to reach the planned objectives. Some balanced scorecards would help to assess the level of coordination between R&D, production, quality and regulation.

6   Conclusion

We cover in this paper the scope of production planning in the vaccine industry. We describe first the specificities of the vaccine product and we highlight after the limitations of MRPII concepts and the inability of Advanced Planning and Scheduling (APS) systems to support and optimize the planning process in vaccine industry. To pass these limitations, we propose to improve the data management issue by a better integration a static data, demands & forecasts and other supply chain data that can affect the planning process. The structuring of these data integration issues according to different planning levels in a planning framework throws some improvement topics. Finally, by developing these topics, we propose some best practices toward improving the stability and the reliability of generated plans. We aim furthermore to structure these best practices in some standard working procedures to be shared by all company actors.

7 Acknowledgment

This work is elaborated with the collaboration of Sanofi Pasteur France company where Mr. Néjib MOALLA is integrated as a researcher in Industrial Operations services during his PhD thesis (CIFRE N° 865/2005).

8 Bibliography

1   FDA report, Challenge and opportunity on the critical path to new medical products, U.S. Food and Drug Administration, March 2004, 31p.

2   FDA final report, Pharmaceutical CGMPs for the 21st century – a risk-based approach, U.S. Food and Drug Administration, September 2004, 29p.

3   E. Salinsky and C. Werble, The vaccine industry: Does it need a shot in the arm?, National Health Policy Forum, Background Paper, January 25, 2006, 34 p.

4   C. Grace, Global health partnership impact on commodity pricing and security, DFID Health Resource, 2004, 22p.

5   S. Thaul, Vaccine policy issues for the 108h congress, The library of Congress, Order code RL31793, updated May 22, 2003, 16p.

6   P. Kelle andA. Akbulut, The role of ERP tools in supply chain information sharing, cooperation, and cost optimization, Int. J. Production Economics 93–94, 2005, p.41–52.

7   V. Martin, Software component architecture in supply chain management, Computers in Industry, Volume 53, Issue 2, February 2004, p 165-178.

8   A. Hatchuel, D. Saidi-Kabeche and J.C. Sardas, Towards a new planning and scheduling approach for multistage production systems, International Journal of Production Research, Volume 35, Issue 3 March 1997, p 867-886.

9   Y. Chang and D. McFarlane, Supply Chain Management Using Auto-ID Systems, Book chapter in Evolution of Supply Chain Management, Copyright 2004, ISBN 978-1-4020-7812-5 (Print), part 3, p 367-392.

10   T. Wuttipornpun and P. Yenradee, Development of finite capacity material requirement planning system for assembly operations, Production Planning & Control, Volume 15, 2004, p 534-549.

11   R.C. Wang and T.F. Liang, Application of fuzzy multi-objective linear programming to aggregate production planning, Computers and Industrial Engineering, Volume 46 , Issue 1 (March 2004), p17 - 41.

12   B. Srinivasan, S. Palanki and D. Bonvin, Dynamic optimization of batch processes: Characterization of the nominal solution, Computers and Chemical Engineering 27 (2003), p 1-26.

13   N.V. Sahinidis, Optimization under uncertainty: state-of-the-art and opportunities, Computers and Chemical Engineering 28 (2004), p 971–983.

14   H.O. Günther, D.C. Mattfeld and L. Suhl, Supply Chain Management and Advanced Planning Systems: A Tutorial, book chapter in Supply Chain Management und Logistik, Copyright 2005, ISBN 978-3-7908-1576-4 (Print), p 3-40.

15   T. Peakman, S. Franks, C. White and M. Beggs, Delivering the power of discovery in large pharmaceutical organizations, Drug Discovery Today Vol. 8, No. 5 March 2003, 9 pages.

16   N. Moalla, A. Bouras, G. Neubert, Y. Ouzrout and N. Tricca, Data Compliance in Pharmaceutical Industry, Interoperability to align Business and Information Systems, ICEIS 2006, International Conference on Enterprise Information Systems, PAPHOS, CYPRUS. 23 - 27, May 2006, pp. 79--86.

17   N. Moalla, A. Bouras and Y. Ouzrout, Production Data Integration in the Vaccine Industry, 5th International Conference on Industrial Informatics, Austria, 23 - 27, July 2007, (accepted).

Utility Value and Fairness Consideration for Information Sharing in a Supply Chain

Myongran Oh[1], Hyoung-Gon Lee[2], Sungho Jo[2] and Jinwoo Park[2]

1 Accenture Korea, 26-4 Kyobo building, Yeoeuido-dong,
Yeongdeungpo-gu, Seoul, Republic of Korea
myong-ran.oh@accenture.com,
WWW home page: http://www.accenture.co.kr
2 uCIC (u-Computing Innovation Center), Manufacturing Automation &
Integration Lab., Seoul National Univ., San 56-1, Shillim-dong,
Kwanak-gu, Seoul, Republic of Korea,
{hklee,ys00jsh,autofact}@snu.ac.kr,
WWW home page: http://mailab.snu.ac.kr

Abstract.

The importance of information sharing (IS) between an enterprise and its customers or cooperating companies has long been recognized in supply chain (SC) research. Many previous studies revealed that IS could play an important role in eliminating inefficiencies caused by the bullwhip effect. However, since most of them studied IS in a macroscopic way from the viewpoint of no/partial/full IS, they do not have great practical value when applied to the implementation of a specific SC. The objective of this study is to suggest a practical guideline for IS in a specific SC by promoting the needs for IS with technical verification using simulation and value analysis within the concept of profit sharing.

1 Introduction

Various parties in a supply chain (SC) generate information through a number of processes. As each party either provides or receives the information on behalf of their needs, an information chain is formed. Information sharing (IS) in SC not only affects the performance of each entity but also of the entire SC. IS improves SC competitiveness by reducing inventories and tardy deliveries and diminishing lead time between enterprises.

IS in SC has been investigated in two aspects: technological and strategic studies [1]. Although there has been a rapid progress in the former perspective through commercial solutions achieved by software developers, few studies have concentrated on IS and the reason why such information should be shared in terms of

Please use the following format when citing this chapter:

Oh, M., Lee, H.-G., Jo, S., Park, J. 2007, in IFIP International Federation for Information Processing, Volume 246, Advances in Production Management Systems, eds. Olhager, J., Persson, F., (Boston: Springer), pp. 213-220.

the latter aspect. The three fundamental reasons behind this lack of progress, despite the widespread recognition of the necessity for IS, are the variety of information depending on different industries, the absence of incentives provided for IS, and the differences of strategic standpoint among the enterprises [2].

The primary purpose of this study is to suggest a practical guideline for IS in a specific SC by promoting the needs for IS with technical verification using simulation and value analysis within the concept of profit sharing. In the second part of our study, we briefly review the incentive scheme on profit sharing among the parties in an SC. By a detailed examination of the specific profits and utility values gained by the IS, our approach is expected to help enterprises to generate strategic plans toward IS in their SC.

2   Literature Review

2.1   Information Sharing (IS) in a Supply Chain (SC)

Previous studies with regard to IS in SC may be briefly divided into two classes: mathematical modeling and survey analysis. Gaonkar and Viswanadham inferred the profits gained from IS by using a linear programming model [3]. The authors assumed an SC based on long-term agreement and classified the IS scenario under certain circumstances into two extreme groups: no IS and complete IS. They were assumed to be based on make-to-stock and make-to-order (MTO), respectively, and the performances of both cases were compared according to their cost. Gaonkar and Viswanadham's study determined the effectiveness of IS in SC and has been extended to prove how the sharing of demand plan affects inventory and lead time in SC by simulation analysis. A number of similar studies extended the IS scenarios into three IS categories: no, partial and full. Narasimhan and Nair's survey research, a representative article among the various survey studies conducted to configure IS in SC, was carried out to test whether IS between retailers and manufacturers affected SC performance [4]. More than 4,500 companies were interviewed for the questionnaire, and the test comprised the following six indexes: market share, return on assets, average selling price, product quality, competitive position and customer service. The study concluded that IS does indeed strengthen SC competitiveness.

However, since most of the aforementioned studies regarded IS in a macroscopic way in terms of no/partial/full IS, they have not had much practical value for the implementation of a specific SC. Thus further study is needed to determine how specific information contributes to SC performance. In addition, most of the previous studies only considered supply side IS, whereas the present study is distinguished by its consideration of both sides of the information flow: the supply side information such as production plan and the demand side information such as demand forecast.

2.2   Profit Distribution Issue regarding Information Sharing (IS)

The problem of profit distribution in IS has come to occupy an important position due to the recent advances in IT technology, since the issue was previously neither recognized nor even technically realizable. "Grove Scheme" is regarded as the exemplary study in this field [5]. The study primarily investigated the profits gained by different organizations according to their performances, and then resolved the distribution of the aggregated profit by giving differentiated incentives. This mechanism assumed the existence of an independent department to adjust and redistribute the overall profit after all other departments report their individual profits. The system was designed so that any party providing distorted information suffers a loss equivalent to their degree of distortion, thus effectively penalizing and thereby preventing such distortion. Feldmann and Muller extended the "Grove Scheme" idea into SC in order to ensure information fairness [6]. Similar to previous research, the authors assumed a third party called "Supply Chain Management" which does not involve or receive any part of the benefits from the SC and merely distributes the benefits.

However, IS in SC may differ markedly from the single organization case, as the parties are legally independent and the expected profits are not equally and precisely measured according to the position in industry held by each specific party. In addition, the basic premise of the third party existing and redistributing the overall profits is unrealistic. This approach faces several problems, including, for example, who assumes the role of the third party, how to solve the ambiguity in the method of profit distribution, and how to get a unanimous agreement on contributions and opportunity cost reductions.

3   Considering the Utility Value and Fairness

3.1   Information Presumption

In this paper we deal with the specified categories of information which contribute to the utility value of SC, rather than simply testing the effect of IS in general. For this reason information that is shared in SC is examined in the following section according to two factors: the information functions and the information flow directions. We restricted the target of interest to manufacturing companies due to the possible wide variation of information type according to industry position.   In order to classify the shared information, the functions of every party in SC are divided into five functions with reference to the SC operation reference (SCOR): plan, source, make, deliver, and return. Next, information types, which are assumed concerning their functions, are filtered according to their potential to affect their SC partners. The presumed information for each function is listed in Table 1.

Purchasing and delivering are the most frequent and visible IS activities with corresponding parties of suppliers and retailers, respectively. On the other hand, information treated in the planning and making functions is provided one-sidedly and potentially shared by manufacturers. We neglected the returning function, since the reverse flow of SC was not a major concern in this study.

3.2 Classifying the Shared Information by Level

The information presumed previously is sometimes shared which affects other parties in SC, but the level of sharing can differ according to their own policies. Prior to assigning the differences, we now reclassify the types of information, as listed in Table 2, according to their flow directions with their generating sources: the supply side information such as production plans and the demand side information such as demand forecast.

Table 1. Presumptions of information shared according to the functions

Functions	Information Presumed
Plan	demand forecasting, launch of new product, current sales figures, phase out of product, promotions
Source	raw material, raw materials on stock, supplier's profile, production plans, purchase plans, purchased list, supplier's goods, supplier's inventory, distribution center, manufacturing plant, delivering vehicles, transportation plans
Make	production plans, manufacturing progress, manufacturing process, bill of material, raw materials, subassemblies, goods on stock, replenishment plans, capacity available
Deliver	customer profile, customer's inventory, sales plan, customer sale, status of orders, customer's credit, contract, goods, distribution center, manufacturing plans, delivering vehicle, transportation plans, delivery difficulties

Table 2. Presumptions of information shared according to the functions

Directions	Supply Side Information		Demand Side Information
Generating Sources	Supplier/manufacturer	Manufacturer	Manufacturer/retailer
Information Flow	supplier ↔ manufacturer		manufacturer ↔ retailer
Details of Information	raw material raw material inventory supplier's profile production plans purchase plans purchased list supplier's goods supplier's inventory distribution center manufacturing plant transportation plans	capacity available demand forecasting production plans manufacturing progress manufacturing process bill of material raw materials on stock subassemblies on stock goods on stock replenishment plans delivery difficulties	customer profile goods on stock demand forecast current sales figures status of orders customer's credit contract launch of new product phase out of product manufacturing plant promotions

Now we classify the level by applying the classifications suggested in Nienhaus et al's survey report [7]. Concerning the demand side information, the importance of demand forecasting and new product launch is relatively high. Concerning the supply side information, delivery difficulties and order status are rather important. We modified each survey result into three levels, as shown in Table 3.

Table 3. Setting the level of information sharing (IS)

Directions	Level	IS
	Level 1	demand forecasting, launch of new product
Demand side	Level 2	Level 1 + current sales figures, phase out of product
	Level 3	Level 2 + goods on stock, production plans, promotions
	Level 1	delivery difficulties, status of orders
Supply side	Level 2	Level 1 + capacity available
	Level 3	Level 2 + goods on stock, production plans

3.3 Considering the Fairness in Information Sharing (IS)

We briefly review the incentive scheme on profit sharing among the parties in an SC. As noted above, however, previous studies faced three problems: unrealistic assumptions on the third parties, ambiguity in distributing the profits, and unanimous agreement on contributions and opportunity for cost reductions.

In our approach, two flow directions in information and profit generation are taken into account in order to describe both manufacturers' and retailers' gains in the form of stock reduction and opportunity cost reduction by introducing a sudden delay of delivery, as depicted in Fig. 1. Since profits cannot be expected according to their own information, they are required, for their own sake, to consider firstly the profit of their cooperating companies. As a result, incentives to share information are given for both manufacturers and retailers to maximize their returns for long-term relationships.

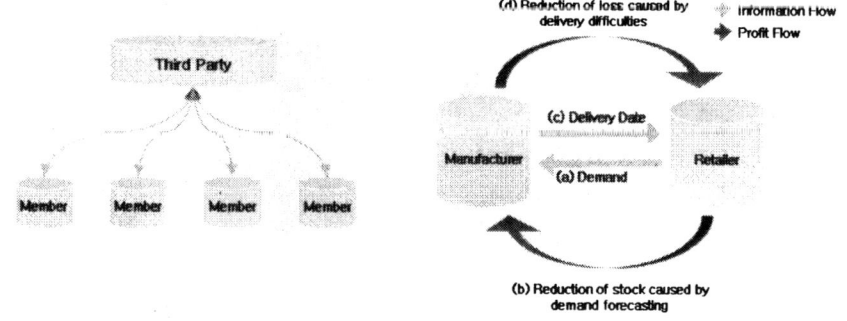

(i) Traditional approach   (ii) Proposed approach
Fig. 1. Proposed approach profit distribution considering fairness in the supply chain (SC)

In the proposed approach, information and profit are mutually provided though twin-side flows. Furthermore, information that decides a company's profit is possessed by the counterpart. For example, as indicated in the figure, when a retailer provides demand information to a manufacturer (a), the latter benefits by stock reduction (b). On the other hand, when the manufacturer sends delivery date to the retailer (c), the latter reduces loss by preventing delivery difficulties (d). Considering that these flows will be repeated between parties having a long term agreement, they will try to increase their overall profits through frequent IS since their own profits are determined by their counterparts.

3.4. Simulation Test

This section tests if specific information listed previously has different utility values according to the level, and determines the valid information to be shared. The following assumptions are made for the test:

- Manufacturers have a long-term agreement with their retailers and manufacture their products in an MTO fashion.
- Demand patterns of retailers include internal and external parameters as well.
- Manufacturers have limited capacities. The priorities regarding the capacity assignments are a) production of delayed deliveries, b) production of currently appointed deliveries, c) production of deliveries those are presumed to exceed capacities in the future, and d) replenishment for safety stock, sequentially.
- Suppliers and customers are regarded as infinite in number, and are not considered as variables.
- Performance regarding the IS is measured by the improvements in service level and average stock status.

A simulation test is conducted with regard to three cases when a manufacturer and retailer share the same level of information. The relationship diagram among the information flow entities is shown in Fig. 2.

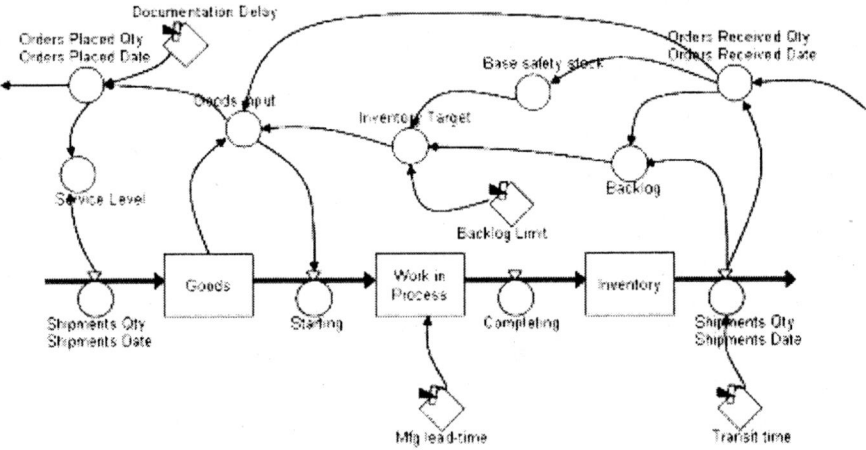

Fig. 2. Relationship diagram among the information flow entities

The test is conducted using the Matlab code. Real data from a company in the beginning of 2005 are applied to generate the customers' orders. Nine product groups comprising 64 products over a 13-week time horizon are considered with 50 repetitions.

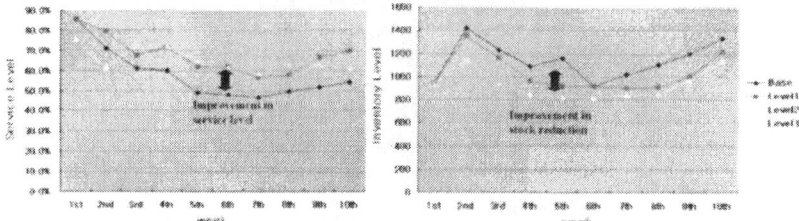

Fig. 3. Variations of service level   **Fig. 4.** Variations of inventory level

Fig. 3 indicates that the service level is improved when information is shared. Especially, the effect of sharing Level 1 information is significant. Fig. 4 also depicts that IS reduces the inventory level and the impact of sharing Level 1 information is clearly evident in this case as well. These results show that some of the information makes a significantly higher contribution to performance than others. Considering the variations led by greater IS along with its level, the overall performance continues to increase but the marginal effect for incremental sharing diminishes, as shown in Fig. 5. This finding indicates that if limited information is allowed to be shared then the company should share the information which brings higher marginal effect.

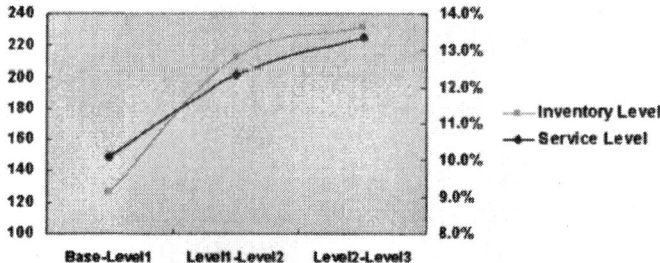

Fig. 5. Incremental improvement through information sharing (IS) by level

Now we examine the profit gained by each party from the above scenarios. The manufacturer reduces the average stock from 0.48 week/year to 0.42 week/year due to the stock information obtained from retailer to alter safety stock and replenish specified stocks that might be in shortage. On the other hand, the profit distributed to the retailer is calculated by the potential reduction in sales loss [volume of goods from manufacturer × improvement in service level], because the benefit might be gained by reducing the loss from delivery difficulties. By anticipating the goods which may be delayed in advance due to the provided information, the retailer sources the products from other manufacturers or adjusts the delivery date with reference to internal stock. The service level with no IS is 57.6%, whereas IS with Level 1 is 67.7%.

4   Conclusion

Two aspects of IS in SC are discussed. First, specified information, rather than the overall information, is tested by simulation study to determine any significant differences in contribution to SC performance. The types of information are re-classified into 3 levels according to their utility values and the direction of flow by modifying the classifications suggested in several previous survey reports. The experimental results support the assumption that certain types of information should be shared among the SC enterprises to improve the overall performance while considering the IS limitation. Secondly, fairness in sharing the profit gained from the improvement is also discussed. Incentives to share the information were naturally given to both parties which have a long-term agreement in MTO production type in our framework. The distinctiveness was based on the need for the parties to consider their cooperating companies' profits for their own sake. By detailed examination of the specific profits and utility values gained by the IS, our approach is expected to help enterprises generate strategic plans toward IS in their SC. Further study is planned to validate our approach for an increased IS scope such as a greater depth of SC and longer time span.

5   References

1. T.M. Rupp, and M. Ristic, Determination and exchange of supply information for co-operation in complex production, Rob. & Auto. Sys., 49(3-4), 181-191 (2004).
2. S. D'Amours, B. Montreuil, P. Lefrancois, and F. Soumis, Networked manufacturing: The impact of information sharing, Int. J. Prod. Econ., 58(1), 63-79 (1999).
3. R. Gaonkar, and N. Viswanadham, Collaboration and Information Sharing in Global Contract Manufacturing Networks, IEEE-ASME Trans. Mech., 6(4), 366-376 (2001).
4. R. Narasimhan, and A. Nair, The antecedent role of quality, information sharing and supply chain proximity on strategic alliance formation and performance, Int. J. Prod. Econ., 96(3), 301-313 (2005).
5. T. Groves, Incentive compatible control of decentralized organizations, in: Direction in large scale systems: many person optimization and decentralized control, edited by Y. Ho (Springer, Berlin, 1976), pp. 149~185.
6. M. Feldmann, S. Muller, An incentive scheme for true information providing in Supply Chains, OMEGA-Int. J. Mng. Sci., 31(2), 63-73 (2003).
7. J. Nienhaus et al., Result of a Study among more than 200 companies, in: Trend in Supply Chain Management, (ETH-Zentrum für Unternehmenswissenschaften, 2003).

Evaluating the Standard Assumptions of Demand Planning and Control

Peter Nielsen and Kenn Steger-Jensen

Aalborg University, Department of Production, Fibigerstraede 16, 9220
Aalborg Oest, Denmark, {peter;kenn}@production.aau.dk

Abstract

This paper investigates customer ordering behavior and compares a particular instance to the standard assumptions of demand planning and control. Experience shows that the performance of these planning processes often do not match what could be expected. Based on a presented case, implications for demand planning performance are inferred. This leads to an analysis tool for diagnosing potential problems in the demand planning approach.

Keywords

Demand planning and control, numerical study, performance impact

1 Introduction

Demand planning systems are a vital part of the business structure of companies. More so in manufacturing companies depending on a demand planning system to deliver feasible plans ensuring the best (or close to) utilization of resources while adhering to material, resource and due date constraints. Today most companies use some variant of the hierarchical demand planning approach presented by Hax and Meal [1]. However the approach's success depends on the conformity to a number of assumptions regarding independence of planning objects' (items, product families, resources) behavior and the ability to aggregate and disaggregate information. This subject has been studied in detail (among others: Axsäter [2]). However no method has been presented that accurately predicts the consequences of deviating from the inherent standard assumptions. This paper suggests a method to evaluate the performance of the demand planning processes and establish the potential impact of deviating from the standard assumptions. The focus is on downstream activities and consequently upstream constraints and impacts are not considered.

This paper contains four distinct parts. First, a literature review, discussing issues within demand planning and control, is presented. Second, results from a study of

Please use the following format when citing this chapter:

Nielsen, P., Steger-Jensen, K., 2007, in IFIP International Federation for Information Processing, Volume 246, Advances in Production Management Systems, eds. Olhager, J., Persson, F., (Boston: Springer), pp. 221-228.

demand patterns, in a company using a mix of ATO and MTS, are presented. Third, a discussion of the implications for demand planning performance and processes is given, before finally a conclusion is reached.

2   The demand planning connection

When looking at demand planning in an manufacturing environment two main aspects naturally need to be taken into account: materials and resources. The relative importance of these two areas will reflect the competitive priorities and market conditions of the particular company. The two are typically manifested through demand planning and subsequently for materials as inventory management. Within inventory management Silver [3] state three statistical assumptions: 1) assuming a particular demand pattern, 2) assuming that parameters of the distribution are known and 3) assuming stationarity. To these assumptions an equally important assumption must be added, namely the assumption of independency of parameters. Although much has changed since Silver statements in 1981, the underlying assumptions have not changed significantly. Demand planning is often implemented as hierarchical demand planning (HDP) [1,4,5]. The main objectives of HDP is to level production output thereby easing allocation of resources and materials. Ideally aggregation naturally levels demand and reduces noise. However to successfully achieve a reduction of noise Sales and Operations Planning must be integrated in the Master Production Scheduling process. In practice only the ideal situation is considered. This coincides with the assumption that if HDP is to work, the aggregate demand plan must successfully be disaggregated to a number of feasible plans [6]. The assumption is a one-to-one relationship in aggregation and disaggregation [1,2]. As a result order size, timing of orders, variance in order size and the number of orders are all assumed to be stochastic and independent. Demand is furthermore often assumed to be seasonal fulfilling one of the main advantages of HDP; namely leveling [5,7]. The interdependency (or lack hereof) between planning parameters is crucial for the performance of the demand planning system in both an ATO and MTS environment.

The major potential pitfall in HDP is of course the ability to aggregate and disaggregate data and plans. To aggregate and disaggregate a dimension of similarity between planning objects is needed. As a result products are typically grouped by similarity of BOM's, while manufacturing resources are grouped by manufacturing capabilities. To justify grouping items in this manner one must assume that they have similar demand patterns and similar use of manufacturing resources [1,5]. However similarities of BOM's does not necessarily mean that the grouped items have similar seasonality (as assumed) or load on resources. The tendency is to tie the underlying data model inexplicitly to materials via the BOM and only connect capacity through a secondary check for feasibility of plans [8]. However this one dimensional approach seems inappropriate in an ATO environment where planning of resources precedes material planning in importance. Another important limitation of HDP is the need to partition demand into time buckets / periods. The size of these time buckets to a large extend determines the degrees of freedom in the demand planning and control system. The underlying assumption is that demand, although stochastic, occurs at a steady rate during the period [8] i.e. that orders are uniformly distributed. Systematic divergence from this assumption gives suboptimal performance of lead times (in ATO) and service level (MTS with multiple replenishments per period) or excessive safety stock (MTS single replenishment per period). This leads to a general problem since the length of the periods is often uncritically chosen as a month due to e.g. the customer credit period, the manner in which forecasts are made or the ability to adjust manufacturing capacity. The demand pattern seldom enters into the equation.

The hypothesis in this paper is that the standard assumptions seldom hold true and that subsequently the demand planning process suffers poorer than expected performance. In this paper an analysis of demand patterns, in a particular case company using a mix of ATO and MTS, is presented. The hypothesis is that the demand pattern in the company deviates significantly from the standard assumptions giving poor performance regarding on-time-delivery, lead times and service level.

3   Results

To investigate whether this hypothesis holds true a method for analysis was developed. The following example focuses on numerically evaluating demand patterns of three products from a major product family over a period of three years. The products are all produced in the same manufacturing cell, but have varying impact on resource load. The products have 80-85% similar components and constitutes approximately 75% of the total demand of the product family. Demand for products 1 and 2 can be assumed to be stationary and normally distributed. Demand for product 3 is somewhat more erratic and does not follow any standard distribution, nor did it exhibit any significant trend. Product 1 is often sold with a combination of other products, making it critical from a demand planning perspective. On the surface a textbook case for the use of HDP. However in practice the company frequently experience problems fulfilling orders on time and unexpected underutilization of resources. To investigate, the following demand pattern parameters were studied: variance of order size within a period, mean order size within a period, the no. of orders per period and the total demand per period. These parameters were checked for correlation within the individual products first. The results are shown in table 1.

The results show that the mean order size and the number of orders for a period are negatively correlated (-0.55) for product 1. This indicates that as the number of orders increase the order size decreases, resulting in more changeovers. This causes problems in the ATO part of the company since a relatively large changeover is incurred between assemblies. Products 2 and 3 show no connection between the number of orders and the mean order size. The standard assumption would be no correlation between the mean order size and the number of orders.

It can be seen from table 1 that the number of orders is positively correlated (0.53 - 0.67) to total demand for all three products included in the study. A positive correlation indicates that as the total demand increases the number of orders also increase. However the parameters are not perfectly correlated, thereby breaking the assumption of a constant ratio of orders compared to total demand. As changeovers are non-trivial this impacts the performance of the aggregate capacity plan.

Table 1. Correlation between different demand pattern parameters.

Product 1	Variance of order size	Mean order size	No. of orders
Mean order size	0.78		
No. of orders	-0.26	-0.55	
Total demand	0.55	0.31	0.53
Product 2			
Mean order size	0.75		
No. of orders	0.13	-0.01	
Total demand	0.66	0.69	0.66
Product 3			
Mean order size	0.77		
No. of orders	0.08	0.21	
Total demand	0.56	0.82	0.67

In an ATO environment the relative size of orders impacts the ability to shift production focus. As a consequence, the variance of order size in a period can be used as a yardstick for the required product mix flexibility. The variance of order size is clearly correlated to total demand and mean order size. Both relationships can be expected, however the correlation to total demand offers some problems. This connection indicates as the total demand increases so does the diversity of order sizes. This would indicate that the manufacturing planning and control system needs greater flexibility when demand is high. This can necessitate a shift from a leveling strategy to a chase strategy. In ATO companies order sizes should not differ between points of time in any systematic manner. To test this, a combined density function for order size and time using the kernel method for multivariate data [9], is developed. The kernel used is the standard bivariate normality density function [9]. To avoid over fitting, the model is fitted to the *relative to mean order size* per day a month as the periods currently used are one month. Subsequently the largest probable relative order size is calculated for each day. Products 1 and 3 show no clear preference for large/small orders at any given time. This is however not the case for product 2. A graph of the largest expected relative order size for product 2 is shown in figure 1.

From figure 1 it can be seen that large orders seem to occur within rather narrow timeframes of the month. In particular days 8-12 and 22-25 tend to be preferred dates for large orders. In an ATO environment lead time is typically an important competitive priority. Since these products are sold as ATO, spikes in the order size stresses the demand planning system, resulting in increased lead times and occasional use of overtime. In practice this means that at certain recurring points of time a chase strategy must be adopted instead of leveling. If the demand pattern was found in a MTS environment instead, choosing an inappropriate inventory management techniques would lead either to poor service level or low inventory turnover. Consider a continuous review technique: If large orders occur at specific times the probability for stock out will be disproportional large at/after these times.

Figure 1. Largest expected relative order size at a given time for different probabilities.

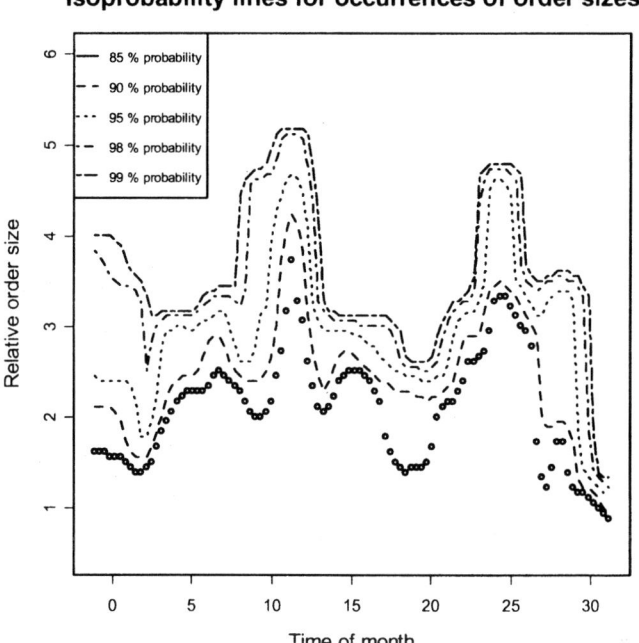

After examining the demand pattern individually, the four parameters were also checked between products, see table 2 for a brief summary.

Table 2. Correlation of demand pattern parameters between products.

Interproduct correlation:	Range of correlation between products:
Variance of order size	0.13 – 0.30
Mean order size	0.24 – 0.39
No. of orders	0.05 – 0.38
Total demand	0.20 – 0.30

From table 2 it is evident that it can be assumed that the demand patterns of the three products are noncorrelated for total demand and variance of order size i.e. they experience no evidence of similar seasonality. This was also supported since a test of autocorrelation of all parameters for all products with a lag of 1-12 periods, showed no significant correlation between periods. However some indication of a connection between the order size and the number of orders per period, does seem to be present. In an ATO environment this gives some problems, since a positive correlation indicate that an increase in the number of orders for one product coincides with an increase in the number of orders for another product. This stresses the manufacturing system and violates the basic assumption of independence that enables leveling.

The findings indicate that the assumption of independence of individual demand pattern parameters is incorrect and subsequently poor performance of the demand planning processes will be expected.

4   Implications

This paragraph includes an overview of the cause and effects between indicators and demand planning assumptions. Subsequently a number of Key Indicators (KI's) and their values, if conforming to standard assumptions of demand planning, is presented. This can be used to diagnose the state of a given demand planning system. Finally avenues of further research are presented.

4.1 Causality

Based on the assumptions of HDP and inventory management techniques in general and the case study, the following steps forms a prudent method for analysis:
1. Identify poorly performing planning areas.
    – These areas will have lower/higher than expected service level or utilization.
2. Identify critical products or resources – i.e. choose the defining dimension for the data model.
    – A multi criteria ABC analysis [10] could be conducted. The criteria could be: materials (volume, cost, similarity of BOM, seasonality, etc.) and resources (capacity, similarity, etc.)
3. Analyze the demand pattern parameters for the individual critical products.
    – Variance of order size, Mean order size, No. of orders and Total demand.
    – Distribution of demand within periods e.g. Kolmogorov-Smirnov test for uniformity of demand rate.
4. Analyze the interaction of demand pattern parameters between critical products.
    – Variance of order size, Mean order size, No. of orders and Total demand
5. Establish where (if any) deviations from the standard assumptions exists.
    – Take corrective actions i.e. adjust the demand planning approach accordingly.
The method presented above is still bound to the data model inherent in HDP i.e. depends on the materials and BOM. The method can be retooled to focus on load of resources, however the data has not been available for this in the current study.

Based on the analysis, a number of KI's for demand pattern parameters can be found. The expected values are presented in table 3.

The KI's presented in table 3, offer the theoretical expected values for correlation between demand pattern parameters for individual items, but not limits on what constitutes a significant deviation from the assumptions. The limits will be situational and depend on the specific interdependencies in a given company. If several of the parameters deviate simultaneously the negative impact on performance is enhanced.

Table 3. Key Indicators for demand pattern parameters and expected values in HDP.

Product 1	*Variance of order size*	*Mean order size*	*No. of orders*
Mean order size	0-1.00		
No. of orders	0	0	
Total demand	0	0	1.00

For the interproduct relationships varying behavior can be expected. If the same seasonality or trend applies for two products the total demand per period will be correlated between products. However if no similar trend or seasonality is present, the demand should be negatively correlated to ensure that leveling is advantageous.

Applying the method involves 1) conducting the analysis and 2) using the results to develop a holistic forecasting and planning method taking the covariance of critical control parameters into account. A demand planning system should be able to handle deviations from the standard assumptions. The consequence is that not only total demand per period needs to be forecasted, but also the characteristics of the demand pattern. Consequently a density function describing several dimension, e.g. order sizes, order frequency, etc., and not just demand rate, should be used, depending on the requirements of the planning process.

An issue that has not been resolved is the impact of using fixed time buckets. The enriched demand pattern models suggested above, still depend on fixed length periods. A potential solution is to use autocorrelation to establish when/whether the demand pattern parameters exhibit systematic behavior. If a systematic ordering pattern is present, then autocorrelation could be used to identify it and subsequently the size of the time buckets can be established. Although it would complicate demand planning further, the path to better demand planning processes could be to dynamically resize time buckets and continually fit demand patterns to these time buckets.

When looking at the suggested method for analysis an issue arises. To get the method to work, a significant amount of data is needed. This means the method in present form only can be applied to products were a substantial ordering history has been established. The dependency on historical data can not be avoided, since it is in the nature of demand planning to – at least to some extend – depend on historical data. Furthermore one can argue that for problems to arise in a demand planning process they must be recurring over time, enabling at least some numerical analysis.

4.2 Further research

To evaluate the impact of deviating from the standard assumptions, simulation studies should be conducted. These would be able to show specific ranges for the KI's presented in paragraph 4.1. This could have a large impact on the practical application of various demand planning techniques, that are often applied without studying the interdependencies of the planning objects.

The implications for practical application are daunting. Imagine demand management with knowledge of which parameters to adjust in customer ordering behavior, i.e. the order sizes, timing of orders, etc.. However it should be noted that further studies are required to validate the results and finalize a descriptive model.

5   Conclusion

HDP is based on a number of standard assumptions regarding demand pattern parameters, focuses on materials and is often applied for MTS. However when looking at an ATO manufacturing environment focus should be on scheduling resources rather than on planning materials. In this paper we presents results indicating reasons for poor performance of demand planning processes. Furthermore suggestions are presented on how to amend these shortcoming of the current demand planning processes. Consequently this paper has a practical impact by giving a method for analysis of demand patterns and subsequently diagnose causes for poor performance of demand planning processes. Secondly the paper offers a theoretical solution for achieving a holistic demand planning approach not only assuming interdependencies among control parameters but including significant occurrences of covariance among parameters. The focus is on downstream influences to the demand planning processes. The presented method can with minor adjustments however be modified to address supplier behavior as well as customer ordering behavior.

References

1. C. Hax and H. C. Meal, Hierarchical Integration of Production Planning and Scheduling, in *Logistics*, edited by M. A. Geisler, vol. 1 of Studies in the Management Sciences, North-Holland/American Elsevier, 1975.
2. S. Axsäter, Aggregation of Product Data for Hierarchical Production Planning, *Operations Research*, 29, pp. 74-756, 1981.
3. Silver, Operations Research in Inventory Management: A Review and Critique, *Operations Research*, 29, pp. 628-645, 1981.
4. G. R. Bitran, E. A. Haas, and A. C. Hax, Hierarchical Production Planning: A Single Stage System, *Operations Research, 29*, pp. 717–743, 1981.
5. C. Holt, F. Modigliani, J. Muth and H. Simon, *Planning Production, Inventories and Work Force*, Prentice-Hall, 1960.
6. S. Axsäter, On the Feasiblity of Aggregate Production Plans, *Operations Research*, 34, pp. 796-800, 1986.
7. Chung and L. J. Krajewski, Planning Horizons for Master Production Scheduling, *Journal of Operations Management*, 4, pp. 389-406, 1984.
8. Vollmann, W. Berry, and D. Whybark, *Manufacturing Planning and Control Systems*, fourth ed., Irwin/McGraw-Hill, 1997.
9. W. Silverman, *Density Estimation for Statistics and Data Analysis*, Chapman and Hall, 1986.
10. Flores and D. C. Whybark, Multiple Criteria ABC Analysis, *International Journal of Operations & Production Management*, 6, pp. 38–46, 1986.

Dynamic Management Architecture for Project Based Production

Akira Tsumaya[1], Yuta Matoba[2], Hidefumi Wakamatsu[2] and Eiji Arai[2]

1   Kobe University, Department of Mechanical Engineering, Graduate
School of Engineering
Rokkodai 1-1, Nada, Kobe 657-8501, Japan
tsumaya@mech.kobe-u.ac.jp
WWW home page: http://www.mech.kobe-u.ac.jp/~tsumaya/index-e.html
2   Osaka University, Division of Materials and Manufacturing Science,
Graduate School of Engineering
Yamadaoka 2-1, Suita, Osaka 565-0871, Japan
{matoba, wakamatu, arai}@mapse.eng.osaka-u.ac.jp,
WWW home page: http://www6.mapse.eng.osaka-u.ac.jp/index-eng.html

Abstract

The production system where production facilities and products are widely distributed like a construction production project causes problems by combining and communicating production facilities mutually. There is a limit of the traditional approaches, where the system obtains the product information from the production facilities. In this paper, the production system architecture which manages the dynamic scheduling and material handling by using parts and packets unification technology is proposed. First, the architecture of the production based production with parts and packets unification technology is proposed. In order to treat both scheduling and material handling, not only the information of "what part" and "when exists" but also "where exists" is used. Then, a pilot system is developed and applied to the case studies about the production scheduling and material handling system in a factory. The results of the case studies show the feasibility of proposed production system using parts and packets unification approach.

Keywords

Dynamic Management, Parts and Packets Unification, Project Based
Production, Scheduling, Material Handling

Please use the following format when citing this chapter:

Tsumaya, A., Matoba, Y., Wakamatsu, H., Arai, E., 2007, in IFIP International Federation for Information Processing, Volume 246, Advances in Production Management Systems, eds. Olhager, J., Persson, F., (Boston: Springer), pp. 229-236.

1   Introduction

A complicated production system is needed to satisfy various requirements to production these days. While it is more and more difficult to manage such various requirements by centralized production management system, the concept of an autonomous distributed production system is proposed as what can adopt to such circumstances. Therefore, numerous researchers have proposed concept of autonomous & distributed production systems in recent years. Many of them focused on product facilities, and constructed intelligent system using network structure, and are available when production facilities and products exist in the limited space like automated factories [1-6].

On the other hand, the production system where production facilities and products are widely distributed like a construction production project causes problems by combining and communicating production facilities mutually. For example, it is difficult to grasp the situation of the production in real time, or the processing performed during the transfer between facilities. In the case of the cooperative work by several different companies, it is more difficult to combine facilities in a network. There is a limit of the traditional approaches, where the system obtains the product information directly but obtains it from the production facilities.

In recent years, RFID tags which can store various data to parts are utilized, and can be used from a viewpoint of cost, size, and memory size. Many application of RFID are introduced and some of them are popular these days [7, 8]. Therefore, the real-time information about various production activities are directly obtained/read/written by using RFID tags which stuck on produced objects.

In this paper, the production system architecture which manages the dynamic production by using parts and packets unification technology is proposed.

2   Concept of Parts and Packets Unification System

For constructing dynamic and flexible production system, it is important to comprehend the state of the product in the production process. To comprehend production state, we need some kind of information. Those are;
— What the product is
— When the product is started or finished in production processes
— Where the product is
— What kind of state the product have

Producing part and packet unification architecture, which is realized by RFID tag attached in the producing part, can store and handle such kind of information [9, 10].

The concept of a dynamic production scheduling system using parts and packets unification technology is shown in Fig. 1. The RFID tag is attached in the producing object, and ID number and the present status are written in it. Status is rewritten according to becoming a unit from materials. The RFID tag transmits ID number, current position, and status through network system whenever it passes through the gate provided in the inside of the factory, the construction site, in the transfer route between them, and the entrance of a warehouse, etc.

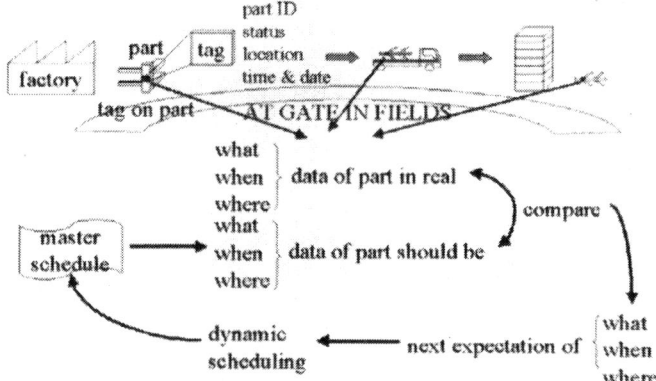

Fig. 1. The concept of a dynamic production scheduling system using parts and packets unified architecture.

As for the production planning system, progress management of the production project is performed by the scheduling system after a master schedule is drawn up. The scheduling system can show the position and status in which each producing object should exist. By comparing with real-time information on each producing object transmitted through network system to the master schedule, a gap between the master schedule and an actual state is detectable. According to the quantity of a gap, re-scheduling is performed when the system judges it is needed.

Using this architecture, a production system can recognize the actual status of all producing parts in real time. The following effects are expected to an autonomous distributed production system.
- Assistance of intelligent facilities (AGV, machine tool, robot, etc.)
- Increasing / free setting of information I/O place
- Improvement of information accuracy in real time
- Improvement of trigger accuracy on dynamic scheduling
- Expansion of trigger range on dynamic scheduling

3 System architecture

The architecture of dynamic production management system using parts and packets unification technology is explained on this section. Figure 2 shows the system architecture in a factory. The network system built by the conventional intelligent manufacturing system has combined production facilities, such as production cells and stations. Meanwhile, in the proposed system, the information that each part have in RFID tag can be exchanged with network system through the gate provided in each production facilities. For example, at the gate provided in the outlet of a processing facility, the status of parts is rewritten according to the production processing.

If the memory capacity of RFID tag will become large in the near future, it is also possible to compose a network system only with RFID tags attached to the parts by storing processing processes and work information in RFID tag, and exchanging information about them with production facilities. However, since the present RFID tag has not satisfactory memory capacity, the proposed system hybridized the network system which combines intelligent production facilities, and parts and packets unification technology. Process and work information are transmitted to production facilities from the scheduling and CAM systems through the network. The part ID in the RFID tag is checked and after processing process, the status of the part is rewritten and is also sent to the CAM system.

Status information in RFID tag changes by the production process such as Fig. 3. Condition of the production process is grasped by the dynamic production management system based on information of parts such as process/delivery start/end time, position, etc. For example, processing time is calculated with process end time and process start time. Cell ability is recognized from this calculated data.

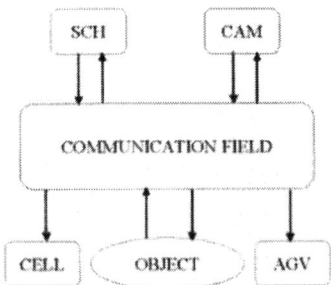

Fig. 2. The concept of a dynamic production system architecture in a factory

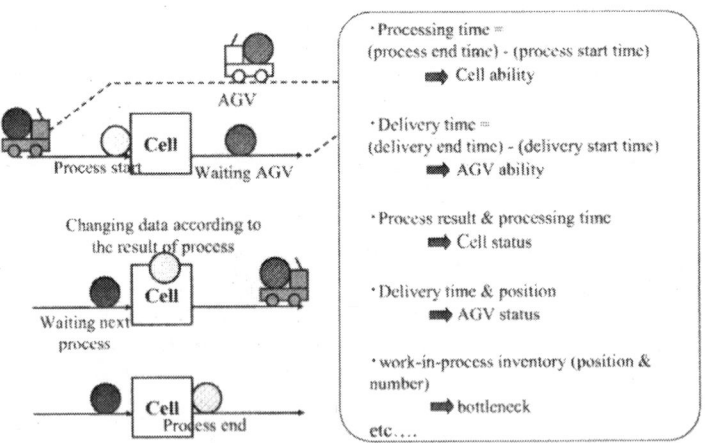

Fig. 3. Parts and packets unified data in a factory

4 Application to Dynamic Scheduling

In this section, an application system of production process within a factory is explained as an example of the dynamic production system. The components of the system are shown in Fig. 4. This example has the following preconditions:

- This factory has following facilities; those are five machining cells, two AGVs (Automated Guided Vehicle), and storage. Five machining cells and storage are arranged in circulate. It is assumed that the conveyance capabilities of two AGV are equal, and it moves in the orbit with one way.
- Previously, the initial value of machining ability on each machining cell is set up. If the ability of machining cell changes during the production, proposed scheduling system calculates new value using the started and finished processing time data obtained from producing objects.
- The order of processing jobs is determined beforehand, and cannot be changed. Part processing which consists of several jobs is prepared. The work data is also provided. According to this work data, each job is given with processing order. Moreover, the processible cells for each jobs are beforehand decided according to the processing type.
- Each producing object has processing geometrical form and volume as data, and machining time is calculated based on these information.
- Scheduling is performed according to the SPT (Shortest Processing Time First) rule which makes the shortest average processing time in the job shop.

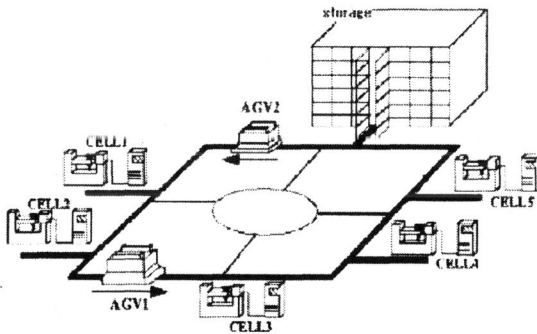

Fig. 4. The components of an application system

Some virtual simulation was performed using the system. Here, the making re-scheduling corresponding to the ability fall of a machining cell is introduced as an example case. This system checks number of waiting parts in front of each machining cell derived from position data of the part, and performs re-scheduling when the number of waiting part is rather than a setting number. On re-scheduling process, the ability of the cell is re-calculated by using information on the processing start time and finish time from latest processed part. The Gantt chart of the simulation is shown in Fig. 5. As these results, the system worked effectively against changing state of production facility.

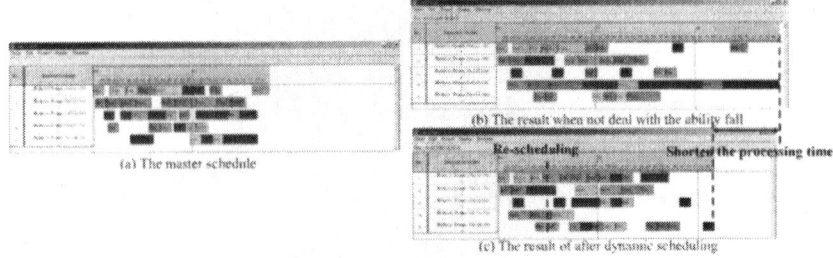

Fig. 5. The Gantt chart of the simulation

5 Application to Material Handling and Scheduling

The system architecture is applied to the management of both material handling and assembly processes. Here, a construction site is taken up as an example case study of the material handling and scheduling for assembly processes.

Figure 6 (a) shows the virtual construct site model of simplified simulated wall construction process on the floor. This model has the following preconditions:

- The wall of the floor is constructed by three companies; A, B, C. Each company makes the wall by using different component materials. There are three component depositories on the floor and each company occupies one of them.
- A horizontal inner wall is constructed by company A. A vertical inner wall is made by company B. A wall surrounding the floor is constructed by company C.
- Each company can construct the wall only when the route between the place of the component depository and the place of the wall are constructed is secured.

Figure 6 (b) shows the snap shot of which warning has gone out when the system detects that it becomes impossible the work of company B because of no access route to the constructing place from the material depository. Like this example, from present position information of component materials and state of the construction progress, it is possible to control the material handling and assembly schedule by checking process before the problem occurred.

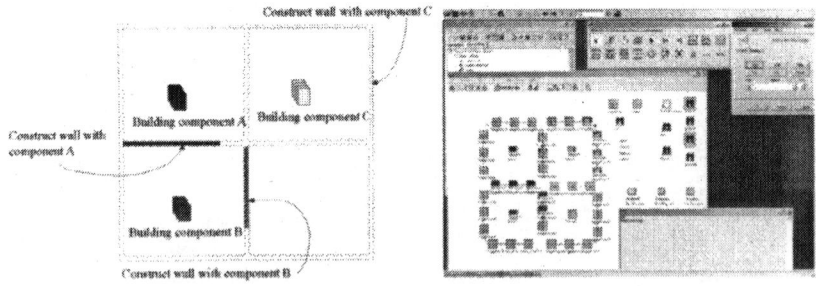

(a) Simulation model   (b) result of the simulation

Fig. 6. Simulation Model and result of the Under Construction Floor of the Building

6 Requirement for System Concept for Construction Production

Construction production differs from general machine production. Parts and units are supplied from different wide-spread fields, the component factories are also distributed geographically, and the different kind of work, such as conveying, processing and assembling, is combined. Moreover, integrate many enterprise which join the construction project is needed. Therefore, it is difficult that an intensive system treats or utilizes each facility's information in such a production style. The parts and packets unified architecture is effective for management in such kind of project based production. Applying the dynamic management architecture for construction production, the following are required.

– Necessity of collaboration of many companies in deferent fields
– Global optimization to avoid conflicts caused by local optimization
– Management of sheared facilities such as cranes
– Correspond to frequent occurrence of operation plan change
– Correspond to frequent occurrence of schedule change

Figure 7 shows the concept of dynamic management architecture for construction production project. Real-time information of parts/units and sheared facilities are gotten from RFID tags through network, and real-time information of workers also obtained from the manager that grasps each worker's status. By comparing with real-time information to the information on the master plan/schedule, a gap between the master plan/schedule and an actual state is detectable. According to the quantity of gap, re-planning/re-scheduling is performed when the system judges it is needed.

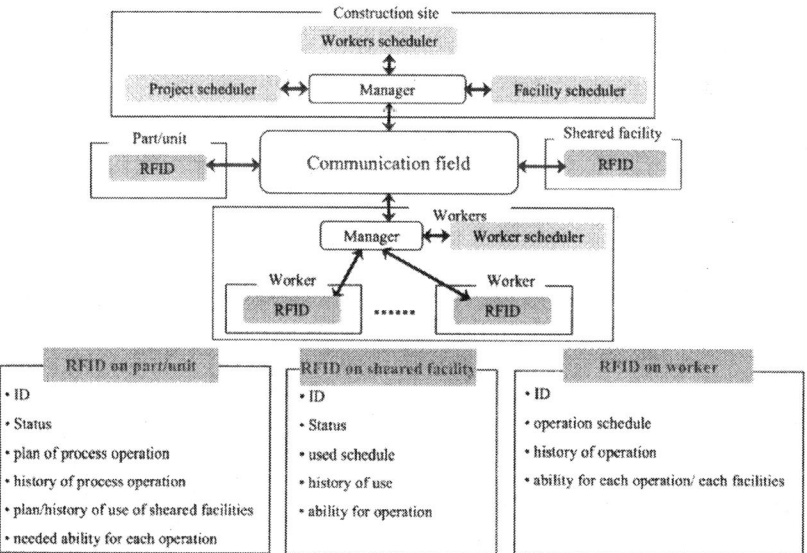

Fig. 7. The concept of dynamic management architecture for construction production

7 Conclusion

The concept and the architecture of the production system with parts and packets unification technology were proposed. In order to treat both scheduling and material handling, not only the information of "what part" and "when exists" but also "where exists" is important. A pilot system is developed and applied to the case studies about the production scheduling and material handling system in a factory and in a construction production site. The results of the case studies show the feasibility of proposed production system using parts and packets unification approach.

References

1   Ueda, K. (1992). An Approach to Bionic Manufacturing Systems Based on DNA-Type Information, *Proceedings of International Conference on Object-Oriented Manufacturing Systems*, pp. 305-308.
2   Ranky, P.G. (1992). Intelligent Planning and Dynamic Scheduling of Flexible Manufacturing Cell and Systems, *Proceedings of 1992 Japan-U.S.A. Symposium on Flexible Automation*, pp. 415-422.
3   Wiendahl, H.P. and Garlichs, R. (1994). Decentral Production Scheduling of Assembly Systems with Genetic Algorithm, *Annals of the CIRP*, Vol. 43, No. 1, pp. 389-396.
4   Fujii. S., Kaihara. T., and Tanaka, M. (1999). A Distributed Virtual Factory in Agile Manufacturing Environment, *Proc. 15th Conference of the International Foundation for Production Research*, II, pp.1551-1554.
5   Sugimura, N., Tanimizu, Y., and Yoshioka, T. (1999). A Study on Object Oriented Modeling of Holonic Manufacturing System, *Manufacturing System*, Vol. 27, No. 3, pp. 253-258.
6   Shirase, K., Wakamatsu, H., Tsumaya, A., and Arai, E. (2005). Dynamic Co-operative Scheduling Based on HLA, *Knowledge and Skill Chains in Engineering and Manufacturing – Information Infrastructure in the Era of Global Communications*, Springer, pp.285-292.
7   Shiibashi, A. (2002). Introduction and Future Development of Suica Non-contact IC Card Ticketing System. *Japan Railway & Transport Review* Vol. 32, pp. 20-27. (http://www.jrtr.net/jrtr32/pdf/f20_shi.pdf).
8   http://www.hitachi.co.jp/Prod/mu-chip/index.html
9   Watanabe, H., Tsumaya, A., Wakamatsu, H., Shirase, K., and Arai, E. (2004). Dynamic Scheduling System with Variable Lot Size Approach Using Parts and Packets Unification, *Proceedings of 2004 JAPAN-U.S.A. Symposium on Flexible Automation (JUSFA2004)*, CD-ROM, JS_038(4 pages).
10   Tsumaya, A., Koike, M., Wakamatsu, H., and Arai, E. (2006). Dynamic Production Management Architecture Considering Preparative Operation, *Pre-proceedings of Advanced Production Management Systems Conference 2006 of the IFIP Working Group 5.7*, pp.289-294.

Fast and Reliable Order Management Design Using a Qualitative Approach

Hans-Hermann Wiendahl
Institute of Industrial Manufacturing and Management (IFF),
University Stuttgart Nobelstrasse 12, 70569 Stuttgart, Germany
Tel: +49 (0)711-970-1968, hhw@iff.uni-stuttgart.de

Abstract

Abrupt and often surprising changes characterize the situation of manufacturing companies. Important for an order management design are those factors which potentially cause turbulence and lead to schedule deviations. The paper describes a method to capture and assess them qualitatively. Based on an analogy to physics, the morphology of turbulence germs is provided. Then, a procedure is described how to successfully transfer approaches for turbulence management to other companies. The last part reflects in detail on the application experience.

Keywords

Manufacturing, management, production planning and control, turbulence

1 Introduction

Manufacturing companies are faced with changing market and supplier conditions, which greatly affect purchase and production logistics. They apply different manufacturing strategies and organizational changes to cope with these turbulent markets. Rapidly changing logistic conditions require fast and appropriate changes to the existing Order Management (OM) system [1]. Many of those in charge respond in an event-driven manner, involving substantial risks: Either so-called "best practices" are adopted without second thoughts, ignoring necessary requirements, or new solution approaches are developed, without considering logistical interactions.

Experience shows that factors which potentially cause turbulence (i.e. dynamics) are the main challenges for OM design. So it must be possible to quickly and efficiently compare the influence of these factors. The paper introduces a novel method of capturing the influence in a qualitative way and transferring OM solution modules successfully from one company to another.

Please use the following format when citing this chapter:

Wiendahl, H.-H., 2007, in IFIP International Federation for Information Processing, Volume 246, Advances in Production Management Systems, eds. Olhager, J., Persson, F., (Boston: Springer), pp. 237-244.

2   Theoretical Fundaments

The research objective of this paper is to compare the logistical requirements that might cause turbulence in an effective and efficient way. An analogy to physics provides the theoretical fundament of the approach.

2.1   Definition of Order Management

Manufacturing companies act in a network of customer-supplier relationships. This requires planning because customers expect reliable delivery promises, while purchase orders need to be placed with the suppliers. The primary order management function is to allocate items, processes and (human and manufacturing) resources to orders in terms of time and quantity [2].

For this purpose, the planning function plans the production flow in advance for a certain period of time, typically based on mean values for lead times. Then, the control function has to implement the plan in the best possible way in spite of changes and disturbances [3]. In doing so, the control function splits up the macroscopic planning results into a microscopic view of single events. Usually, such a plan can never be executed precisely. Therefore – similar to the quality control approach – a tolerance for permissible deviations has to be specified [4].

2.2   Turbulence

The definition of turbulence in OM follows physics [5]: Turbulence exists if individual values (microscopic view) deviate significantly from the representative value (macroscopic view of an aggregated value) [6]. To detect a significant deviation, it takes tolerances defining allowed deviations between individual values and the representative, i.e. typically the mean value. These conditions limit the prediction of an individual value by a mean-value based planning approach for reaching the required reliability. As a result, controlling turbulence takes two views [6, 7]:

- Objectively, turbulence can be identified by significant deviations – i.e. there are individual values deviating more than allowed from the mean value, Fig.1a.
- Subjectively, turbulence means to have objective turbulence not under control. Comparing required to actual reliability – i.e. requirements compared to capabilities – shows if significant deviations are under control, Fig. 1b.

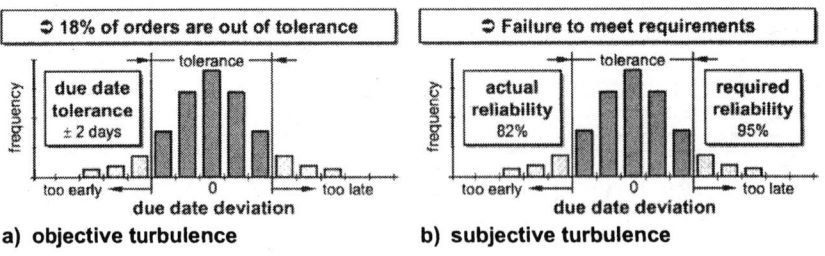

Fig. 1. Objective and subjective turbulence (example of due date deviation)

3   Turbulence Germs in Order Management

Applying these findings to OM operation, it is necessary to identify the turbulence triggers. The so-called turbulence germs that affect planning and control can be clustered in five groups:

- Relevant for planning are variations, fluctuations and plan adaptations.
- Tolerances define the allowed deviation from planned values. They judge the relevance of a turbulence germ and represent a cause cluster of their own.
- Control deals with unexpected deviations that jeopardize the promised delivery.

Note that turbulence germs do not necessarily lead to plan deviations. The relationship between requirements and capabilities determines their relevance, Fig. 2b.

OM covers the processes Source, Make and Deliver [8]. Setting them against the five cause clusters will result in the morphology of OM turbulence germs [9], Fig. 2.

Variations

Variations represent heterogeneous requirements within the same period. The relation between variations resulting from heterogeneous requirements (delivery times, lead times, and replacement times) and permissible deviation (planning tolerance) defines the relevance of turbulence germs.

From a delivery point of view, heterogeneous customer requirements complicate planning: If they lead to heterogeneous order lead times, they will violate the basic assumptions of the successive planning concept MRP II with its mean-based lead time planning approach. If variations exceed planning tolerances, the smooth order flow is disrupted, resulting inevitably in plan deviations. From a make point of view, the necessity of large batches for technological reasons can also create a critical heterogeneous order flow. Heterogeneous replenishment conditions (e.g. long-lead items) have a similar effect on sourcing.

Cluster	Source	Make	Deliver	
Variation	Het. sourcing conditions • times • quantities	Heterogeneous order flow • batch creation • heterogeneous order mix	Het. delivery requirements • times • quantities	
Fluctuation	Fluctuations in sourcing • times • quantities	Fluctuations in production • fluctuating order mix • fluctuating availability	Fluctuations in demand • times • quantities	Planning
Adaptation	Short material life cycles	Short technology cycles	Short product life cycles	
	Plan adaptations supplier • deadline • quantity	Adaptations in production • ongoing modif. order mix • ongoing modif. availability	Plan adaptations forecast • deadline • quantity	
Inconsistency	Sourcing tolerances ⟷	Production tolerances ⟷	Delivery tolerances	Tolerance
Deviation	Quality deviations material	Quality deviations product	Changes to final product	
	Unreliable suppliers • deadline • quantity	Unexpected stoppages • machine breakdown • staff shortage	Order changes • deadline • quantity	Control

Fig. 2. Morphology of turbulence germs in order management

Fluctuations

Fluctuations represent heterogeneous requirements within different periods. Main requirements are the variability of demand in terms of total quantity (delivery view) or order mix (make view): the more varying the quantities in relation to the available quantity flexibility in manufacturing, the greater the turbulence. Fluctuating sourcing conditions (times or quantities) have a similar effect.

Plan adaptations

Planning should anticipate future. Rolling planning approaches are used to deal with changing delivery and supply situations for the same planning period. That means that depending on the point of time in which the forecast is made, the quantity of the planned period changes. Frequently changing circumstances complicate planning, e.g. short product life cycles or forecast deviations concerning deadline and quantity (deliver), as well as short life cycles of materials and purchased parts or ongoing plan adaptations by the supplier (source). From the make perspective, short technology cycles and ongoing modifications of order mix or resource availabilities (e.g. staff, machine, tools) can cause turbulence.

Inconsistent tolerances

Logistical requirements also affect the interface between planning and control. From a delivery point of view, the requested delivery tolerances are of particular relevance. For instance: In case of ship-to-stock, it could be sufficient if the delivery arrives anytime within a specified week, while just-in-time deliveries directly to the assembly line must arrive on the exact day or even shift. For differing external requirements, the valid internal planning tolerance must be decided on beforehand. Heterogeneous production or sourcing tolerances complicate planning and control in a similar way.

Deviations

Deviations from planned values occur when unexpected events after order release necessitate interventions. From a delivery perspective, changes of orders in terms of end product, order quantity or deadline require an intervention by control. From a make perspective, a distinction between unexpected quality deviations, staff shortages and machine breakdowns makes sense. During sourcing, unexpected material deviations (technical specification) and unreliable suppliers can result in missing parts.

Turbulence Profile

To analyze the impact of turbulence germs in practice, it is advisable to generate a turbulence profile. The investigation is based on a qualitative evaluation of the germs. Application experience from approx. 40 cases shows the practical advantages of a workshop-based procedure to evaluate the germs by technical experts:
1. Explain the theoretical foundation of turbulence and its consequences for OM.
2. Select the relevant germs and judge them individually (by each technical expert).
3. Aggregate the results to create a company-specific turbulence profile, discuss the judgment (similarities, differences) and reasons for the impact of each germ.

4   Case Study

The application is described for a company unsatisfied with its logistical performance. To improve logistics, a two-step procedure was agreed on:

- Firstly, develop an OM design concept. Derive solution blocks and combine them into a consistent approach for OM design, if necessary based on a quantitative assessment of the root causes.
- Secondly, transfer appropriate solution blocks from other companies. Compare the turbulence profiles. Discuss the concept's validity, identify transferable solution blocks and transfer experience.

The following description reflects main discussion points and the OM design consequences for planning and control for one exemplary area.

4.1   Initial Situation

The mechanical forming area is characterized by job-shop production with 30 capacity-relevant work centers and 250 workers; work-in-process of 1,800 orders with 11,000 operations in production; simple product structures (1-3 levels). Fig. 3 shows the investigated turbulence profile for this area; on average, planning germs are rated higher: Heterogeneous delivery quantity requirements and demand fluctuations dominate here. From a control point of view, unreliable production processes and deviations from the required material specification are rated highest, since the company works at the cutting-edge of manufacturing technology.

Of course, such a qualitative evaluation does not replace quantitative analyses but reduces effort by focusing on important points. Here, quantitative results confirm the qualitative evaluation and highlight two aspects: The planning approach disregards the manufacturing complexity (changing demand, job shop production) and in addition, planners fix the schedule (firm order quantity and production operation dates) as early as possible. As a result, the control activities after order release cannot implement the fixed schedule within the agreed tolerances.

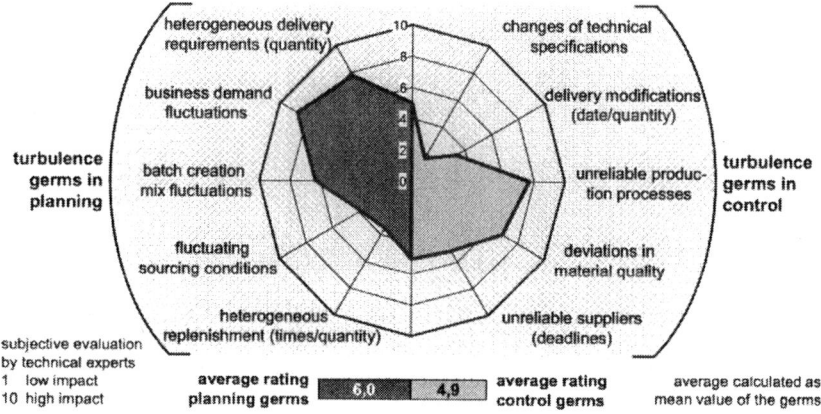

Fig. 3. Turbulence profile company 1, area mechanical forming

4.2   Requirements and Solution Blocks for Planning

Today, planners schedule in a semi-automated way based on the MRP approach; they are supposed to check capacity to avoid an overload. Unrealistic production plans lead to the following questions: Firstly, what level of detail is adequate for planning – i.e. is the detailed scheduling of operations still necessary? Secondly, can the MRP approach with its infinite capacity planning algorithm still be used or is a finite planning approach advisable?

Demand and mix fluctuations get high ratings by the experts (Fig. 3). Detailed quantitative analyses verify that these germs create bottlenecks at different times and places for this job-shop production environment. At first, the idea was to redesign the factory into product segments based on simple material flows. Economic reasons, however, did prevent this lean alternative (due to high investment in new machines). To check for possible bottleneck situations at the work centers, a higher planning complexity must be accepted. This requires the scheduling of operations.

In addition, the capacity planning approach needs to be redesigned: On the one hand, heterogeneous delivery requirements are rated high. Customers order different quantities, resulting in heterogeneous production lot sizes with different capacity demand. On the other hand, the available capacity flexibility is high but differs for each work center in time and quantity. Complex material flows complicate the precise temporal forecast of capacity demand, resulting in high WIP fluctuations; quantitative analyses and discussions with the foremen prove this thesis. A 'double scheduling' approach produces an adequate capacity forecast horizon:

- Purely manufacturing on demand requires infinite flexibility. So, the infinite scheduling run quantifies the capacity demand (quantity and period).
- A realistic plan takes into account finite capacities (and material availability). So, the finite scheduling run forecasts the probable due dates of the orders.

The results of both scheduling runs have to be compared. Deviations from order or resource view point out bottlenecks and indicate the need for planner interventions.

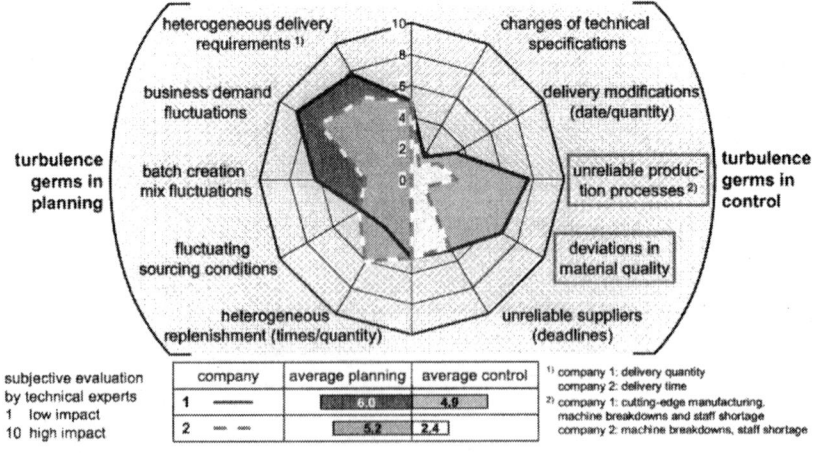

Fig. 4. Comparison of turbulence profiles (company 1 and company 2)

4.3   Requirements and Solution Blocks for Control

Control functions corresponding to the designed planning functions are:
- WIP-oriented order release checking availability of physical material to damp the 'unreliable suppliers' germ and support order execution according to plan.
- Recording of 'sequence discipline' to monitor execution behavior.

More important is another topic: From a control point of view, the above-mentioned 'unreliable processes' and 'material quality deviations' germs get the highest ratings. This advises regular rescheduling, i.e. non-fixed operation dates.

Internal experts fear that this prevents successful software-based OM: Experience shows that ongoing plan changes undermine plan's acceptance. Such a discussion reflects the classical area of conflict between plan stability and dynamics:
- On the one hand, planning should anticipate future. The turbulence analogy shows: The less plans change the more stability – and reliability – they create.
- On the other hand, planning should be up-to-date. If demand changes or a process disturbance occurs, it makes sense to update the old plan. But the turbulence analogy implies that fast changing plans lose people's acceptance.

In this area of conflict, each company has to find its own position. Based on the results, a comparison with two other companies' successful implementations should help to clarify the arguments.

4.4   Comparison of Turbulence Profiles

The companies handle turbulence differently: The first executes a fixed schedule with high flexibility; the other achieves acceptance with a not-fixed schedule.

Company 2 produces equipment assemblies. The turbulence profile shows that a highly reliable production allows fixed schedules, Fig. 4. Central planner focus on planning germs and foreman use local flexibility to deal with the control germs. A strict check for material availability supports execution according to plan.

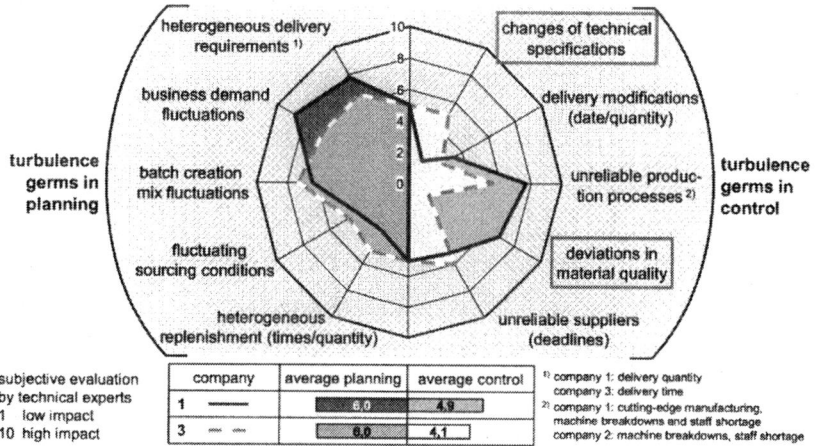

Fig. 5. Comparison of turbulence profiles (company 1 and company 3)

With the turbulence profile of Company 3 (plant manufacturer), the rating of control germs is higher, Fig. 5. People achieve the plan acceptance in two ways: Firstly, they distinguish between promised, planned and actual dates, visualize the bottlenecks and indicate where planners should take action. Secondly, they split up information into a loading list for 2-3 days (individual orders for work preparation) and the demand information for 4-6 weeks (capacity forecast per week).

The project team decided to integrate the company 3 solution into their concept. It fits their requirements and proved practicable in a comparable situation.

5   Summary

Turbulent markets require a fast, effective and efficient order management alignment. The proposed method supports this goal in various ways: It allows the structured qualitative analysis of requirements, derives consistent solutions, and encourages the discussion of logistical interactions and their effects on OM design and operation. The results enable an effective comparison of the logistical requirements and solution blocks from different companies. The outcome accounts for the chosen design solution and hints at possible software tools.

Of course, a qualitative evaluation does not replace quantitative analyses. But the effort for data collection, analysis and verification can be substantially reduced. Experience from about 40 case studies proved the practicability of the approach.

6   Acknowledgements

The results are part of the ongoing research project "Model-based Order Management Design for discrete manufacturing" which is funded by the German Research Foundation DFG (WI 2670/1).

7   References

1. H.-P. Wiendahl, S. Lutz, Production in Networks, *CIRP Annals* **51/2**, 573-586 (2002).
2. H.-H. Wiendahl, Changeability in Production Planning and Control : A Framework for Designing a Changeable Software Tool, *Advances in Production Management Systems: Modeling and Implementing the Integrated Enterprise*, edited by IFIP 5.7 (2005).
3. W. J. Hopp, M. L. Spearman, *Factory Physics* (Irwin, Chicago, 1996).
4. J. Juran, B. Godfrey (ed.), *Juran's Quality Handbook* (McGraw-Hill, New York, 1999).
5. B. S. Massey, *Mechanics of Fluids* (Van Nostrand Reinhold, London, 1998).
6. H.-H. Wiendahl, N. Roth, E. Westkämper, Logistical Positioning in a Turbulent Environment, *CIRP Annals* **51/1**, 383-386 (2002).
7. H. Mintzberg, That's not Turbulence, Chicken Little, It's Really Opportunity, *Planning Review* 11/12, 7-9 (1994).
8. Supply Chain Operations Reference Model (May 12, 2007); http://www.supply-chain.org.
9. H.-H. Wiendahl, Turbulence Germs and their Impact on Planning and Control – Root Causes and Solutions for PPC Design, *CIRP Annals* **56/1** 2007 (to be published).

Achieving Agility of Supply Chain Management through Information Technology Applications

Yi Wu and Jannis Angelis
Warwick Business School, University of Warwick, Coventry, CV4 7AL,
UK
Yi.Wu05@wbs.ac.uk:
Jannis.Angelis@wbs.ac.uk

Abstract

Agility in supply chains is critical for competitive advantages as it helps to explore and exploit opportunities in fast changing markets. Firms are increasingly dependent on information technology (IT) for supply chain management as a competitive tool to facilitate such agility. However, little research has been done on the role of IT on supply chain agility. The paper aims to address this gap by further investigating how IT applications affect supply chain agility. We propose that IT infrastructure integration, consisting of data consistency and cross-functional application integration is critical to achieve agility as various integration processes in agile supply chains can be hampered by fragmented IT infrastructures which enable information flow and coordination activities across function units and network partners. We further illustrate IT infrastructure integration impacts on agility in various operational dimensions, such as speed, flexibility across the supply chain.

Keywords

Information technology infrastructure, Supply chain agility, Flexibility, Responsiveness

1 Introduction

Agility in the supply chain is critical for competitive advantages as it helps to explore and exploit opportunities under time-to-market pressures, and seeks to provide prompt response to customer requirements at an acceptable cost. It has

Please use the following format when citing this chapter:

Wu, Y., Angelis, J. J., 2007, in IFIP International Federation for Information Processing, Volume 246, Advances in Production Management Systems, eds. Olhager, J., Persson, F., (Boston: Springer), pp. 245-253.

gained the significant attentions from both academics and practitioners currently [1-3] and it has been a main objective for leading companies [4].

Firms are increasingly dependent on information technology (IT) for supply chain management as it can help to reduce cost, shorten product life cycle and increase information visibility across supply chains [5]. IT is also recognized as an important role in supply chain agility, as Breu *et al* [6] ague that '*information systems are seen to assume a fundamental role in developing agility*'. Particularly, in pursuing of supply chain agility, various integration processes can be hampers by fragmented IT infrastructure which enables information flow and coordination activities across functional units and network partners [7].

Despite the critical role of IT that has been identified in supply chain management [3, 8], theoretical and empirical research pertaining to the impacts of IT infrastructure on supply chain agility has been limited. We draw the concepts from related literatures on IT infrastructure, supply chain agility and supply chain operations to develop key constructs and relationships. We focus on the critical factors of IT infrastructure for supply chain agility which consists of various integrations, and the impact of IT on agility. More specifically, we address the following questions: 1) What are the values of applying IT in supply chain agility? 2) How does IT infrastructure impact supply chain agility?

The paper is organized as follows. Section 2 presents a review of literature on supply chain agility and the relationship of IT with agile capabilities of supply chain management, while section 3 proposes our framework and relevant propositions. The paper closes with theoretical and managerial implications.

2 Literature Review

Supply chain management (SCM) is moving from vertical integration to horizontal integration, involving inter-firm integration and extensive outsourcing to achieve efficiency [9]. Within the time-based competition, agility is becoming important as it is about '*customer responsiveness and mastering market turbulence*' [9]. Goldman *et al.* [10] identified four basic dimensions of agility: enriching customers, cooperating to enhance competitive advantages, organizing to master change and uncertainty and leveraging the impact of people and information. The definition provides a basic conceptual view with the relevant elements of agility, stressing the responses to changes and capturing changes as opportunities [11].

In the context of supply chain, agility lies in the same theoretical premises as agile manufacturing [12]. More specifically, Aitken *et al.* [13] propose a three-level model with key principles to agile enterprises from rapid replenishment, lean production and organizational agility to individual action. In the research of van Hoek et al.[9], four dimensions of agile supply chain have been identified, which are customer sensitivity for a customer oriented supply chain; virtual integration to leverage information across supply chains; process integration to master changes through focusing on core competencies and network integration to coordinate with partners[1,9]. Many studies have shown that the integration of supply chains at a multi-enterprise level can create competitive advantages and improve the overall performance [17, 18].

In the case of automotive industry, the supply chain is characterized by complexity, uncertainty and heterogeneity [15]. A large number of studies illustrate the strategic importance of agile manufacturing [11, 16]. Specifically, in the automotive industry supply chain which is heavily dependent on the whole supply chain, a single agile manufacturing enterprise may have difficulties responding rapidly to changing market requirements due to limited resources [15].Employing an IT system that links customers and suppliers at various stages through real time communication and information exchange may enable innovative and cost-effective product design [17].

IT plays a key role on the various integration processes as synchronizing suppliers in the network by providing real time information [17]. Therefore, through the agile capability to realize operation on actual demand, information should be instantly available through information sharing and exchange and organizations are designed for maximum efficiency during integration processes [14]. A bulk of literature has addressed the benefits of IT on SCM from direct operational benefits to the creation of strategic advantages [19, 20]. Levary [19] illustrates several implications such as product cycle time. Meanwhile, suppliers may provide better collaboration with an OEM due to the IT applications [10]. In particular, when agility is needed, IT is of particular importance [21].

However, the integration processes in supply chains can be hampered by the fragmented IT infrastructure, which enables information flows and coordination activities across functional units and network partners. A well-integrated IT platform is not only the individual physical parts as it requires the standards for integration of data, applications, and processes to realize the information flow [22, 23]. The two construct important to IT infrastructure are data consistency which should enable the process integration [20] including the information flow by defining key entities to realize information sharing, and cross-functional application integration which enables the management of the supply chain-related processes and realizes the ability to interface with supply chain applications among partners in real time [23].

3 Research Model and Propositions

In agile supply chain, customer sensitivity emphasizes customers and markets, including customer-focused logistics and rapid response. Supply chains are becoming demand-driven rather than forecast-driven in order to effectively respond in real-time demand. With IT infrastructure integration, supply chain partners can capture data on demand, thus leading to customer-focused supply chains [1]. Thus we propose that,

P1a IT has a positive impact on responding to changes in production and services
P1b IT has a positive impact on the responsiveness of processing market demands on new products

It is critical to leverage the strengths and competencies of partners to realize fast responsiveness to market requirements [1], which is called network competition. This can be interpreted as a company's dependability on its partners. This research regards dependability as coordinating with partners while focusing on their own competencies through network integration. Thus we propose that,

P2 IT increases the degree of dependability among partners in the supply chain

Process integration is related to uncertainty across the supply chain, placing emphasis on self-management teams so that core modules of products can be delegated. Alliances among various suppliers, manufacturers and customers will be inevitable [24].Therefore, while focusing on their own competencies, companies are much more likely to increase product variety and improve the ability to handle orders with special requirements. Thus we propose that

P3 IT improves product and volume flexibility along the supply chain

Virtual integration emphasizes on leveraging people and information along the supply chain. Information gathering and dissemination [26] are two important attributes towards organizational learning, a key dimension for SCM, given the complex and often dynamic nature of SCM [27]. Due to IT enhanced connectivity, individuals can share their own interpretations of information to make consensus-focused development more efficient. Thus we propose that,

P4 IT positively impacts on information acquisition and dissemination of organizational learning.

The model in Figure 1 indicates the operations impact of IT on agile capabilities of the supply chain.

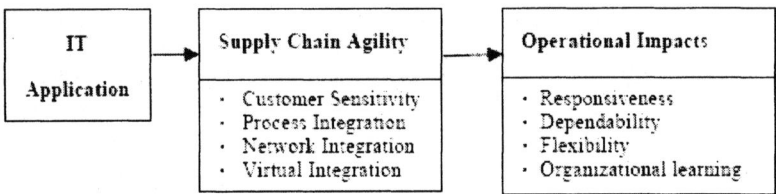

Fig. 1. Conceptual Framework

4 Case Study

We conducted multiple case studies in Chinese automotive companies, where firms pursue effective operations to meet rapid growth and demanding customers in which supply chain performance plays a crucial role on sustaining competitive advantage. Semi-structured interviews were deployed in order to gain an in-depth understanding of the broader organizational patterns of information technology applications and their relation to supply chain agility, and a survey used to test propositions.

4.1 Case Background

The studied companies are operating in a volatile and competitive market as customers are increasingly demanding with changing needs which is relevant to Goldman et al [10] description of agility.

The OEM (A) is one of the biggest auto manufacturers in China with its own R&D. The company is facing the multi-market competition from Japanese, Korean, American and European auto makers. However, the company seeks for the emerging market and can lunch new models with good quality and low cost in short time based on market demands. It has a close relationship with its suppliers on the product design and development.

The 1st tier supplier (B) is a manufacturer of automotive seating systems for various auto makers. According to the CEO, the automotive industry is driven by cost and there are increasing pressures and constant changes in customer requirements. At present, the company hopes that the OEM can provide the technology to realize real time communication and information exchange in order to improve the overall supply chain performance.

The 2nd tier supplier (C) is manufacturing the foams for the automotive seating system and has the competitive advantage on the production processes in China, although it is a relatively new entrant (2 yrs). They have a close link with its customers across this chain, as it has already involved in the new vehicle model R&D process in OEM. On the other hand, it has already realized the inventory share with some customers through IS integration to establish a close relationship with customers at the strategic level, according to the CEO.

4.2 IT Applications in the Companies

Table 2 provides details of the IT employed in each company. In Company A, they develop ERP cooperated with third party. IT is based on the standard ERP solution with the features presenting their own business characteristics. The data are consistent across departments to ensure information flow. Accordingly, Company B is using a standard ERP system. However, due to the lack of management skills and relevant knowledge, finance department needs to entry all data manually. Therefore, it increases the data inaccuracy in the business process. As for Company C, due to the business increasing in terms of customers, they find that MRP can not fully support the business.

There is only one way connection from Company A to B through the platform A established for supplier management to delivery monthly, weekly and daily planning so that Company B can produce their manufacturing schedule and procurement planning. Company A is considering of building up a real time information exchange system and share with suppliers, but the problem is whether suppliers would like to share their information.

Only email communication set up among Company B and C, which constrains overall performance. The CEO of Company B acknowledged the company's need for a closer electronic interaction with the suppliers to increase overall competitiveness.

Table 1. IT applied in the company

Company	Application	Description	Type
OEM –A	MySAP	ERP Solution	Commercial off the shelf
1st Supplier—B	UF	ERP Solution	Commercial off the shelf
2nd Supplier –C	No Name	MRP Solution	In house software development

4.2 Operational Impact of IT

The interviewees from the above three companies concluded that the impacts of IT applications are significant on the operational dimensions. As one senior manager from logistics department described, 'without IT, all the operation processes will be in a mess'.

- Responsiveness

In Company A and C all departments share information so as to make the fluent information flow so as to improve the responsiveness of the change and the process from downstream/upstream. Considering the lack of knowledge on IT applications, the company found that without integrated IS, it threats data accuracy which affects the responsiveness of the business processes. The CEO of Company B agreed that through the portal of SRM they have coordinated planning mechanisms and process alignment across the corporate boundaries due to the IT connectivity. However, the demand information is restricted to suppliers; suppliers can not get any information directly from end-customers to schedule their manufacturing. Therefore, because information on end-customer requested change is delayed from the OEM to the suppliers, the latter have limited time to re-schedule their daily manufacturing. The findings indicate that companies are moving towards a more integrated supply chain in order to have the competitive advantages of the overall supply chains, but so far at a slow rate.

- Dependability

Under the dynamic and changing markets, it is difficult for companies be responsive. Coordination is becoming critical. As in this case, suppliers are increasingly involved on collaborative working product design, and can focus more on their competencies.

However, results from the survey show that IT does not have a great impact on a single supplier on each sourced product as most interviewees pointed out that IT facilitates the connection between partners through the information share process, such as planning and inventory. For example, suppliers are planning their manufacturing schedules based on OEM plans. With instant access to the OEM's manufacturing execution schedule (which is updated every two hours) it is easier for the suppliers to operate an outbound JIT system and have lower inventory levels than before the IT integration. Being in an industry driven by cost, company A engages into several collaborations with various suppliers for each product sourced.

- Flexibility

The findings suggest that IT has a great impact on the realization of product and volume flexibility, because of the visibility IT provides. As the CEO from Company B remarked, the flexibility of volume, products and delivery time is really important. However, the company needs to have the ability to handle non-standard orders at the shop floor. Therefore, the company is working on improving operations flexibility.

- Organizational learning

Given high information availability among partners, it is imperative to achieve technology integration throughout the supply chain by adopting EAI or EDI systems to provide employees with high information availability.

4 Discussion and Conclusions

The exponential increase of corporate investment on IT suggests a strong impact of IT applications on reshaping and improving firm capabilities. However, it has been recently argued that IT cannot create value in a vacuum, as illustrated by the significant failure of IT firms to create sustained growth in high technology markets [29]. In this paper, we investigate indirect links of IT and corporate value creation. In doing so, we focus on supply chains, and more specifically on supply chain agility. Our research is motivated by the importance of IT on supply chain agility in complex manufacturing environments such as the automotive industry. We develop a conceptual model to address the theoretical gap of IT, supply chain agility and value creation.

This paper has several theoretical and practical implications. First, we extend the current literature of IT and supply chain agility by investigating the impact of IT and the possible ways of realizing value. Moreover, we highlight the role of IT as a platform for various integration processes in an agile supply chain and the direct link among IT application, agility and operational drivers.

Second, we lay the theoretical ground for applying IT to achieve supply chain agility. This paper illustrates how IT applications affect specific operational drivers on achieving agility. Furthermore, we provide an integrated perspective of IT and supply chain agility and illustrate how to leverage IT applications along agile supply chains to improve its responsiveness, dependability and flexibility.

In terms of management practice, we found that the succesfull application of IT systems across the supply chain it is not dependent on the technology adopted throughout per se, but rather the continuous engagement of trading partners. As one senior manager points out "*sometimes it is not technology problems, but that the suppliers do not want to share their inventory*". Furthermore, our case study findings suggest that the information flow across the supply chain was limited. We suggest that low information transfer among partners may impede several operational capabilities, such as flexibility. Last, we found a misalignment between existing IT management practices and applications being implemented. Put it differently, we suggest that managers should steer considerable effort towards efficient data integration and analysis in order to gain from the implementation of complex IT systems. Under this stream, we raise managerial attention to further training of efficient IT management.

References

1. M. Christopher, The Agile Supply Chain-Competing in Volatile Markets, *Industrial Marketing Management*, 29 (1), 37-44(2000).

2. J.B.Naylor, M.M. Naim, and D.Berry, Leagility: Integrating the Lean and Agile Manufacturing Paradigms in the Total Supply Chain, *International Journal of Production Economics*, 62 (1/2), 107-118 (1999).

3. D. J. Power, A.S. Sohal, and S. Rahman, Critical Success Factors in Agile Supply Chain Management, *International Journal of Physical Distribution & Logistics Management*, 31(4), 247-265 (2001).

4. M. L. Fisher, What is the Right Supply Chain for Your Product? *Harvard Business Review*, 75(2), 105–116 (1997).

5. R. R. Levary, Better Supply Chains through Information Technology, *Industrial Management*, 42(3), 24-30(2000).

6. K. Breu, C. Hemingway, M. Strathern, & D.Bridger, Workforce Agility: The New Employee Strategy for the Knowledge Economy, *Journal of Information Technology*, 171, 21–31(2001).

7. A. Barua, P. Konana, A.B. Whinston, and F.Yin, Assessing Net-Enabled Business Value: An Exploratory Analysis, *MIS Quarterly*, 28(4), 585-620 (1995).

8. A. Gunasekaran,and E.W.T. Ngai, Information Systems in Supply Chain Integration and Management, *European Journal of Operational Research*, 159, 269-295 (2004).

9. R. I.Van Hoek, A. Harrison, and M. Christopher, Measuring Agile Capabilities in the Supply Chain, *International Journal of Operations & Production Management*, 21(1/2), 126-147(2001).

10. S.L. Goldman,R.N.Nagel, and K. Preiss, *Agile Competitors and Virtual Organisations* (Van Nostrand Reinhold, New York, 1995).

11. H.Sharifi, and Z. Zhang, a Methodology for Achieving Agility in Manufacturing Organization: An Introduction, *International Journal of Production Economies*, 62,7-22(1999).

12. H.S. Ismail, and H. Sharifi, (2006), A Balanced Approach to Building Agile Supply Chains, *International Journal of Physical Distribution &Logistics Management*, l36 (6), 431-444 (2006).

13. J.Aitken, M. Christopher, and D. Towill, Understanding, Implementing and Exploiting Agility and Leanness, *International Journal of Logistics: Research and Applications*, 5 (1), 59-74(2002).

14. Y.Y.Yusuf, A.Gunasekaran, E.O.Adeleye, and K. Sivayoganathan, Agile supply Chain Capabilities: Determinants of Competitive Objectives, *European Journal of Operational Research*, 159, 379-392(2004).

15. H.Q.Xu, C.B.Besant, and M.Ristic, System for Enhancing Supply Chain Agility through Exception Handling, *International Journal of Production Research*, 41(6), 1099-1114(2003).

16. K.Cheng, D.K.Harrison, and P.Y. Pan, Implementing of Agile Manufacturing-An AI and Internet Based Approach, *Journal of Materials Processing Technology*,76, 96-101(1998)

17. M. Christopher, *Logistics and Supply Chain Management: Creating Value-adding Networks* (Prentice Hall, London,2005).

18. R.E.Spekman, and K.JW.N. Myhr, An Empirical Investigation into Supply Chain Management: A Perspective on Partnership, *Supply Chain Management*, 3(2), 53-67(1998).

19. R.R.Levary, Better Supply Chains through Information Technology, *Industrial Management*, 42 (3), 24-30(2000).

20. T.W.Malone, J.Yates, and R.L. Benjamin, Electronic Markets and Electronic Hierarchies, *Communications of the ACM*,30 (6), 484-497(1987).

21. S. Vickery, C. Droge, and R. Markland, Dimensions of Manufacturing Strength in the Furniture Industry, *Journal of Operations Management*, 15,317-330(1997).

22. J.W.Ross,Creating a Strategic IT Architecture Competency: Learning in Stages, *MIS Quarterly Executives*, 2(1), 31-43 (2003).

23. A. Rai, R. Patnayakuni, and N.Seth, Firm Performance Impacts of Digitally Enabled Supply Chain Integration Capabilities, *MIS Quarterly*, 30(2), 225-246(2006).

24. H. Lee and C. Billington, Managing Supply Chain Inventory: Pitfalls and Opportunities, *Sloan Management Review*, 33 (3), 65-7(1992)

25. M.Christopher, and D.R.Towill, Supply Chain Migration from Lean and Functional to Agile and Customized, *Supply Chain Management: An International Journal*, 5 (4), 206-213(2000).

27. M.J.Tippins, and R.S. Sohi, IT Competency and Firm Performance: Is Organizational Learning a Missing Link?, *Strategic Management Journal*, 24, 745-761(2003).

28. G.T.M.Hult, R.F.Hurley, L.C. Giunipero, and Jr. E.L. Nichols, *Organizational Learning in Global Supply Management: a Model and Tests of Internal Users and Corporate Buy*ers, *Decision Sciences*, 31 (2), 293-325(2000).

29. T.C.Powell, and A. Dent-Micallef, Information Technology as Competitive Advantage: the Role of Human, Business and Technology Resources, *Strategic Management Journal*, 18(5), 375-405 (1997).

PART IV

Modelling and Simulation

Supply Chain Management Analysis: A Simulation Approach of the Value Chain Operations Reference Model (VCOR)

Carlo Di Domenico[1], Yacine Ouzrout[2], Matteo M. Savinno[1]
and Abdelaziz Bouras[2]
[1] University of Sannio – Dept. of Engineering, Piazza Roma, 21 – 82100
Benevento, Italy
{matteo.savino, carlo.didomenico}@unisannio.it
[2] LIESP Laboratory – University of Lyon (Lyon 2)
160 Bd de l'université
69676 Bron Cedex, France
{yacine.ouzrout, abdelaziz.bouras}@univ-lyon2.fr

Abstract

The impact of globalization and worldwide competition has forced firms to modify their strategies toward a *real time* operation with respect to customer's requirements. This behavior, together with the possibilities of communication offered by the up to date Information and Communication Technologies, moves the top management toward the concept of *extended enterprise* in wich a collaborative link is established among supplier, commercial partners and customers. When the information flows involves each agent of the chain, from suppliers to the final distribution centres, the extended enterprise becomes a *virtual firm*, which can be defined as a set of stand-alone operational units that acts to reconfigure itself as a value chain in order to adapt to the business opportunities given by Market. The present work is intended to verify the effective quantitative advantages given by the introduction of the *Value Chain* concept into the supply chain management through a simulation approach. The paper, after a description of SCOR and VCOR methods, makes a comparison between the two methods by the implementation of a simulation approach which point out the main additional requirements that are added to the VCOR model for its implementation.

Keywords

Supply Chain Management, Simulation, SCOR and VCOR Modeling,

Please use the following format when citing this chapter:

Di Domenico, C., Ouzrout, Y.,Savinno, M. M., Bouras, A., 2007, in IFIP International Federation for Information Processing, Volume 246, Advances in Production Management Systems, eds. Olhager, J., Persson, F., (Boston: Springer), pp. 257-264.

1   Introduction

Supply chain Management is one of the focus areas for companies and researchers over the last two decades. It deals with cost effective way of managing materials, information and financial flows from point of origin to the point of consumption to satisfy customer requirements. For a long time each member of the supply chain was considered as isolated, while nowadays, with the new ICT tools, companies are strictly integrated and in a *Supply Chain Dynamics*, A positive aspect of this dynamic is the effective collaboration which may lead to higher performance. A negative aspect is the independent decision making, which may create various delays and aggravate the forecasting error. The present paper deals with supply chain modeling and simulation, with discrete event approach, of the two main methods: SCOR and VCOR. Our main research interest in this work is to clarify the critical factors for minimizing the negative effects of supply chain dynamics and to gain insight on how to effectively manage them.

2   Supply Chain and Value Chain Management

A Supply Chain can be defined as a system network that provides raw materials, transforms them into intermediate commodities and/or in finished goods and distributes them to the customers through a delivery system [1]. The aim is to produce and distribute the right quantities, to the right locations, at the right time, while reducing costs and maintaining a high level of service.

Supply chain management (SCM) is recognized as a contemporary concept that leads in achieving benefits of both operational and strategic nature. SCM is concerned with smoothness, economically driven operations and maximizing value for the end customer through quality delivery. The limitations are mainly due to the fact that SCM as a concept does not extend far enough to capture customer's (end user) future needs and how these get addressed [2].

Another important theory that can be defined as strategic in the context of SCM is the concept of Value Chain Management. The Value Chain was described and popularized by Michael Porter in his 1985 best-seller, Competitive Advantage: Creating and Sustaining Superior Performance. Porter defined the "Value" as the amount that buyers are willing to pay for what a firm provides, and he conceived the "Value Chain" as the combination of generic activities operating within a firm, activities that work together to provide value to customers.

The huge importance of focusing on the customer has forced the integration of the optimization techniques of the SCM, Customer Relationship Management (CRM) and Product Lifecycle Management (PLM) (figure 1):

• PLM is a strategic business approach that helps enterprises to achieve its business goals of reducing costs, improving quality and shortening time to market, contemporarily innovating its products, services, and business operations. A definition can be: "*a strategic business approach that applies a consistent set of business solutions in support of the collaborative creation, management, dissemination, and use of product definition information across the extended enterprise from the product concept to the end of its life, integrating people, processes, business systems, and information.*" [3].

- The CRM is the creation, the development, the palimony and the optimization of long period relationships more profitable among consumers and firm. The success of CRM is based on the understanding of the consumers' needs and desires, and it fulfils setting such desires to the center of the business, integrating them with the firm's strategy, the people, the technology and the business process [4].

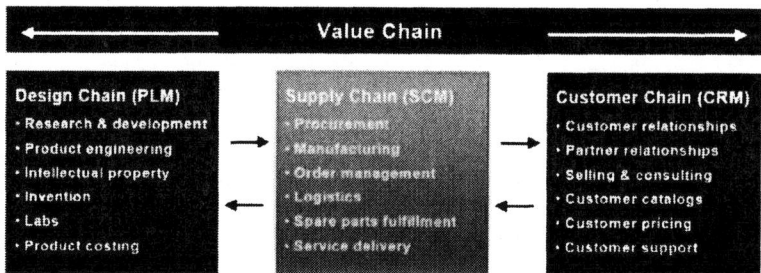

Fig. 1. Integration of PLM, SCM, and CRM from www.process-wizard.com

3   SCOR vs VCOR modeling

The VCOR model was developed from the perspective of being a value chain framework which supports and enables companies to integrate their three critical domains: Product Developments, Supply Network Integration and Customer Success, using one reference model to support the vision of an integrated Value Chain (VC). To achieve this goal VCOR uses a "process based common language". The main objective of this model is to increase the performance of the total chain and support the current evolution; for that, it proposes three main modeling layers:
- *Strategic Level:* The top level of the model includes all the high level processes in VCs and is represented through the Process Categories Plan–Govern–Execute.
- *Tactical Level:* The second level of the model contains "abstract" processes decomposed from the Strategic Level to implement and fulfill the strategic goals set in the top level of the model hierarchy.
- *Operational Level:* The third level of the model represents specific processes in the value chain related to actual activities being executed. In a VC perspective this is the level where fine-tuning occurs.

In order to measure the performance of the chain, the Supply Chain Council and the Value Chain Group (VCG) [8] have introduced different metrics and KPI to test supply chain reliability, responsiveness, flexibility, costs and efficiency in managing assets. According to the definition of the VCG a metric is "*a quantifiable variable that reflects a specific state of business performance during process execution within a strategic value chain context*". In VCOR a metric is characterized by different features: Metric Name, Definition, Priority, Metric Class & Sub-Class, Formula, Input Requirements, Dimension, Calculation Rules, etc. We will use these metrics in the simulation model to analyze the dynamic and the performance of the processes.

4   The simulation of SCM and PLM processes

Nowadays in literature there are few examples of simulated SC; moreover the number sensitively decreases if we consider the examples that apply the SCOR model [5][6][7]. This section presents first the advantages of simulation in comparison to the analytical and mathematical models and finally the realization of a generic architecture to simulate the VCOR model [8].

A SC is a dynamic, stochastic and complex system composed by a lot of actors; so in rare occasions it is possible to analytically model it. The benefits in using simulation in SCs can be summarized as follows [9][10]: Capacity to capture data for analysis; Decrease the risks inherent to changes in planning; Investigate the impact of innovations; Investigate relations between suppliers and other actors; Rationalize the number/size of order lots; Investigate opportunities to decrease the varieties of product components and standardize them throughout the SC...

4.1   Problem Description

The aim of this section is to illustrate the steps of our work, starting from the SCOR model simulation to the VCOR's one. We started from the level 1 of SCOR following a top-down approach. Once defined the Macro-Processes involved in each element of the chain, we choose their configuration and we depict the level 3 elements implicated in the work. To describe this model we use the following example of SC (figure 2).

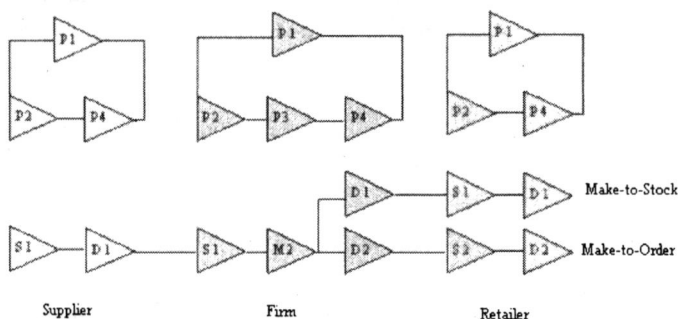

Fig. 2. Supply Chain Configuration

The Firm has a production process that transforms raw materials in finished goods using a *Make process* in its configuration. The other members, like Retailer and Suppliers, have only package processes that are included in the *Deliver process*. For some products, the production follows the rule of the *Make-to-Stock*, while for the others a *Make-to-Order* production is assigned. The simulation tool is based on the orders management; when any member needs a quantity of products it becomes a client and sends an order signal to its providers. An order is characterized by its identifier, quantity/type of product, timestamps and status (ordered, delivered...).

In this context, the necessary SCOR Level 3 elements have been realized in ARENA blocks and then gathered and organized in ARENA sub-models in order to set up the SCOR processes (figure 3).

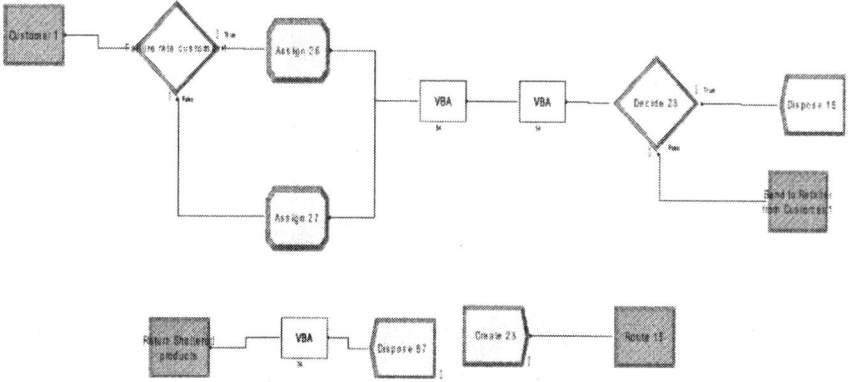

Fig. 3. Supply Chain Level 2: a sample of ARENA Implementation

4.2   SCOR and VCOR Simulation

We describe in this section the implementation of a sample actor of the chain: the retailer. It is composed by a *Source* and a *Deliver* process with two configurations: *Make-to-Stock* and *Make-to-Order*. An Arena block is implemented with a rule in which each request made by the consumer automatically becomes an order of the Retailer to the Firm. In the next figure the elements of the *Source process* are shown from the level 3 of SCOR. The *Schedule Product Deliveries S2.1* has the role to check the inventory levels stored in the database. If the effective level is under an "*s*" value, this module sends an order to the Firm, writing the order information's in the Firm's Demand database. The *Receive Product S2.2* receives the commodities send by the Firm and changes the order status to "Delivered". The *Verify Product S2.3* element checks the incoming products and *Transfer Product S2.4* transfers them in the inventory. The Retailer Inventory database is updated and the Customers demand can be satisfied. These generic processes based on SCOR are used for each actor of the SC (retailer, supplier, customer...) and implemented in the ARENA model.

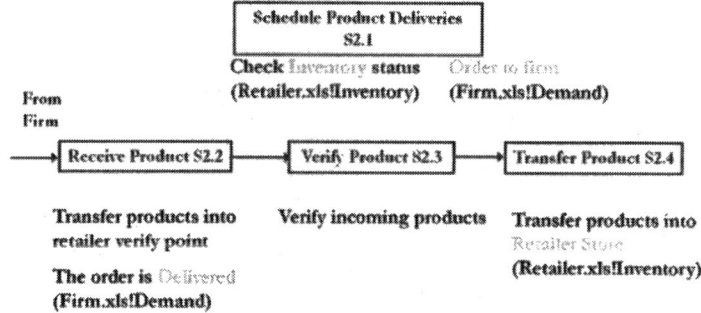

Fig. 4. Retailer SCOR Source process

In a second phase, the specifics elements of VCOR processes: Market, Research, Develop, Sell and Support processes have been added in the ARENA model:

a. Customer Behavior model

The VCOR simulation needs a model for consumer behavior; we have adopted the Adaptive Learning Model [11] that consists in a mathematical model which is able to simulate the satisfaction level of the consumer. According to this model the customers express their level of satisfaction through a vote for each product. This vote decrements in an exponential way with the time (product's end of life), and increases if the support service, introduced in the VCOR model, is able to resolve the problems of the clients. It is possible thanks to an "Innovation Factor" in the mathematical model that is enabled when an old product is modified to meet the market's demands. Our consumer behavior model is adapted from the Hopkins's one and is formalized as:

$$x_i(k+1) = (1-\alpha)\cdot x_i(k) + \alpha \cdot u_i(k) \quad (1)$$

where:

$x_i(k)$: the vote of the customer i, at instant k.

α: forgetting factor $(0<\alpha<1)$

$u_i(k)$ is the user input function

In our model ui(k) has the following expression

$$u_i(k) = f(k)\cdot I_n + \xi \cdot s + \beta \cdot \Delta p + \delta \cdot d + \phi \cdot q + \eta \cdot x_{j\neq i}(k) \quad (2)$$

where:

s: indicates if the last request of support was correctly satisfied (1) or not (-1)

ξ: takes into account how important is the quality of service for the customer.

In our application's case a measurement system will be adopted with respect to the firm's customer satisfaction requirements.

b. New Retailer configuration

With the application of VCOR model, the Retailer configuration defines a new process. The Support element has the objective to solve the Customers problems when a delivered lot is defective (figure 5).

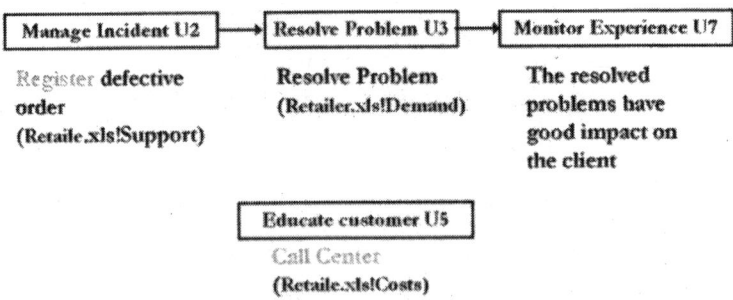

Fig. 5. Retailer VCOR Support process

The *Manage Incident U2* has the role to register in the Support database the defective order, the quantity and type of products returned, the Customer identifier and the timestamps. The *Resolve Problem U3* resolves the problem writing a new order in the Retailer Demand database with a quantity equal to the number of damaged products. Once that order is delivered the *Monitor Experience U7* registers the operation. It also modifies a variable in the Customer Behavior model that increments the vote. The *Educate Customer U5* simulates a Call Center; it has the important goal to decrement the percentage of defective products caused by bad installation decrementing the value of a variable in a Customer ARENA module.

c. Firm configuration

The Firm has four new processes derived from PLM and CRM of the Value Chain Management. The Market, Research and Develop processes belong to PLM while the Sell process to the CRM. With these processes a company tries to analyze the market and take consequently strategic decisions, like the restyling of a product or the acquisition of new technologies. Since it is very difficult to reproduce a complex market analysis or a restyling of a product, the ARENA blocks simulate these events in terms of required time and associated costs. The *Analyze Market M1* module periodically checks the Customer satisfaction level in the Retailer Demand database and if the satisfaction goes down under a specified threshold activates the *Architect Solution M4* module. This block finds the product type that has the least number of sales and decides to adapt it. In order to change a product, the Firm needs to modify its production line with the introduction of new technology. The *Introduce Technology* changes the production process in the *VCOR Build Product B3* (figure 6)

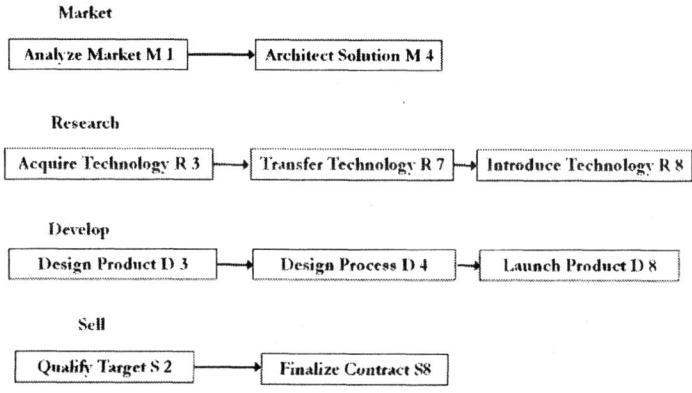

Fig. 6. Firm Market, Research and Develop processes

The *Develop process* materially changes the Bill Of Material of the product and it is responsible to launch the product. The D8, in fact, sets the variable "Innovation Factor" in the Customer Behavior model and, as consequence, the consumer vote considerably increases. The Sell process tries to identify the clients in the market in order to develop relationships and proposals. The *Qualify Target S2* classifies the Clients by the basis of their priority and determines which of them can be supplied by the Firm. It calculates, in fact, the total number of products required by the Clients and finalizes the contracts (S8) that don't exceed the fixed percentage of the dedicated production capacity.

5   Conclusion

This paper has discussed the study of the two most famous models used to implement the concepts of Supply Chain Management and Value Chain Management: SCOR and VCOR, through a simulation approach. Starting from an analysis of their standard architectures, using a top-down approach, a simulation tool has been developed under the ARENA software. Once implemented the SCOR model on an example of supply chain, the tool was extended to VCOR.

As confirmed by some experimental results made to validate the simulator, the adoption of VCOR model is a huge financial and organizational stress, but these efforts are fully repaid by the benefits in term of quality of service, market extension, competitiveness, flexibility, "quick response", innovation and other features essential to survive in the Global Market. A future work could complete the implementation of templates about all the processes of VCOR model, extending the flexibility of the tool realized, with a validation test on a real industrial case study to obtain a robust simulation tool.

6   Bibliography

[1] M. Christopher, "Logistics And Supply Chain Management: Creating Value-Adding Networks" Ed. Prentice Hall; ISBN-13: 978-0273681762; (2004)

[2] A.S. Al-Mudimigh, M. Zairi, A.M. Ahmed. "Extending the concept of supply chain: The effective management of value chains" IJPE; Ed. Elsevier Vol.3; (2004)

[3] CIMdata: Product Lifecycle Management *www.CIMdata.com*

[4] Wen-Bao Lin "The exploration of customer satisfaction model from a comprehensive perspective" Expert Systems Application Vol. 33(1) (2007)

[5] Hermann, Lin and Pundoor "Supply Chain Simulation modeling using the Supply Chain Operations Reference Model". Inderscience Publicher; IJSCM Vol. 2 (2003)

[6] P. Bolstorff R. Rosenbaum. "SC Excellence: A Handbook for dramatic: improvement using SCOR Model" Ed. AMACOM; ISBN: 978-0814407301; (2004)

[7] J. Geunes, P.M. Pardalos, and H.E. Romeijn. "SCM: Models, Applications and Research Directions ", Kluwer Academic Publishers, ISBN 1-4020-0487-7 (2002)

[8] Value Chain Group. "EMEA Webpage" *www.value-chain.org*

[9] Boucher, Kuhn and Janke. "Integrated Modeling and Simulation of Inter-Organizational Business Processes". (2003)

[10] H. Akkerman, "Emergent Supply Networks: System Dynamics Simulation of Adaptive Supply Agents"Proceeding of the 34th Hawaii ICSS'01. (2001)

[11] Ed Hopkins "Adaptive Learning Models of Consumer Behavior" Levine's Bibliography - UCLA Department of Economics (2006)

An Integral Model for Mapping Variant Production in Supply Chains

Sören Günther and André Minkus

ETH Zurich, Center for Enterprise Sciences (BWI)
Kreuzplatz 5, 8032 Zurich, Switzerland
sguenther@ethz.ch, aminkus@ethz.ch
WWW home page: http://www.lim.ethz.ch

Abstract

The capability to efficiently manage product-variety is nowadays a critical success factor for many companies. However, existing models still lack a suitable support for mapping and analyzing variant productions. The paper contributes to this area with a comprehensive but practicable approach. Product families are defined through common attributes with differentiating characteristics. Product structures and process plans of variants belonging to the same product family are represented using generic "plan skeletons". It is shown how the approach can thereby reduce modeling efforts and enhance the clarity of the resulting model. Furthermore, linking such a model to a simulation allows for assigning performance indicators to the product family's attributes instead of many product variants. Variant-induced costs can thus be disclosed as additional costs compared to a base product for each characteristic.

Keywords

Modeling, Simulation, Product Variety, Supply Chain Management

1 Motivation

The number of product variants offered by companies has strongly increased within recent years. This becomes obvious by an example of the automobile industry: whereas in 1914 the slogan of Henry Ford applied "you can have your car in any color, so long it is black", in the years 2003 and 2004 only two of the 1.1 million Mercedes A-class-cars produced in the Raststatt plant were identical [1]. This development can be observed across all industries. In a European-wide evaluation of studies about the future of production the increasing product variety was identified as one of the main challenges for the coming years [2].

Please use the following format when citing this chapter:

Günther, S., Minkus, A., 2007, in IFIP International Federation for Information Processing, Volume 246, Advances in Production Management Systems, eds. Olhager, J., Persson, F., (Boston: Springer), pp. 265-272.

The reasons for this trend are manifold. Often companies try to meet the growing individualization of customer needs by differentiation strategies and thereby to gain a competitive advantage. Many enterprises hope that with sinking or stagnating sales they can secure the turnovers by offering additional product variants and special features to less attractive market segments. Furthermore, the increasing internationalization requires an adaptation of products to regional distinctions (e.g. different voltages and/or frequencies).

Hayes and Wheelright stated already in 1979: "Coping with product variety forces a manufacturing firm to confront a fundamental trade-off – the increased revenue that can result from more variety versus increased costs through the loss of economies of scale." [3]. Efficiently managing and controlling the product variety thus has turned out to be a critical success factor. According to Ramdas [4] two fundamental types of approaches can be differentiated: on the one hand the containment and/or decrease of product variety during the product development, on the other hand the efficient management of existing variants in the production network. Given the complexity of today's production and logistics networks the utilization of models can provide an important decision support for the latter approach. First, mapping the reality in a structured and clear model simplifies the understanding of the numerous interdependences, especially across company borders. In addition, with the help of such models alternative scenarios can be analyzed and evaluated. However, this demands for robust, tangible and efficient approaches, which are able to model the value adding network and to support variety-related decisions.

2 State of the art in modeling

In scientific literature numerous authors deal with modeling and analyzing production and logistics networks. Beyond that, a vast number of commercial software solutions is offered under the designation "supply chain management (SCM) software". Those solution claim to support the planner in the decision making process (see for example [5]).

Although different kinds of modeling approaches have been developed and applied (e.g. network models or reference models) many authors consider simulation models as the most promising technique. This is mainly due to the fact that they are able to map dynamic changes, stochastic events, and thus uncertainties [6]. Terzi and Cavalieri state: „Among these quantitative methods, simulation is undoubtedly one of the most powerful techniques to apply, as a decision support system, within a supply chain environment." [7]. The Eureka study already mentioned confirms that industry also expects simulations to become very important especially when dealing with a high product variety [2]. However, one of the main problems when using simulations is the large number of necessary input parameters, resulting in a high model complexity and large efforts for gathering those data. In addition, often specific knowledge is required, e.g. simulation languages [8].

Modeling concepts can be divided into resource-oriented and product-oriented approaches. It is notable that almost all models belong to the first category, which means that first resources (i.e. suppliers, machines, storages etc.) are modeled and afterwards the material flows between these resources are defined. Mainly this is due to the fact that most of those approaches have evolved from modeling and simulation of rather "isolated" manufacturing systems on shop floor level. The focus thereby is on the production resources and their performance (e.g. utilization, maintenance intervals, down-times etc.). These approaches, however, are lacking a sufficient support of the fundamental idea of supply chain management, which is the alignment of the entire value creation towards the final customers: "the supply chain (...) encompasses every effort involved in producing and delivering a final product or service, from the supplier's supplier to the customer's customer." [9]. Even more, in hardly any of the models the particularities of product variants are considered. Instead, most of the approaches model variants of the same product family similar as completely different products. The well-known and beneficial concepts of product and process configuration are not used in this context (see e.g. [10]).

Among the few product-oriented models is the "Product Chain Decision Model" which utilizes the concept of Petri Nets [8]. Here "place nodes" represent product components, "transition nodes" model process steps. The "Supply Chain Analyzer" - a simulation software developed by Nienhaus - is based on the product structure, too [11]. However, both approaches are not capable of modeling product variants in an appropriate and efficient way.

3   The Modeling Approach

3.1   Different Views on a Supply Chain

In order to map the complexity of production networks with conflicting objectives integral models – as opposed to singular models such as flow charts – have to be used. They incorporate different views on the system to be analyzed, i.e. a product view, a process view, and a resource view. Only the combination of these views can answer the question "Which things *(product)* are processed by whom *(resource)* in which order and according to which rules *(process)*?" (following [11]).

3.2   Product View

The modeling approach to be presented is primarily based on the product view. Therefore, the model belongs to the "product-oriented" approaches. When modeling a variant production entire product families rather than single products should be considered. According to APICs product families are defined as a group of products (also called "variants") with similar features or functions, similar product structures, and a high percentage of same processes in the process plan [12]. Product families are designed as such during the product design phase and – if necessary – expanded throughout their life cycle [13]. A variant is a specific product of a product family and can be derived from the model of the product family ("product configuration"). The variants within a product family can be distinguished by the different characteristics of common attributes, e.g. an attribute "color" with the characteristics "green", "yellow", and "red". Inversely, for a certain variant all characteristics of the attributes of the product family are definite.

Consequently, building the supply chain model starts with modeling a product family and defining its attributes and characteristics. For every product family (0..n) attributes each with (1..m) characteristics need to be defined resulting in $\prod n_i$ (for i=1..m) potential variants. For this set of variants rules can be created which describe all valid or exclude all invalid combinations of product characteristics, e.g. "IF (material = leather) THEN (color = black OR beige)". If the model is connected to a discrete event simulation these rules can dynamically be changed during the simulation run in order to e.g. model product phase-ins and phase-outs. Figure 1 shows an example for a product family with three attributes and the resulting set of variants.

Name	Color	Voltage	Type	Valid
All	All	All	All	
Variant_1	black	220V	Standard	Valid
Variant_2	silver	220V	Standard	Valid
Variant_3	red	220V	Standard	Valid
Variant_4	black	120V	Standard	Invalid
Variant_5	silver	120V	Standard	Invalid
Variant_6	red	120V	Standard	Valid
Variant_7	black	100V	Standard	Valid
Variant_8	silver	100V	Standard	Valid
Variant_9	red	100V	Standard	Valid
Variant_10	black	220V	Deluxe	Invalid
Variant_11	silver	220V	Deluxe	Invalid
Variant_12	red	220V	Deluxe	Valid
Variant_13	black	120V	Deluxe	Valid
Variant_14	silver	120V	Deluxe	Valid
Variant_15	red	120V	Deluxe	Invalid
Variant_16	black	100V	Deluxe	Invalid
Variant_17	silver	100V	Deluxe	Valid
Variant_18	red	100V	Deluxe	Invalid

Fig. 1. Modeling a product family with its attributes and characteristics

This concept of modeling a product family has important advantages, especially when using the model as input for a discrete event simulation. In the course of the simulation performance indicators can be assigned to few attributes of a product family instead of having to analyze a large number of product variants. This way e.g. the effects of offering additional features to the customer (in terms of more characteristics such as more colors or even in terms of new attributes) can easily be determined. For each characteristic variant-induced costs can be disclosed as additional costs compared to a base product. Furthermore, the modeling effort can be reduced by a generic representation of the product family, resulting in a clear and comprehensible model. This will be explained further in the following.

The product structure – and the structure of the product family respectively – forms the basis of the entire supply chain model. It is described through a maximum bill of materials which can be determined by combining the bill of materials held at various companies. As the model currently only allows for convergent product structures, these can be represented in a tree-like structure. The "leafs" located to the left represent the inputs (e.g. raw materials or parts purchased at suppliers which are not relevant to the analysis). These are then processed and assembled resulting in the final product to the right. For each of the components in the tree the required input quantity has to be defined. Depending on the attributes of the product family the product structure (i.e. the bill of materials) of the variants may differ. Hence, the information needs to be represented in a generic way. From this the product structure for a certain variant can be generated during the so called product configuration.

The product structures of all variants within a product family are therefore represented in a so called *plan skeleton*. The plan skeleton is an ordered graph with its nodes depicting product components and process steps. Different from a common bill of materials comprises ramification nodes with associated decision rules which describe in which situation a specific component is to be chosen. This allows for an efficient and concise representation of similar plans while at the same time keeping redundancy low. This concept will be further explained in the following chapter.

3.3   Process View

The process view is the secondary view on the supply chain. It observes the system with respect to the order in that elementary functions (e.g. manufacture product, store product) are carried out. In the model on hand it is derived by adding process steps to the plan skeleton which defines the product structure.

Similar as for the product structure the processes may also differ depending on the variant's characteristics. The process configuration then selects the process steps needed to manufacture a certain product variant from a predefined set of processes and puts them into a feasible sequence. Plan skeletons also allow for representing this process-related knowledge of a product family. To further ease the supply chain model only three different types of process nodes are used within the plan skeleton: production, assembly, and storage processes.

- Production nodes define a production process. Thereby the model doesn't differentiate between different types of production processes but uses processes with generic parameters to describe production planning and control, lead time, as well as material and production costs.
- Storage nodes define a storage process. Again, to simplify the model only one type of storage is used, no matter which kind of material is stored (raw materials, in-process inventories or finished goods). Parameters relate to production planning and control, lead time, material management and storage costs.
- Assembly nodes unite parallel production paths. An assembly node can also have only one predecessor. For each of the predecessor paths an input quantity is defined, i.e. the number of components per unit of the super ordinate part, into which the component is built.

In addition, ramification nodes (so called "decision nodes") can be inserted to define parallel paths of the graph. Each of those paths can again comprise any number of process nodes or even other decision nodes. The parallel paths are merged in a "join node". Using the decision nodes a maximum process plan can be modelled which comprises all process plans of the different variants within the product family. To each decision node an attribute of the product family is assigned. Furthermore, the characteristics of that attribute are allotted to the paths. This is done in a subtractive process, which means that one path is set as default for all characteristics and each characteristic that is assigned to another path is subtracted from the default set. Here the advantage over resource-oriented approaches becomes especially apparent as those approaches require the definition of each individual manufacturing process for every product variant which is often quite time-consuming and leads to a high complexity of the model. A simple example for a plan skeleton is given in Figure 2.

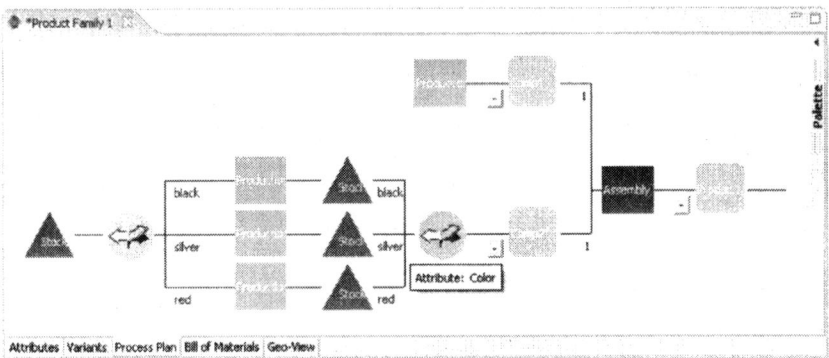

Fig. 2. Example for a simple plan skeleton with one decision node.

From the plan skeleton of a product family different views can be generated by filtering the plan skeleton with respect to certain characteristics: The bill of materials can be derived by hiding all process steps. It can be displayed either for an individual or for several product variants. A process plan for a certain variant can be deduced by omitting the decision nodes and only showing the paths referring to the characteristics of the product variant. It is also possible to display the process plan for several selected variants. In this case all necessary decision nodes of the attributes of the selected variants are included.

3.4   Resource View

The modeling approach especially aims at small and medium sized enterprises. Such companies are often involved into supply chain networks with legally and financially independent partners. Consequently, all of the partners should be able to understand and use the model although each of them has a different understanding of the roles within the value chain, e.g. the supplier of one partner might be the customer of another partner. Hence, in the present model suppliers, OEMs, and customers are not differentiated but all represented by a class *company*. The supply chain model focuses on the value-added chain up to the final product; therefore it doesn't include the distribution network. Retailers and distributors are not taken into account.

A company stands for an owning entity of the supply chain. Each company can comprise several *sites* representing the premises the company owns. They are defined by their name and a geographic location. A site manufactures or assembles finished or semi-finished products from raw-materials and/or sub assemblies for which *machines* are needed. In addition, a site may comprise one or more *storages* to store raw materials, work in-process, and finished goods.

These production network objects can then be combined with the plan skeleton. Thereby, the resource view becomes the tertiary view in the supply chain model. The site responsible for performing a process step is assigned to the respective process node in the plan skeleton. If the site is assigned to a production or assembly node then the usage of a machine within that site is assumed. If it is assigned to a storage node then a storage location of that site is accessed. Of course a site can be assigned to multiple nodes mapping the different process steps which are performed in that plant. Again, the difference to resource-oriented approaches becomes apparent: The proceeding is exactly the opposite from those approaches, which assign the material flows needed to produce a product to a network of resources.

3.5   Modularization of the Model

There is a major difference between product-oriented and resource-oriented models: In product-oriented approaches the product view is the primary view and the process and resource view are dependent views (secondary and tertiary); for resource-oriented approaches this is exactly the opposite. Therefore, in the latter case inheritance as one of the main features of object-oriented modeling is often used to provide a modeling library for supply chain resources such as site, machines etc.

Using a product-oriented approach even allows for going a step beyond this. As the entire model is based on the product structure, any component node within a plan skeleton can be "exported" together with its subtree as plan skeleton of a new product family (e.g. a module or subassembly). All attributes assigned to the decisions nodes within the subtree become attributes of the new product family, and the assignment of production resource to nodes is inherited, too. This module can then be inserted as component node into the plan skeletons of other product families. For example first the product family "drilling machine" is modelled. As the same motor is used for other product as well it is exported as new product family. It can then e.g. be inserted into a plan skeleton of the product family "chipping hammers" without having to model all processes and resources again. This way the model can easily be modularized and even a library of reusable components can be developed.

4   Conclusions and Outlook

The modeling approach presented in the previous chapter has been implemented in a prototypical software solution using the object-oriented programming language Java 1.5 (the Figures 1 and 2 were taken from this software). This allows for a 1:1 representation of the supply chain model and ensures platform independency. A main requirement is to keep the software small – though comprehensive – so that it can be run on any standard personal computer or laptop. Furthermore, it should guide the user and thereby require no specific IT knowledge or modeling expertise.

First applications of the prototype in industry proved that the product-oriented approach is well suited for modeling variant productions. Modeling efforts could be reduced especially through inheriting entire components together with the resources assigned to them. Also, the clarity of the model was increased remarkably.

In a next step the supply chain model will be connected to a discrete event simulation. The results of that simulation can then again be displayed along the product structure. The representation of the simulation results in the same structure that is underlying the model facilitates the interpretation. The user can thus directly and visually grasp bottlenecks and potential for improvement along the product creation process without having to analyze complex tables or figures. Finally, the decision support for variant-related problems will benefit from the generic representation of product families through their attributes and characteristics as mentioned earlier.

5   References

1. N.N., Wahnsinn mit Methode: Zögerlicher Kampf gegen Variantenvielfalt, *Automobil-Produktion*, January 2005, 38-43 (2005).
2. E. Schirrmeister, P. Warnke, C. Dreher, Untersuchung über die Zukunft der Produktion in Deutschland: Sekundäranalyse von Vorausschau-Studien für den europäischen Vergleich, Abschlussbericht des Eureka-Factory-Projekts Informan 2000+ (Fraunhofer ISI, Karlsruhe, 2003).
3. R.H. Hayes, S.C. Wheelright, Link Manufacturing Process and Product Life Cycles, Harvard Business Review, January-February 57(1), 133-140 (1979).
4. K. Ramdas, Managing Product Variety: An integrative review and research directions, Production and Operations Management 12 (1), 79-101 (2003).
5. F. Laakmann, K. Nayabi, R. Hieber, Market Survey 2003 - Supply Chain Management Software. Detailed investigation of supply chain planning systems. 2nd edition, edited by scm-CTC (scm-CTC, Stuttgart, 2003)
6. S. Jain et al., Development of a high-level supply chain simulation model, Proceedings of the 2001 Winter Simulation Conference, 1129-1137 (2001).
7. S. Terzi, S. Cavalieri, Simulation in the supply chain context: a survey, Computers in Industry 53, 3-16 (2004).
8. J. Blackhurst, T. Wu, P. O'Grady, PCDM: a decision support modeling methodology for supply chain, product and process design decisions, Journal of Operations Management 23, 325-343 (2005).
9. SCOR, Supply Chain Operations Reference (SCOR) Model, Version 8.0 (Supply Chain Council, Pittsburgh, 2006).
10. K. Schierholt, Process Configuration – mastering knowledge-intensive planning tasks (vdf Hochschulverlag, Zürich, 2001).
11. J. Nienhaus, Modeling, Analysis and Improvement of Supply Chains – A Structured Approach (Dissertation No. 15809, ETH Zürich, 2004).
12. APICS, Dictionary 11th Edition (APICS – The Association for Operations Management, Alexandria, VA, 2004).
13. P. Schönsleben, Integral Logistics Management – Operations and supply chain management in comprehensive value-added networks, 3rd edition (Auerbach Publications, 2007).

Integrating Activity Based Costing and Process Simulation for Strategic Human Planning

Takayuki Kataoka [1], Aritoshi Kimura [1], Katsumi Morikawa [2] and
Katsuhiko Takahashi [2]
1 Department of Information Systems Engineering,
Faculty of Engineering, Kinki University,
Takaya Umenobe 1, Higashi-Hiroshima, Hiroshima, 739-2116, JAPAN
{kataoka, kimura}@hiro.kindai.ac.jp
WWW home page: http://www.hiro.kindai.ac.jp
2 Department of Artificial Complex and System Engineering,
Graduate School of Engineering, Hiroshima University,
4-1 Kagamiyama 1-chome, Higashi Hiroshima, 739-0857, JAPAN

Abstract

This paper presents a method of integrating activity-based costing (ABC) and
process simulation in human planning. Our studies have already proposed a
method of integrating ABC and process simulation in business process
reengineering (BPR) and showed a case study of a chemical plant. Some
studies have also already showed some examples of various aspects in
manufacturing systems. Although a large number of studies have been made
on product-mix/machine loading or scheduling, little is known about human
planning. In this paper, effective BPR methodologies to achieve dramatic
improvements in business measures of workers' skills and costs based on ABC
are discussed. First, two important tools: process simulation method and ABC
analysis that can be customized by organizations for their own BPR are shown.
As these tools have been separately used, a unified approach of process
simulation and ABC analysis for process redesigns based on simulation
Secondly, a method of process simulation design is shown. It is repeated to
consider working ratio and running time of resources. Thirdly, a method of
ABC analysis that can be customized by organizations is shown. It
automatically shows the data that has been gathered from many sources. By
utilizing the data, the process simulation is implemented, and the result of
simulation gives the data to ABC analysis. Lastly, this paper shows a case
study in BPR and the effectiveness of our method.

Keywords
Activity-Based Costing (ABC), Human Planning, Process Simulation

Please use the following format when citing this chapter:

Kataoka, T., Kimura, A., Morikawa, K., Takahashi, K., 2007, in IFIP International Federation for Information
Processing, Volume 246, Advances in Production Management Systems, eds. Olhager, J., Persson, F., (Boston:
Springer), pp. 273-280.

1   Introduction

Recently, there has been a great deal of interest in business process reengineering (BPR). Hammer and Champy [1] defined reengineering as "the fundamental rethinking and radical redesign of core business processes to achieve dramatic improvements in quality, cost, and cycle time". Earlier literature on BPR focused only on the principles of BPR and provided us with cases of successful BPR projects that address only the "what" and "why" questions [2]. However, there are few comprehensive and methodological approaches to conducting BPR projects to answer to the "how" question. Therefore, this paper focuses on development of comprehensive BPR methodologies. This paper first describes the process simulation method and the activity-based costing (ABC) analysis, which are likely considered as the most effective methods and tools in BPR methodologies. The former is effective to analyze whether the newly designed process can meet performance standards. "What if" questions related to process performance may be asked. The latter is a very useful tool when cost saving is the major goal of a BPR project. However, both are usually used separately in a BPR project. It can be seen clearly that rather than using these methods separately, it is better to integrate both methods in doing reengineering process. Therefore, this paper proposes a new methodology, ABPCSS (Activity-Based Process and Costing Simulation System), which explicitly unified both process simulation method and activity-based costing analysis. At last, this paper also gives a simple case study of strategic human planning about the use of the unified approach.

2   BPR Methods and Tools: Process Simulation Method and Activity-Based Costing(ABC) Analysis

There are several methods and tools for process modeling and analysis. Among them, the process simulation method and the ABC analysis are especially effective to the understanding of complex business processes.

2.1   Process Simulation

Strictly speaking, simulation is the use of a model as a basis for exploration and experimentation. Like all modeling approaches, simulation is used because it is cheaper, safer, quicker, and more secure than using the real system itself. The idea is that the model becomes a vehicle for asking "what-if" questions. That is, the simulation model is subject to known inputs and the effects of these inputs on the outputs are noted.

Whatever the simulation modeling approach adopted, a simulation model must be able to mimic the changes that occur, through time, in the real system. The majority of applications of dynamic simulation methods in management science seem to use discrete event models [3].

The simulated process is summarized into four steps:

    1. Specify model.

2. Build model.
3. Simulate model.
4. Use model.

Each step is conducted as follows:

Step 1: Specify model.

Describe the objectives of the system to be simulated, which is in discrete condition where the situation change as time goes. The following things are described:

· Entity: tangible component of the system.
· Class: grouping of similar objects that share similar features or attributes.

Moreover, to describe how entity changes as time goes, the following things are also described:

· State: one for a period of time that is occupied by entities.
· Event: a state at which the entity changes state.
· Activity: one that the entity is changing state and this takes time to happen.
· Process: chronological sequence of activities.

Step 2: Build model.

In order to build a simulation model, we must understand the logic of the system to be simulated in terms of the entities and their interactions. In order to do that, an activity cycle diagram (ACD), one of the description methods, was made. An ACD only has two symbols: active state and dead state, as follows:

· Active state: one whose time duration can be directly determined at the event which marks its start.
· Dead state: one whose duration cannot be so determined but can only be inferred by knowing how long the active states may last.

Step 3: Simulate model (by VIMS (Visual Interactive Simulation and Modeling System)).

The usual way in which a VIMS model is developed is to begin with a blank background screen and then to pace icons on the screen to represent the major components of the system. These icons are then linked together by drawing lines on the screen to form a type of network, which captures the logical interactions between the entities of the system.

Step 4: Use model.

The use of the model involves the making of runs and the interpretation and presentation of the outputs. When simulation results are used to draw inferences or to test hypotheses, statistical methods should be employed.

2.2 Activity-based costing (ABC)

ABC is a technique that measures the cost of activities. Costs that are generally treated as overhead are allocated to activities identified in the process model. ABC can provide insight into overhead.

There are 5 steps in an ABC analysis, as follows:

1. Analyze activities.
2. Gather costs.
3. Trace costs to activities.

4. Establish output measures.

5. Analyze costs.

We describe each of these steps in detail here [4]:

Step 1: Analyze activities.

This step conducts process decomposition, call "bill of activities", in the ABC effort. Then, activities are classified as primary, secondary, required, or discretionary. A primary activity is one that directly supports the organization's mission. Secondary activities support primary activities. Activities that the organization must be performed are required, while discretionary activities are ones that are truly optional.

Step 2: Gather costs.

The next step gathers costs for each activity. Collecting costs could be the most difficult aspect of conducting ABC because of lack of historical data for existing activities. The overhead cost of producing products or delivering service often represents a larger percentage of total costs than material, labor, or machinery.

Step 3: Trace costs to activities.

Cost is traced to an activity based on the charges for each activity. Resources used by activities, and hence their costs, are traced to activities. The total input cost of each secondary activity is allocated to the primary activities it supports. Total input cost for each activity can be calculated at this stage.

Step 4: Establish output measures.

Each activity may have more than one output. One primary output must be identified and it should be quantifiable. The unit cost of an activity is calculated by dividing the total input cost of the activity by its primary output volume.

Step 5: Analyze costs.

Cost analysts can use cost driver analysis to identify activities or factors that influence the cost and performance of other activities. The cause for high cost should be treated at its source. Cost data can be analyzed based on cost elements (i.e., labor, material, service, and supplier) in order to compare alternative designs to the baseline. Non-value-added activities are targets for elimination or improvement.

3   A Unified Approach: Integrating Process Simulation and Activity-Based Costing (ABC)

In general, it is better to use some methods and tools as one unit than use them separately. Even though there have been a lot of BPR methods and tools developed, research concerning a unified approach is scarce [5]. Therefore, we try to develop a new system methodology for BPR: an Activity-Based Process and Costing Simulation System (ABPCSS), which unified the largely used BPR methodologies: the process simulation method and the activity-based costing analysis. We explain this below. The primary components of ABPCSS are:

· Database system: This is a set of related files in the ABPCSS. Resources, activities, and cost objects are stored in the files.

· Activity-based process simulator: This uses the simulation model to forecast the performance of reengineered processes.

· Activity-based costing simulator: This calculates the detailed cost of resources consumed by activities and the cost of activities consumed by the cost objects within the existing and reengineered processes.
· Graphical user interface: This allows user to communicate with any component of the ABPCSS.
  Figure 1 shows data flow among each system component.

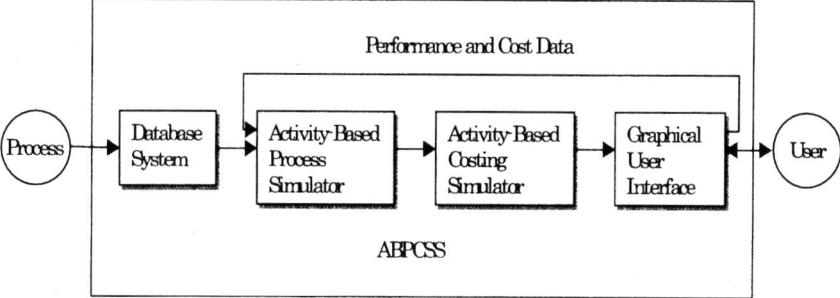

Fig. 1. A framework of activity-based process and costing simulation system (ABPCSS)

The process data and the bill of activities can be fed into the activity-based process simulator to analyze the performance of an existing process, and then performance and cost data can be fed into the activity-based costing simulator in order to analyze the cost of existing process. Then, the performance and cost data can be used to compare the reengineered process with the existing process. Design of the new process may be revised, and then re-evaluated with the ABPCSS until a finalized process is determined.

4   A Case Study: Small-Scale Cellular Manufacturing Operation

In order to illustrate the above procedures, we will take up the reengineering problem of the small-scale cellular manufacturing operation, which is made by using Arena, one of VIMS. This is conducted by a number of workers, which execute various jobs involving collaborative operations. Therefore, how to allocate single and collaborative operations to appropriate number of workers becomes an important topic in the equipment factory.

4.1   Modeling for Simulation

First, a simulation model concerning operations for such small cellular manufacturing workers is expressed, where the objective is to find improvement points. Preconditions for small cellular manufacturing operation discussed here, are as follows:
· The kinds of materials: 3
  Materials 1: average working time: 50 minutes, transportation time: 10 minutes

Materials 2: average working time: 75 minutes, transportation time: 12 minutes
Materials 3: average working time: 40 minutes, transportation time: 8 minutes
· The number of cells: 5
· The number of processes in each cell: 4
· Each process needs the specified skills
· About processes in each cell:
1st process: the skill must be more than level 6, 2nd process: more than level 5, 3th process: more than level 4, 4th process: more than level 6
· Organization: fixed 5 workers (A,B,C,D,E)
· Each worker's skill consists of 10 degrees:
Worker A: skill level 8: 28 dollars/hour, Worker B: skill level 7: 28 dollars/hour
Worker C: skill level 7: 28 dollars/hour, Worker D: skill level 8: 28 dollars/hour
Worker E: skill level 8: 28 dollars/hour
· Processing time per a day is 8 hours and a half (including break time: 30 minutes)
· Each worker is fixed in each cell

4.2   Implementation and Analysis

The average value of 10 working days is considered as simulation results. WIP (work-in-process) is started from zero. If it exists on the end of a working day, it is made on the next day.
Case A: 5 workers is fixed in each cell
The results of simulation are shown in Table 1 and 2. Based on these results, "Total Cost", "Idle Cost", and "Number Out" are focused to search for better results.

Table 1. An example of simulation results

Total Cost	Busy Cost	Idle Cost	Number Out
35,519	22,473	13,047	140

Table 2. Instantaneous Utilization

Worker A	Worker B	Worker C	Worker D	Worker E
0.9699	0.5699	0.9824	0.9154	0.9773

4.3   Optimization simulation

Preconditions for optimization simulation discussed here, are as follows:
· Organization is classified in 4 groups
Group 1: Main Staff, G2: Main Staff2, G3: Sub Staff, G4: Sub Staff2
· Working place of each group
Main Staff : 2 processes in Cell 1 ~ 3, Main Staff2: 2 processes in Cell 4 or 5
Sub Staff : 2 processes in Cell 1 ~ 3, Sub Staff2: 2 processes in Cell 4 or 5
Case B: how many temporary workers of 8 working hours per a day should be added as the optimal simulation?

· Optimized procedure
  Step 1: Maximum of "Number Out"
  Step 2: Minimum of "Total Cost"
  Step 3: Minimum of "Idle Cost"
  (The maximum number of added workers is 12)
Step 1: Maximum of "Number Out".
  These results are implemented by "Out Quest for Arena". Figure 2 shows the process of optimization. The optimal value is indicated at about 30 times. However the result of 204 products is lead by adding 4 temporary workers, it is not only one. Therefore, optimization of "Number Out" is added to the precondition, and the minimum of "Total Cost" and "Idle Cost" are similarly implemented by optimization simulation.

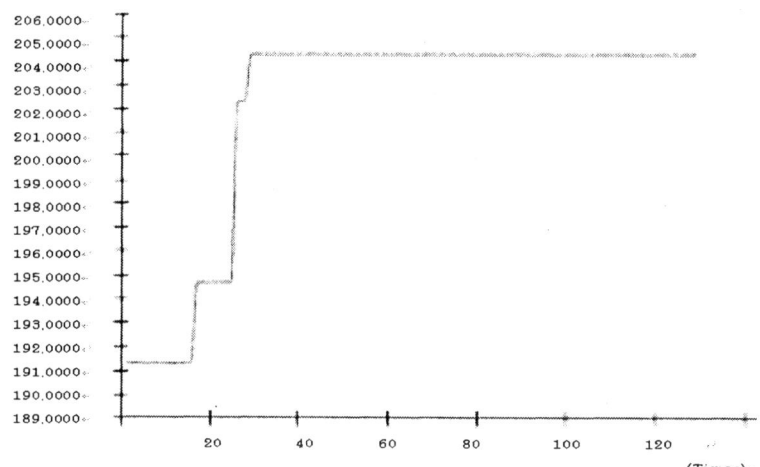

Fig. 2. Optimal solution for Step 1

Step 2: Minimum of "Total Cost"
  The optimal value is indicated at about 30 times.
· Additional worker: 1
· Number Out: 204
· Total Cost: 34,841
· Idle Cost: 4,329
Step 3: Minimum of "Idle Cost"
  This result shows that "Idle Cost" in "Total Cost" is 1/8. Additionally, compared with *Case A*,
· Number Out: 64 (increase)
· Total Cost: 678 (decrease)
· Idle Cost: 9,396 (decrease)
  As a result, *Case B* is better. Especially, it shows a great reduction in "Idle Cost". It can become even 1/3.
Case C: how many workers should be added and how long should they be employed as optimal simulation if additional members are 4 part-time workers?

As the result, the value of 166th simulation is the minimum. The result shows the following.
- Additional workers: 3 (1st worker: 2 hours, 2nd: 2 hours, 3rd: 4 hours)
- Total Cost: 35,344
- Idle Cost: 4,346
- Number Out: 205

There are some differences from "Total Cost" in compared with *Case B*. The result of *Case B* is better than that of *Case C*. In other words, the result shows that to add temporary workers who can work for 8 hours is better than to add part-time workers can work for 4 hours in this case.

5   Conclusion

Recently, in order to survive, companies cannot win the severe competition without doing reformation to adapt the change. The methodology proposed here fulfills the important role in executing BPR. This paper has focused on process simulation method and ABC analysis, and proposed a unified approach into which they are integrated. This paper has also presented a case study, in order to improve the effectiveness of the unified approach.

In present, we are collecting some successful example of BPR application and in the same time, doing some research to develop new methodologies, which apply information technology to BPR. This has been done, as we want to consider how to mix information technology with BPR to measure the efficiency of the organization after changing the organization itself or its way of work.

6   References

1. Hammer, M. and J. Champy., Reengineering the Corporation: A Manifesto for Business Revolution, (Harper Collins Publisher, Inc, New York, 1993).

2. Chen, M., BPR Methodologies: Methods and Tools, in D.J. Elzinga, et al. eds, Business Process Reengineering, Kluwer Academic Publisher, 187-211 (1999).

3. Pidd, M., Tools for Thinking: Modeling in Management Science, (John Wiley & Sons, Inc, New York, 1999).

4. Brimson, J.A., Activity Accounting: An Activity-Based Costing Approach, (John Wiley & Sons, Inc, New York, 1991).

5. M. Ozbayrak, M. Akgun, and A. K. Turker., Activity-Based Cost Estimation in a Push/Pull Advanced Manufacturing System, International Journal of Production economics, 87,49-65 (2004).

A Methodology for Modeling a Quality Embedded Remanufacturing System

Youngseok Kim, Hong-Bae Jun, Dimitris Kiritsis, and Paul Xirouchakis
EPFL, STI-IPR-LICP, ME B1, Station 9, 1015 Lausanne, Switzerland,
youngseok.kim@epfl.ch
WWW home page: http://licpwww.epfl.ch

Abstract

The uncertain quality of used products highly affects the performances of remanufacturing systems and quality of remanufactured products. Hence the quality issues of used product cannot be neglected in remanufacturing. To apply the quality concept into remanufacturing system control or simulation, individual management of each product is required. To this end, a multi-agent approach can provide good solutions. The first step in applying the quality concept with the multi-agent approach is an effective modeling of a remanufacturing system. This study proposes a methodology for modeling a quality embedded remanufacturing system (QRS) with two layers. The first layer represents elements in a remanufacturing system as it is. The representation also contains the information of a multi-agent structure for a QRS. The second layer expresses the status of each product and resource agent, and their relationships to manage the multi-agent system. As modeling tools to support the proposed methodology, directed graphs and Petri-Nets are used. A case study is introduced to show an application of the proposed methodology.

Keywords

Remanufacturing, Modeling methodology, Agent, Petri-Nets, Quality

1 Introduction

Used products are basic elements in remanufacturing. The uncertain quality of used products and components/parts which are disassembled from the used products highly affect the performances of remanufacturing systems and the qualities of remanufactured products [1]. Hence quality issues cannot be neglected in remanufacturing. To reflect the uncertain quality status of parts (for simplicity, in this study 'part' is a general term for a used product, a disassembled component/part,

---

Please use the following format when citing this chapter:

Kim, Y., Jun, H.-B., Kiritsis, D., Xirouchakis, P., 2007, in IFIP International Federation for Information Processing, Volume 246, Advances in Production Management Systems, eds. Olhager, J., Persson, F., (Boston: Springer), pp. 281-288.

and a reassembled product) to the remanufacturing system control or simulation, each part has to be managed individually. Until now, it has not been easy to manage each part individually considering its characteristics, because there was no way to identify each part in a remanufacturing system. But emerging product identification technologies like radio frequency identification (RFID) make it possible.

This study focuses on the modeling methodology for a quality embedded remanufacturing system (QRS). The modeling is important because all analysis, management, control, and simulation works are carried out based on the modeled information. To manage parts and resources individually in a remanufacturing line, this study applies a multi-agent framework where each part and resource are regarded as an agent, respectively. The agent approach has been recognized as an effective solution to overcome the traditional approaches' limits in abilities of the expansion, reconfiguration, maintenance without shut down, and so on [2]. Defining each part as an agent can also enrich the individual controllability of each part, because remanufacturing systems usually handle multiple kinds of products.

The remaining of this study is as follows. Section 2 reviews previous related research. Section 3 represents necessary modeling elements in a remanufacturing system. Section 4 proposes a modeling methodology and proper modeling tools. Finally, section 5 introduces a case study to show how the proposed methodology can be applied to a real case.

2   Previous Related Research

Most previous research on remanufacturing systems focus on the disassembly phase which highly involves quality uncertainty and parts composition diversity of used products. They usually borrow ideas of graph representation from research on assembly modeling tools. Demello and Sanderson [3] analyzed optimal assembly/disassembly sequences with AND/OR graphs, in which all possible components and parts are marked as nodes and assembly/disassembly relationships as arcs. Zussman and Zhou [4] proposed disassembly Petri-Nets (DPN). They formally imported the concept of disassembly operation quality into their tool by inserting quality test transitions right after each transition for a disassembly action.

Modeling tools for simulation or control of manufacturing/remanufacturing systems have been considered. In the simulation or control, their logic and information are usually managed based on system status; a simulator collects and analyzes information of status change of manufacturing/remanufacturing systems, and a controller decides the next status of the system depending on its current status. Such systems can be regarded as a kind of an automaton system. Petri-Nets are widely used for representing these automata systems because of their well known efficiency in managing system status. For example, ElMekkawy and ElMaraghy [5] used timed Petri-Nets (TPN) to develop a rescheduling algorithm. Jeng et al. [6] handled a resource sharing with process nets with resources (PNRs). Furthermore, as a more generalized simulation tool, CIMPACT developed software, SIMAS [7], which enables a simulation of manufacturing/remanufacturing systems from the perspective of processing time.

Among previous research on modeling, extended two-level colored Petri-Nets (XCPN) [8] can be considered as a suitable modeling tool for a remanufacturing system modeling because of the following reasons. First, the modularization concept of the tool enables separated modeling of resources and products. Second, the color, inherited from colored Petri-Nets (CPN), and time characteristics make stochastic simulation possible, which is necessary in simulating the stochastic characteristics of remanufacturing systems. However, a small modification is required because the XCPN lacks the characteristics for handling the quality dependent part routing.

As explored above, although there are many research works on modeling tools, there is lack of modeling tools for remanufacturing system simulation or control incorporating the quality concept. To our knowledge, there is no literature that explains an overall methodology of modeling remanufacturing systems. To overcome these limitations, this study proposes a structured modeling methodology and an overall outline of suitable modeling tools for the methodology.

3   Modeling Elements

To represent a QRS well, necessary elements have to be extracted. To this end; we apply a top-down approach. In this approach, we inspect necessary elements composing the QRS, and their sub elements until the bottom level.

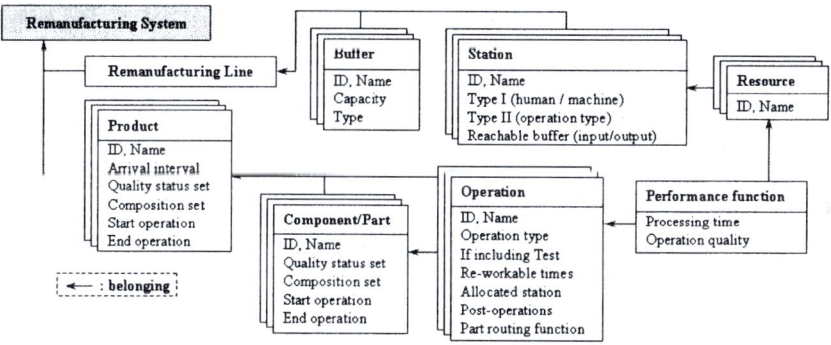

Fig. 1. Elements in a quality embedded remanufacturing system (QRS).

Fig.1 shows the extracted elements and their attributes. Two main elements in a remanufacturing system are a remanufacturing line and products which flow on the line. The line comprises buffers and work stations which have some resources. A remanufacturing system can handle multiple kinds of products which are composed of some parts. Each product and disassembled part is assigned with operations to be processed.

To simulate or control a remanufacturing system, performance related information of each operation is required, e.g. processing time and operation quality. Such performance indices are affected not only by an operation but also by a resource which is in charge of the operation, because even the same operation can show different performance depending on a resource's type, degree of expertise, degree of fatigue, and so on. Hence the performance information does not belong to one element but to the combination of an operation and a resource.

This study defines the quality as an abstract level attribute, because the quality characteristics are very different from product to product and cannot be generalized. Although there are other elements and attributes that are not mentioned above like warehouse, part transfer vehicle, life-time of a resource, and so on, to keep the model compact and simple, relatively unimportant elements and attributes are not considered in this study.

4   Modeling Methodology

4.1 Two-layered model: intuitive model and status-relationship model

The extracted necessary information should be modeled from two kinds of perspectives: a business-side and a system-side (Fig.2). In general, field experts who have only business-side knowledge encounter difficulty in understanding system-oriented models because the information from the system-side perspective are too bizarre and complicated. Hence the system should be modeled in an intuitively understandable way from the business-side perspective. It helps field experts and system analysts to communicate with each other, which leads to improving the accuracy of the modeled information.

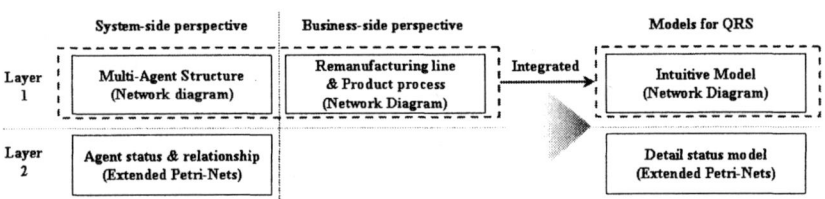

Fig. 2. Structure of Models for QRS.

From the business-side perspective, the above extracted necessary elements should be modeled as they are in the real world. Therefore, it is preferable to divide the business-side model into two kinds of models: a remanufacturing line model and product-process models, where work stations and buffers are modeled with their characteristics in a remanufacturing line model and operations of each product with disassembly relationships are modeled in the product-process models. For linking two models, information on resource allocation to operations and operation time and quality performance should be added.

From the system-side perspective, we propose a two-layered structure, where the information representing the structure of a multi-agent framework for a QRS is modeled in the first layer, and the status of agents and their relationships are expressed in the second layer. This information can be surely expressed in an integrated model. However, the complexity of such an integrated model is very high, which leads to ambiguity in constructing the multi-agent structure.

Although modeling from two kinds of perspective is preferable, this study suggests integrating the model from the business-side perspective and the first layer model from the system-side perspective into one model (Fig.2), because they are similar in terms of the structure and contained information. In consequence, a two-layered model should is proposed for a QRS: an intuitive model and a status-relationship model.

An intuitive model is composed of two parts like the model from the business-side perspective: remanufacturing line model and process model of products which flow on the line (product-process model). The relationship information between the two models should be added to integrate the two parts. Therefore the intuitive model can be defined as 4-tuples:

$$\text{Intuitive model} = \{M_R, M_P, \lambda_{cs}, \Delta_{qt}\}$$

where M_R is a remanufacturing line model, M_P is a product-process model, λ_{cs} is a resource allocation information, and Δ_{qt} is a quality and time performance information depending on the combination of operations and resources.

A status-relationship model can be derived from XCPN, because it already has many features that we need as we explained above. The model can be defined as follows:

$$\text{Status-relationship model} = \{PN^T{}_{XCPN}, Q, P_t, P_q, C\}$$

where $PN^T{}_{XCPN}$ is a XCPN without time related feature, Q is quality information of parts, P_t and P_q are time and quality performance information, and C is correlation information between operation time and quality. With existing features in XCPN, we can model the status of agents and their relationship.

The two models must support graphical notation. Modeling only with formal definition may cause much confusion for modelers as the system size increases. For this reason, we recommend modeling an intuitive model with a directed graph for a remanufacturing system in which overall time distribution of an operation and overall quality performance of a resource are expressed. For the status-relationship model, graphical notation of XCPN can be used with addition of symbols related with quality dependent part routing.

4.2 Necessity of conversion method

As we discussed in section 4.1, a two-layered model is preferable for applying a multi-agent approach to a QRS. An intuitive model has the basic structure information of a remanufacturing system. The status-relationship model has the supporting logic and characteristic information. It means that the latter should be constructed based on the information of the former, and the consistency between them has to be kept.

To avoid inconsistency between them, it is necessary to develop a method that converts an intuitive model into a status-relationship model. The status-relationship model should contain additional information such as possible status and synchronization protocol of each element type, which do not exist in an intuitive model. Therefore the conversion method should provide such information. Hence the 'conversion' in this research means not only the change of expression but also the addition of pre-acquired information. For example, a resource agent should be able to have the status of waiting jobs, carrying a job on, examining an operation quality, and so on.

4.3 Framework of QRS modeling and multi-agent system construction

With the two-layered model and conversion method mentioned above, the real world information of a remanufacturing system can be modeled, and the information can be used for the construction of a multi-agent remanufacturing system in the virtual world. Fig.3 shows an overall framework for the modeling of a QRS and the constructing a multi-agent remanufacturing system based on the model.

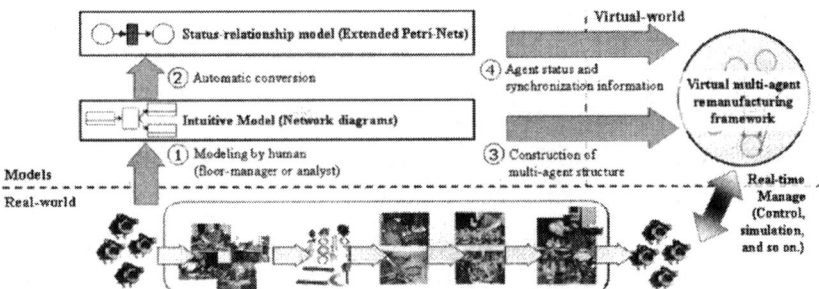

Fig. 3. Framework of modeling and constructing a multi-agent remanufacturing system.

The modeling and construction of a multi-agent remanufacturing system based on the model information can be done in 4 steps as follows:

Step 1. Create an intuitive model;
Step 2. Create a status-relationship model based on the conversion method;
Step 3. Construct a multi-agent structure in the virtual remanufacturing system;
Step 4. Load detailed status and synchronization information to each agent.

If the objective of the modeling is to control a remanufacturing system, each element in the constructed virtual remanufacturing system should be synchronized to the corresponding element in the remanufacturing system. But just for the simulation work, the synchronization is not needed. Instead of that, the virtual remanufacturing system must contain the module which can emulate real world situation and manage time.

5   Case Study

For the better understanding of the proposed methodology, this section introduces a remanufacturing line example.

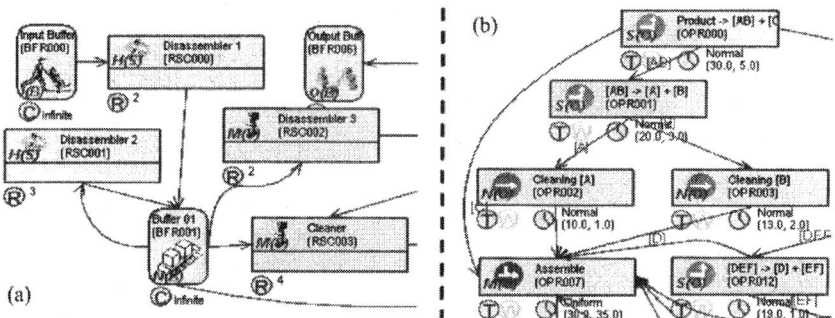

Fig. 4. Intuitive model; a part of (a) remanufacturing line and (b) operations of a product.

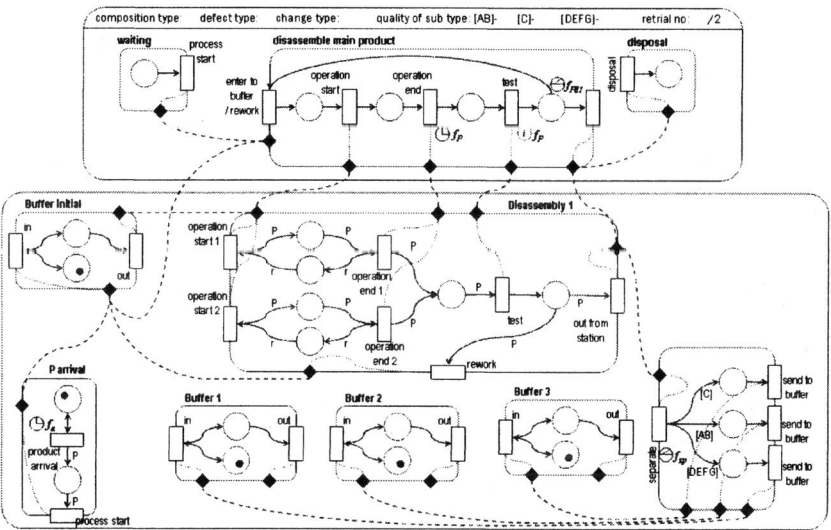

Fig. 5. A part of status-relationship model converted from the intuitive model in Fig.4.

Fig.4 is an example of an intuitive model. Fig.4(a) represents a part of the remanufacturing line and Fig.4(b) represents a part of the process of a product that flows on the line. It is designed by a directed graph which has various node types depending on element types such as resource, buffer, and operation in process. Fig.5 shows a part of status-relationship model, which is converted by the pre-defined conversion method. In this figure, only partial buffers and work stations in the arrival point of a used product in the remanufacturing system are expressed. It is designed based on the XCPN notation. Some features which are not included in XCPN are added to represent dynamic quality and part routing. In the model, each workstation, buffer and operation is expressed as a module, and some modules are added for the agent life-cycle management.

6 Conclusion

In this research, we have proposed a methodology for modeling a QRS. In the methodology, a two-layered modeling approach has been suggested and the necessity of a conversion method has been explained. The proposed modeling framework enables us to represent a remanufacturing system well, considering its characteristics. Furthermore, the extracted necessary elements are helpful for enhancing the correctness of models. In spite of these contributions, there are still some limitations in this study: First, formal definition with graphical notation for intuitive model and detail specification for extended Petri-Nets should be done. Second, quality related features are preferable to be detailed and generalized based on analysis of many case studies.

References

1. V.D.R. Guide, M.E. Kraus, and R. Srivastava, Scheduling policies for remanufacturing, *International Journal of Production Economics* 48(2), 187-204 (1997)

2. W.M. Shen, Q. Hao, H.J. Yoon, and D.H. Norrie, Applications of Agent-based Systems in Intelligent Manufacturing: An Updated Review, *Advanced Engineering Informatics* 20(4), 415-431(2006).

3. L.S.H. Demello, and A.C. Sanderson, And/or graph representation of assembly plans, *IEEE Transactions on Robotics and Automation* 6(2), 188-199 (1990).

4. E. Zussman, and M.C. Zhou, A methodology for modeling and adaptive planning of disassembly processes, *IEEE Transactions on Robotics and Automation* 15(1), 190-194 (1999).

5. T.Y. ElMekkawy, and H.A. ElMaraghy, Real-time scheduling with deadlock avoidance in flexible manufacturing systems, *International Journal of Advanced Manufacturing Technology* 22(3-4), 259-270 (2003).

6. M Jeng, S.L. Xie, and M.Y. Peng, Process nets with resources for manufacturing modeling and their analysis, *IEEE Transactions on Robotics and Automation* 18(6), 875-889 (2002).

7. SIMAS II, CIMPACT (2000), available at http://www.cimpact.ch.

8. D. Sakara, Modeling and scheduling of remanufacturing systems, Ph.D. dissertation, STI-IPR-LICP, EPFL, Lausanne, Switzerland, 2006.

Towards a Reference Model for After-Sales Service Processes

Elena Legnani, Stefano Ierace and Sergio Cavalieri
University of Bergamo, Department of Industrial Engineering
Viale Marconi, 5 - I – 24044 Dalmine, Italy
e-mail: (elena.legnani, stefano.ierace, sergio.cavalieri)@unibg.it

Abstract

In the last years, given the high market pressure and the increased competition in several industries, the search for new business opportunity is focusing on service activities. After-Sales (AS) service has become increasingly important as a source of differentiation and market share for manufacturers and resellers, as well as a strategic driver for customer's retention. These changes often call for a new conceptual definition of the product-service bundle marketed to the final customer and for a thorough revision of the logistical and organizational configuration of the whole service chain. It comes out the necessity to design appropriate processes and to have a general and shared definition of their structure. Aim of the paper is to suggest a model which provides a common configuration of the assistance processes according to a framework that links the different typologies of assistance with the product service strategies offered by companies. Some case studies have been considered in order to validate the proposed framework and the model.

Keywords

After-Sales service, product service strategy, business process, performance measures

1 Introduction

Market competition has recently pushed companies to seek for other forms of competitiveness different from a traditional cost leadership strategy. The widening of geographic horizons due to the availability of communication technologies, the more pressing and specific requests of the final customers and the downfall of technological barriers are inducing manufacturing firms to look for new sources of competitive advantage (Thoben et al., 2001). In this context, companies are shifting from a product centric view to a more innovative customer centric one: their business role cannot be considered ending up with the transactional undertaking of product sale. They must strive their efforts in ensuring a long-lasting and stable relationship

Please use the following format when citing this chapter:

Legnani, E., Ierace, S., Cavalieri, S., 2007, in IFIP International Federation for Information Processing, Volume 246, Advances in Production Management Systems, eds. Olhager, J., Persson, F., (Boston: Springer), pp. 289-296.

with the final customer through the overall product life-cycle by providing a value-added portfolio of connected services. A bundle of tangible and intangible components extends the physical functionalities of the core product in order to fulfill the increasing requests of customization, uniqueness perception and variety demanded by the final customer (Rispoli and Tamma, 1992, Thobeh et al., 2001).

The AS service embraces a wide concept which spans several aspects: it is commonly referred to as either customer support, product support, or technical support and service (Goffin and New, 2001) and it has been acquiring a strategic role within companies' business. At first, it is a source of differentiation and revenue generation: for example, the profit margins are often higher than the ones obtained with the product sales and it may generate at least three times the turnover of the original purchase during a given product lifecycle (Alexander et al., 2002). Moreover, it affects the definition of the product service mix offered to the customer and the physical and organizational configuration of the overall logistics chain. Finally, it is also a power marketing force for promoting the brand of a company.

Since such a radical change of the strategic vision of AS service requires the design or the thorough reengineering of the service business processes (Hammer and Champy, 1993), their common understanding, their inherent activities, related performance metrics and best practices should be considered and properly assessed. The present paper attempts to fill this necessity by proposing a model which aims at providing a comprehensive mapping of the AS service processes, with particular regard to the assistance ones. The paper is organized as follows: §2 is a literature review about the existing models used to map AS activities, §3 reports a framework which links a product service strategy perspective to a technical support perspective and suggests a model to map the assistance processes while §4 concerns the case studies used to test and validate both the framework and the model. Some conclusions and further developments are finally reported.

2   Literature review

A business process reference model aims at providing support for the solution of practical problems in reality and its purpose is the generalization and standardization of the processes and structures of a system: it integrates the concepts of business process reengineering, benchmarking and process measurement into a cross functional framework. With regard to the AS domain, what is really missing is a reference model which provides a clear and comprehensive definition of the processes and the activities making up a service chain.

Normally, the natural approach is to apply models from the manufacturing sector to the service one (Ellram et al., 2004). The most known *product-based models* adopted to describe service chains have been developed by Lee and Billington (1995), who analyze the flow of goods among suppliers, manufacturers and customers within an uncertain environment, Croxton et al. (2001), whose model conceptualizes a supply chain that includes the business processes, the management components and the structure of the chain and the Supply Chain Council (2006) which proposes the Supply Chain Operations Reference (SCOR) model to map the supply chain processes and their related metrics and best practice.

On the other hand, focusing on *service-based models* specifically thought to represent service activities, in literature there are several approaches. They originate mainly in marketing and industrial engineering (Donabedian 1980; Edvardson, Olson 1996) or in the quality management area (Meffert, Bruhn 1995; Bullinger, Haischer, Renner 1994) and there are also a number of matrixes, describing how various characteristics of a service relate to each other (Johansson, 2003). However, it comes out that they have mainly a strategic approach and none of them focuses on the link between the supporting service strategies and the related processes structure.

As a result *product-service models*, which consider both tangible and intangible aspects as well as a strategic and an operational view, are required and their relating activities and performance metrics should be considered and properly defined. An attempt towards this necessity is the After Sales Chain Operations Reference (ASCOR) model proposed by Cavalieri et al. (2006) which has a limited applicability since it is focused just on a typology of assistance service (on-site support defined in section 3 of this paper) and the Customer Chain Operations Reference (CCOR) model defined by the Supply Chain Council (2006) which is still in an early development phase.

Table 1 reports a comparison among the quoted models considering their strength and weakness aspects from the service perspective. Despite the product-based models which can be roughly adapted to the service area, considering the models specifically designed for service, it turns out that the service-based models have mainly a strategic view while the current product-service models (such as ASCOR and CCOR) have an operational view which considers the processes associated to the service activities and their relating performance metrics.

Table 1. Comparison among the models from the service perspective

	Model	Support for service	Weakness for service
Product – based models	Lee and Billington (1995)	It manages uncertainty and it utilizes capacity levels and flexibility versus inventory.	Services cannot be inventoried. The model cannot easily address the differences in quality of services.
	Croxton et al. (2001)	It utilizes a process view to meet uncertain demand.	It fits the product and component flow of goods.
	SCOR model (2006)	Services are process driven.	The processes of make, deliver and return do not fit services.
Service – based models	Donabedian (1980), Edvardson et al. (1996), Meffert et al.(1995), Bullinger et al. (1994)	They are mainly focused on service strategies.	They are not focused on service processes and do not consider an operational perspective. Performance metrics are not provided.
Product-service models	ASCOR (2006)	It is focused on assistance processes and their related performance measures.	It is limited since it considers just a typology of assistance support and do not focus on other service activities. It does not consider service strategies.
	CCOR (2006)	It is focused on all the service activities within the customer interaction and their related performance measures.	It considers a reactive support neglecting a proactive perspective.

3   Towards a reference model

3.1   Aligning product service strategies and assistance processes: a framework

As highlighted from the literature review and shown in the Table 1, there is a lack of product-service models which associate a service strategic view with a more process-operational one. In this section, a framework which aligns the product service strategies with the required assistance supports has been created in order to facilitate companies in identifying their position and individualizing the key assistance processes to manage.

Focusing on the service strategies associated to a product, Lele (1997) states that any product can be assigned to one of four AS service segments. Considering low and high fixed costs (which occur regardless of the duration of equipment downtime) vs low and high variable costs (which change according to the duration of equipment downtime), these strategies are classified as follows: *disposable*, refers to products which do not cost much and when they are broken they are normally not repaired (like small household appliances and inexpensive office equipment); *repairable*, refers to products which need an assistance support but do not need to be repaired as soon as the problem occurs (like home PCs); *rapid response*, refers to products which need to be repaired as soon as the problem occurs (like office PCs); *never fail*, refers to products where a failure is not an acceptable option for customers (such as medical equipment, airplanes, etc).

Regarding the processes to perform, Cavalieri and Corradi (2002) identify different typologies of support according to the service level offered, the level of involvement of the customer and the sustained costs. The supports can be:

- *indirect support* where the company provides an appropriate documentation to the customer who is able to autonomously perform the diagnosis, identification and application of the solution;
- *remote support* where the customer autonomously sorts out the problem with the help of an expert;
- *off-site support* where the company collects the faulty product through its assistance channel, it repairs and gives it back to the customer;
- *on-site support* where the customer is not able to solve the problem and needs the help of an expert who solves the problem at the location where the customer and the problem currently reside.

Next to these traditional forms of support, in the last years new ones have also been proposed: they are called *proactive supports* and are forms where the repair service or upgrade are scheduled by the company which provides the product. There are predictive maintenance techniques which help to determine the condition of the equipment in order to predict when an assistance support should be performed. There also forms of *customised support* that are developed according to specific customer requirements.

In Figure 1 an alignment between the product service strategies defined by Lele and the assistance supports is illustrated.

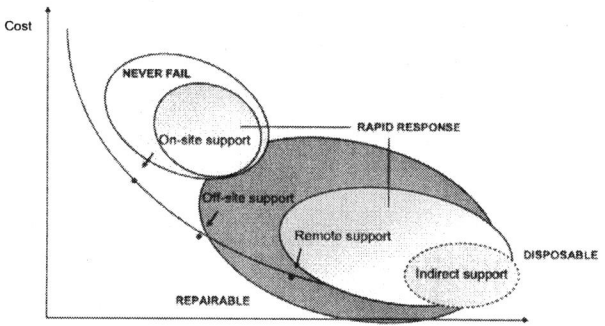

Fig. 1. Products and service support alignment

According to this framework, it results that:

- the never fail group embraces an on-site support since it normally refers to cumbersome products, with a very high capital value, that need a skilled technical support and thus repaired at the customer place;
- the rapid response group refers to products which need to be fixed promptly since they have high variable costs. On-site, remote or indirect supports are mainly performed;
- the repairable group includes indirect, remote or off-site supports since time bonds are not pressing. Even if products belonging to this group have a very high value, a repair is demanded but not necessarily immediate because the associated variable costs are not considerable;
- the disposable group may require an indirect support even if the low value of the associated products generally do not lead to any assistance support.

The proactive and the customised supports have not been considered within the work developed in this paper: this could lead to a further development of the model afterwards reported.

3.2. Mapping assistance processes

According to the product service strategies above mentioned, companies need a standard model to refer to while revising their AS processes which maps in detail how these processes can be performed and evaluated through the use of suitable indicators. This section reports in detail the processes associated to the assistance supports previously defined according to the same semantic structure of the Supply Chain Operations Reference (SCOR) model formalism. In order to provide an exhaustive and reliable structure, the model has been designed considering different scenarios coming from interviews in some companies and the suggestions and expertise of several industrial people. Furthermore, the ASCOR model and the CCOR model have been analyzed and integrated.

To understand the model structure, the processes are reported according to their assistance typology and, as an example, an insight of the off-site support process and its related performance metrics is depicted.

Table 2. Assistance processes: definitions and metrics

Indirect support	Remote support	On-site support	Off-site support
-Receive Inquiry/Request -Route Request -Identify Solution -Propose Solution -Release Solution to Customer -Close Request	-Receive Inquiry/Request -Route Request -Identify Solution -Propose Solution -Distribute Solution -Release Solution to Customer -Close Request	-Receive Inquiry/Request -Route Request -Schedule On-Site Assist -Propose solution -Obtain Materials -Repair Product -Dispose Materials -Close Request	-Receive Inquiry/Request -Route Request -Propose solution -Repair Product -Dispose Materials -Close Request

Off-Site support process definition: "repair product"

- *Repair Product*: the process of preparing, decomposing the product, replacing the part and re-assembling the product at the store/service center.

Off-Site support performance metrics related to "repair product" process

- Annualized Service Event Rate: n° of service calls per system per year;
- Customer Commit Resolution Time met: % of time a customer problem/question is resolved within the agreed upon time;
- First Time Fix Rate: % of time the problem was fixed during the first contact with the customer;
- Repair Product Total Cost: Process costs; it includes direct and indirect cost.

4 Case studies

In order to validate the soundness of the proposed framework and to define the key assistance processes to map, three companies have been used as case studies. Figure 2 shows how product service strategy affects the processes management: each section of the diagram is filled with a different shade of grey according to the emphasis that each company gives to a specific process.

Company 1 performs a never fail strategy. It has three different manufacturing and commercial branches, it is present in more than 50 countries in the world and it provides products and service for folding carton, corrugated board and flexible materials markets. It has a widely range of machines, plants and spare parts. Since it has to provide a rapid and timely intervention, analysing its peculiar assistance processes, it comes out that the most important ones are typical of the on-site support, like the scheduling, the material obtaining and the product repair.

Company 2 refers to a rapid response strategy. It is an American multinational society involved in the computer science industry and it operates both in the hardware and in the software market. It offers facilities and IT service, personal computer, access equipment and solutions for imaging and printing. The example tested is about the support requested by a company operating in the express service field which needs an assistance support for 24 hours a day and for 7 days a week. In this case, it has been observed that as soon as the request is received it is fundamental to identify a rapid solution and correctly release it to the customer. This implies that an on-site support is often required; however, just in case of easily solvable failures, remote and indirect support are performed directly by the customer with the help of a company expert.

Company 3 is an example of repairable strategy. It is an Italian company which produces motorbikes. It has 8 plants in the world and it is present in more than 50 different countries. In this case the service strategy does not require a very timely support, thus the company mainly offers an off-site support together with a remote and indirect support. This has been proved considering that for the company the most important processes are accomplished at the service center, like the solution proposal and the product repair.

The following table summarizes the results carried out from the analyzed case studies.

Fig 2. Case studies findings

| | Company 1 | Company 2 | Company 3 |
	Never Fail	Rapid response	Repairable
On-site support	Schedule on-site	Schedule on-site	Schedule on-site
	Obtain materials	Obtain materials	Obtain materials
	Repair product	Repair product	Repair product
Off-site support	Route request	Route request	Route request
	Repair product	Repair product	Repair product
	Dispose materials	Dispose materials	Dispose materials
Remote support	Route request	Route request	Route request
	Identify solution	Identify solution	Identify solution
	Release solution	Release solution	Release solution
Indirect support	Identify solution	Identify solution	Identify solution
	Propose solution	Propose solution	Propose solution
	Release solution	Release solution	Release solution

5   Conclusions and further developments

This paper emphasizes the lack of suitable product-service models which consider both tangible and intangible aspects related to the AS area. For this reason, a framework which links product service strategies with assistance supports is proposed. Moreover, though in literature processes associated with the AS service have already been defined, their common understanding and a shareable structure should be properly defined. Thus a reference model has been suggested in order to map assistance activities. Summarizing, the proposed work could allow enterprises to: i) relate more coherently their AS strategy to their assistance operational processes and ii) identify the key processes to manage.

Further developments of this research could lead to the definition of a more complete standard reference model. On the one hand, the work could be addressed to additionally develop the assistance processes, including the proactive and customised supports and more specific performance metrics. On the other hand, the model could be enlarged in order to map all the AS activities linked to the interaction between the customer and the service provider. In this sense it is worthwhile mentioning the current research efforts within the Supply Chain Council, aiming to develop a framework encompassing the customer centred perspective (CCOR model), the process-service designer perspective (DCOR, Design Chain Operation Reference model) and the product-service supply chain management (SCOR model).

References

Alexander W.L., Dayal S., Dempsey J.J., Vander Ark J.D. (2002), 'The secret life of factory service centers', The McKinsey Quarterly, No.3, pp. 106-115.

Bullinger H.J., Haischer M., Renner T. (1994), 'Qualitätsmanagement und Zertifizierung fördern den Unternehmenserfolg (in German), Computerworld Schweiz, Nr. 23/94, S. 1-4.

Bundschuh R.G., Dezvane T.M. (2003), 'How to make after sales services pay off', The McKinsey Quarterly: The Online Journal of McKinsey & Co., No.4 (http://www.mckinseyquartely.com).

Cavalieri S., Brun A., Ierace S. (2006), 'After-Sales Processes in the SCOR model: a proposal and an empirical application', 14th Working Seminar on Production Economics, Vol.3, pp. 25-36.

Cavalieri S., Corradi E. (2002), 'L'evoluzione del servizio di assistenza del post-vendita: modelli di supporto, aspetti logistici ed opportunità IT (in Italian)', Atti del XXVIII Convengo Nazionale ANIMP, Spoleto 25-26 Ottobre.

Croxton K.L., Garcia-Dastugue S.J., Lambert D.M., Rogers D.L. (2001), The Supply Chain Management Process', The International Journal of Logistic Management, pp. 17-32.

Donabedian A. (1980), 'The definition of quality and approaches to its assessment. Explorations in quality assessment and monitoring', Health Administration, Ann ArborlMichigan.

Edvardson B., Olson J. (1996), 'Key concepts for new service development', The Service Industries Journal, April, Vol.16, No.2, pp. 140-164.

Ellram L.M., Tate W.L., Billington C. (2004), 'Understanding and managing the service supply chain', The Journal of Supply Chain Management, pp. 17-32.

Goffin K., New C. (2001), 'Customer support and new product development', International Journal of Operations and Production Management, 21 (3), 275-301.

Hammer M., Champy J. (1993), 'Reengineering the Corporation: A manifesto for business revolution', Harper Collins Publishers NY.

Johansson P., Olhager J (2003), 'Industrial service profiling: matching service offerings and processes', International Journal of Production Economics, 89, 309-320.

Lee H., Billington C. (1995), 'The evolution of Supply Chain Management Models and Practice at Hewlett-Packard', Interface, Vol.25, No.5, pp. 42-63.

Lele M. (1997), 'After-Sales service-necessary evil or strategic opportunity?', Managing Service Quality, Vol.7, No.3, pp. 141-145.

Meffert H., Bruhn M. (1995), 'Dienstleistungsmarketing: Grundlagen-Konzepte-Methoden (in German)', Wiesbaden (Gabler).

Rispoli M., Tamma M. (1992), 'Beni e Servizi, cioè prodotti (in Italian)', Sinergie, n.28.

Supply Chain Council (2006), 'Supply Chain Operations Reference model, version 8.0', 'Customer Chain Operations Reference model', 'Design Chain Operations Reference model', www.supply-chain.org.

Thoben K.-D., Jagdev H., Eschenbaecher J. (2001), 'Extended Product: Evolving Traditional Product concepts', Proceedings of the 7th International Conference on Concurrent Enterprising: Engineering the Knowledge Economy through Co-operation, Bremen.

Analysis of the Human Role in Planning and Scheduling via System Dynamics

Katsumi Morikawa and Katsuhiko Takahashi
Graduate School of Engineering, Hiroshima University,
1-4-1, Kagamiyama, Higashi-Hiroshima 739-8527, Japan
TEL: +81-82-424-7704, FAX: +81-82-422-7024
{mkatsumi, takahasi}@hiroshima-u.ac.jp

Abstract

A system dynamics model is proposed to analyze the human role in planning and scheduling. Based on the interview research results planning and scheduling activities in manufacturing companies are investigated. By selecting important elements in planning and scheduling, a model of the human role is developed assuming a simple manufacturing environment. Two modes of the model is examined in the simulation experiments, and the results indicate the importance of the human role in smoothing the workload by look-ahead decisions and in reducing the uncertainty by collecting information and coordinating planning and scheduling activities.

Keywords

Planning, scheduling, human, system dynamics

1 Introduction

Manufacturing companies have been facing with the world-wide severe competition. Selecting and introducing a suitable integrated production planning and management system is an important strategic decision to improve the productivity and cost competitiveness. Although many companies have already introduced such planning systems successfully, we found that introducing an entirely new planning system without understanding manufacturing conditions and current roles of the human planner and scheduler may lead to unsatisfactory consequences [1]. Most manufacturing organizations still require human contributions (Jackson *et al.* [2]), and much research effort has been devoted to investigating difficulties in real-life planning and scheduling tasks (for example, Stoop and Wiers [3]), and analyzing/

Please use the following format when citing this chapter:

Morikawa, K., Takahashi, K., 2007, in IFIP International Federation for Information Processing, Volume 246, Advances in Production Management Systems, eds. Olhager, J., Persson, F., (Boston: Springer), pp. 297-304.

modeling their activities (Crawford *et al.* [4], MacCarthy *et al.* [5], and MacCarthy and Wilson [6]). These studies emphasize the importance of human contributions, but such importance is not recognized properly in some companies we have investigated. Qualitative and quantitative researches are still required to develop and introduce successful production planning systems that work with humans collaboratively.

The aim of this study is to propose a model of production planning and scheduling activities of humans, and analyze the human role through simulation experiments. In making the model it is difficult to select detailed elements and then formulate quantitative equations between them. Therefore this study focuses on macro-level elements and develops the model using system dynamics. Detail of the model is explained after discussing the planning and scheduling activities considered in this study.

2   Production Planning and Scheduling

Defining the meaning of planning and scheduling is not so simple. McKay and Wiers [7] have classified the tasks of planner, scheduler, and dispatcher based on a long period of investigation of companies. This study implicitly assumes several small or medium scale companies interviewed [1] and define the planner and scheduler, and their tasks as follows.

The planner decides daily production plans often using the production planning system. The production plan generally defines the product types to be produced, the production quantities of these products, and due dates of them. The planner cannot always use the production plan produced by the production planning system because of the following reasons; (i) the planning system cannot consider all of the capacity constraints of the shop floor in general, (ii) information within the planning system is sometimes different from the actual conditions, (iii) the planning system cannot handle some information such as expected future orders appropriately. However, except for making very simple products, the information system is an important tool that can process complicated information effectively. Thus the planner generates production plan by obtaining necessary information from the shop floor and the sales department, by using interactive planning/re-planning function of the system, and finally by modifying the obtained plan manually if such modifications are needed.

Based on the upper planner's decision, the scheduler decides detailed production schedules. Typical decisions include worker allocation, machine allocation, and job sequencing. These decisions often need a deep knowledge about the shop floor conditions and also the processing requirements of orders. Therefore, a foreman often plays the role of the scheduler. In make-to-order companies producing different products everyday, it is very difficult to estimate the workload of each order acccurately. The foreman often can give good estimation based on his (her) long experience. When making detailed schedules, sophisticated optimization techniques are seldom used in practice. Instead, the foreman often makes schedules based on simple calculations using spread sheet software and his (her) experience.

The human role fairly depends on the manufacturing environments. The planner of companies producing a limited number of product types repetitively by assembling parts with relatively simple operations can produce detailed production plans using the production planning system. In such companies, the human role is mainly to handle unexpected events such as machine failures, delivery delay, and so on. On the other hand, if the processing time of operations inevitably fluctuates, or planned production lead time is fairly long and thus it is often required to modify the plans, the automatic planning is difficult and thus the adjustment of plans by the human planner and scheduler must be important.

3   A Model of the Human Role via System Dynamics

In this section we propose a model of the human role in planning and scheduling based on system dynamics. Several authors have already proposed several production planning and inventory control models via system dynamics. Two examples are a domestic manufacturing company illustrated by Coyle [8], and a supply chain model of Intel by Gonçalves et al. [9]. Although these models are highly helpful in developing our model, this study has focused on the condition-dependent actions of the planner and scheduler, and the collaborative decision-making between them at a deeper level. To our limited knowledge, there seems no similar system dynamics-based approach for modeling the planner and scheduler.

Fig. 1 shows the developed causal loop of the decision activities of planner and scheduler under a hypothetical manufacturing environment based on the activities described in section 2. The upper area of the figure corresponds to the planner's decisions, and the lower area to the scheduler's decisions. The boundary of these two areas is not rigid and may depend on manufacturing environments. Needless to say actual planners and schedulers consider many other factors and make several other decisions. The objective of the proposed causal loop is not to provide a comprehensive model, but rather to focus on typical decision making activities and illustrate their relations.

The primal task of the production planner is to select items to be produced and decide their production quantities. In the proposed model all orders are measured in hours, and the planner decides the order release rate (RR) of the received orders (RO). The system normally needs a fixed order process time ($D1$) to release the received orders. Therefore, the feasible release rate of orders is $RO/D1$. However, the planner also considers other factors. First is the workload of the shop floor (WL). If the current workload of the shop floor is relatively high, it may be better to reduce the release rate based on the level of the current workload. The model assumes that the desired workload (DWL) of the shop floor is given by the regular production time (REG) multiplied by the planned lead time (PLT). If the shop floor is overloaded, i.e., $WL > DWL$, and the feasible release rate is greater than the regular production capacity, it may be better to reduce the release rate. On the other hand, if the feasible release rate is less than the regular production capacity, and the workload of the shop floor is less than the desired level, the planner may go to the sales department and collect the information about expected future orders. Let EO denote the workload of the expected orders of the next working day. If *EO* is greater than the regular production capacity, a candidate look-ahead procedure is to expedite required processes for the current received orders and to prepare for the expected future orders immediately. These activities can increase the feasible release rate to some extent, and thus it is expected to reduce the future overtime production. Based on the above idea, the release rate is defined by the following equation:

$$RR = \begin{cases} FRR - \dfrac{WL - DWL}{D2}, & \text{if } WL > DWL \text{ and } FRR > REG, \\[2ex] \dfrac{RO + EO - REG}{D1}, & \text{if } WL < DWL \text{ and } FRR < REG \text{ and } EO > REG, \\[2ex] FRR, & \text{otherwise.} \end{cases}$$

where *D2* means a constant value representing the length of time to resolve the overload.

The planned lead time is a coordination parameter between the planner and scheduler. We assume that both the planner and scheduler know the possible shortest lead time (*SLT*). However the shortest lead time often incurs difficulties in making detailed schedules within the regular production time. If the planned lead time is longer than the shortest time, the scheduler can make the following decisions; reduce the number of set up operations by combining the same or similar orders, adjust the processing sequence to improve the equipment utilization, allocate orders to workers considering the skill level of each worker, and so on. The degree of freedom of scheduler's decision increases with the increase of planned lead time. This relation is expressed by the efficiency term in the figure because we assume that the planner can produce efficient schedules using the planning freedom. The actual production time is largely determined by the production rate and efficiency of the schedule, but partially affected by the uncertainty involved in the environment. If the actual overtime production time is relatively large when compared with the expected time, the scheduler requests the increase of the planned lead time to hedge against the uncertainty. However, selecting a long planned lead time beyond a certain level is unacceptable from the planner's viewpoint because of higher work-in-process inventory and longer response to customer demand. Therefore deciding the best planned lead time considering these conflicting factors is an important role of the planner and scheduler.

The decision of production rate (*PR*) is the primal task of the scheduler. A normal production rate (*NPR*) can be defined as follows: $NPR = WL/PLT$. In some manufacturing environments, the scheduler has a certain degree of freedom in deciding the production rate. For example, the scheduler may be able to add workload of 30 minutes in addition to *NPR* by making a good schedule, or shift a partial amount of workload to the next working day. This type of freedom is named adjustable range (*AR*) in Fig. 1. For example, *AR*=5% means the scheduler can adjust the workload up to $0.05 \cdot REG$. Fig. 1 also shows the information flow from order release rate (*RR*) to production rate (*PR*). If the information exchange between the planner and scheduler is conducted properly, the scheduler can know the current release rate in addition to the current workload level. Based on the above discussions, the production rate is assumed to be defined as follows:

$$PR = \begin{cases} REG, & \text{if } NPR \geq (1 - AR) \times REG \text{ and } NPR < REG \text{ and } RR > REG, \\ REG, & \text{if } NPR \leq (1 + AR) \times REG \text{ and } NPR > REG \text{ and } RR < REG, \\ NPR, & \text{otherwise.} \end{cases}$$

The meaning of the above equation is straightforward. For example, the top equation means that if the normal production rate is slightly lower than the regular production capacity, and the release rate is increasing, then increase the production rate to the regular production time. Such decisions are expected to reduce the cumulative overtime production.

A remaining important equation is the following balance equation between the inflow, stock, and outflow of workload:

$$\frac{dWL}{dt} = RR - PR$$

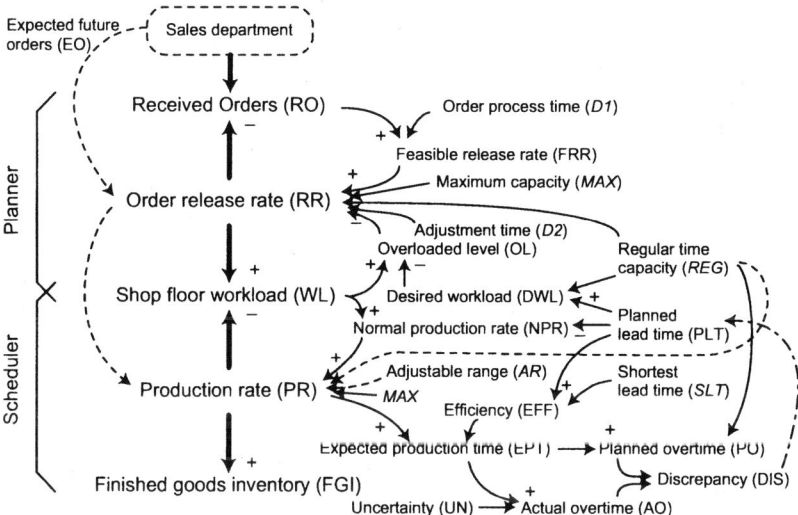

Fig. 1. Causal loop diagram of planning and scheduling activities

4   Simulation Experiments

Based on the causal loop diagram shown in Fig. 1, two modes of the simulation model were developed; simple mode and full mode. The simple mode was composed of elements shown in Fig. 1 except the relationships shown in broken lines. On the other hand, the full mode included all relationships shown in Fig. 1. However, both modes excluded the directed broken arc started from *DIS* (discrepancy) to *PLT* (planned lead time). This relation was considered separately in the simulation experiments.

The following values of the input parameters were used in all simulation runs: $REG=8$ (hours), $MAX = 10$ (hours), $D1 = 4$ (days), $D2 = 3$ (days), $SLT = 5$ (days). Fluctuating demand was generated randomly within the interval [5, 12] (hours) in each day. Three adjustable ranges (AR), 0%, 5%, and 10%, were prepared, and the level of uncertainty was selected from the following values; 0%, 2.5%, 5%, 10%, 15%, and 20%. When the planned lead time (PLT) was 5 days, there was no efficiency improvement. If PLT was 5.5 or 6, the efficiency improvement was 2.5% or 5%, respectively. The length of the simulation was 250 working days, and all random events were generated using the same seed in all simulation runs.

The primal performance measure is the cumulative overtime production in hours. Utilizing the available regular production time as much as possible, and avoid the overtime production if possible is often considered as an important planning goal. Fig. 2 shows the cumulative overtime of the simple and full modes under several uncertainty levels and adjustable ranges when the planned lead time was 5.5 (upper graphs) or 6 (lower graphs). As we assumed that longer planned lead time enables higher production efficiency, the cumulative overtime increased when the planned lead time was 5.5. We can interpret the figure as follows: (i) Increasing the level of uncertainty increased the cumulative overtime production. (ii) The full mode was effective in reducing overtime production when compared with the simple mode. (iii) In full mode, increasing the adjustable range also reduced the cumulative overtime. (iv) The difference of cumulative overtime between the simple mode and the full mode with $AR=0$% expresses the effectiveness of releasing workload considering the future expected orders. The results indicate that the planner's look-ahead releasing decision reduced the cumulative overtime. (v) When the planned lead time (PLT) was 5.5 and thus the efficiency was 2.5%, the differences in overtime between two modes and between different adjustable ranges in full mode were both larger when compared with the results of $PLT=6$. This means that the full mode is an effective model in utilizing the limited available capacities. In other words, decisions of planner and scheduler highly affect the cumulative overtime production especially under higher workload conditions. Fig. 2 also indicates that the performance of the simple and full modes was relatively stable when the level of uncertainty was small. This means that the importance of the planner and scheduler is enhanced when the shop floor condition is relatively unstable and uncertain. Namely, reducing the uncertainty by collecting up-to-date information and coordinating planning and scheduling activities is an important role of the planner and scheduler.

The average workload level is also an important performance measure when deciding the planned lead time. Fig. 3 shows the effect of planned lead time on the cumulative overtime and the average shop floor workload. Bold lines correspond to the full mode, and thin lines to the simple mode. As the average workload was nearly the same for both modes, one broken line is used in the figure. We assumed that the production efficiency increases when the planned lead time is increased. Therefore the cumulative overtime decreased for increased planned lead time. On the other hand, the average shop floor workload increased, and the rate of increase was proportional to the regular time capacity. This behavior can be explained by the structure of the causal loop model shown in Fig. 1. Another important role of the planner and scheduler is to adjust the planned lead time adequately by considering the trade-off between the average workload and the expected overtime. Although the proposed model does not have an explicit function to describe this type of decision, the importance of selecting the appropriate length of the planned lead time is partially explained by the simulation results.

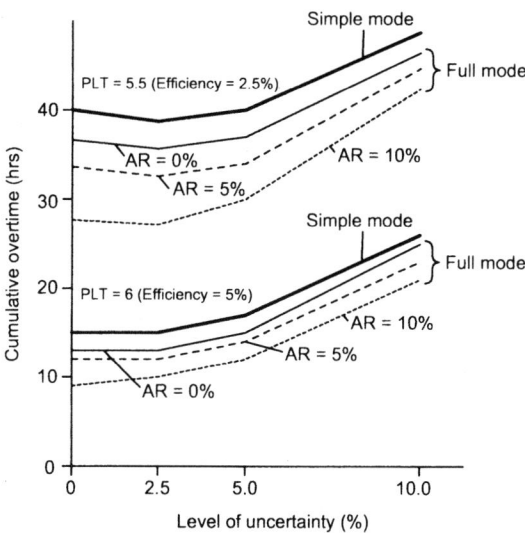

Fig. 2. The cumulative overtime of simple and full modes

5   Conclusion

This paper describes planning and scheduling activities in manufacturing companies based on the interview research conducted previously, and then proposes a system dynamics-based model of the human role in planning and scheduling. The proposed model involves the condition-dependent actions of the planner and scheduler, and the collaborative decision-making between them at a deeper level. Simulation experiments using two modes of the proposed model illustrate the importance of the human role in smoothing the workload by look-ahead decisions and in lowering the uncertainty by effective information exchange. Enhancing the model considering other important activities is needed in future study. In addition, validation and refinement of the model based on actual data is also an important remaining work.

Ackowledgment

This research was partially supported by the Japan Society for the Promotion of Science with a Grant-in-Aid for Scientific Research, No. 19510149.

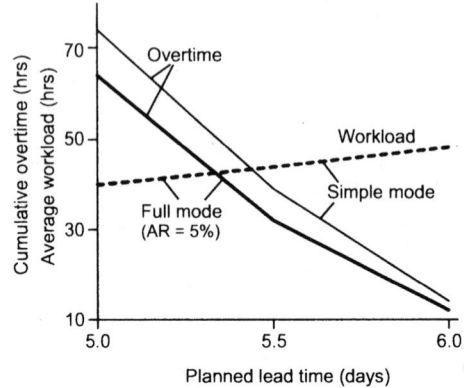

Fig. 3. The cumulative overtime and average workload varying the planned lead time when *UN*=2.5%

References

1. K. Morikawa and K. Takahashi, Modeling planning and scheduling tasks based on interviews, *Proceedings of the 7th Asia Pacific Industrial Engineering & Management Systems Conference*, December 17-20, Bangkok, Thailand, 1657-1661 (2006).
2. S. Jackson, J. R. Wilson, and B. L. MacCarthy, A new model of scheduling in manufacturing: Tasks, roles, and monitoring, *Human Factors*, 46(3), 533-550 (2004).
3. P. P. M. Stoop and V. C. S. Wiers, The complexity of scheduling in practice, *International Journal of Operations & Production Management*, 16(10), 37-53 (1996).
4. S. Crawford, B. L. MacCarthy, J. R. Wilson and C. Vernon, Investigating the work of industrial schedulers through field study, *Cognition, Technology & Work*, 1(2), 63-77 (1999).
5. B. L. MacCarthy, J. R. Wilson, and S. Crawford, Human performance in industrial scheduling: A framework for understanding, *Human Factors and Ergonomics in Manufacturing*, 11(4), 299-320 (2001).
6. B. L. MacCarthy and J. R. Wilson (Editors), *Human Performance in Planning and Scheduling* (Taylor & Francis, London, 2001).
7. K. N. McKay and V. C. S. Wiers, Planning, scheduling and dispatching tasks in production control, *Cognition, Technology & Work*, 5(2), 82-93 (2003).
8. R. G. Coyle, *System Dynamics Modelling: A Practical Approach* (Chapman & Hall, London, 1996).
9. P. Gonçalves, J. Hines, and J. Sterman, The impact of endogenous demand on push-pull production systems, *System Dynamics Review*, 21(3), 187-216 (2005).

A Conceptual Modeling Technique for Discrete Event Simulation of Operational Processes

Henk Jan Pels and Jan Goossenaerts
Technische Universiteit Eindhoven, Faculty of Technology Management,
Postbox 513, 5600 MD, Eindhoven, The Netherlands, h.j.pels@tue.nl

Abstract

A formal modeling technique, based on colored timed Petri net and UML static structure modeling languages is used to teach students to model their business process problem as a discrete event system, before they build a working simulation model in a simulation tool (in our case Arena). Combining Petri net and UML static structure diagrams, one can build an abstract, well defined and complete model. This model enables the simulation analyst to make an unambiguous, complete and yet easily readable model of the target operational process. The two most important classes of decisions that are reflected in the conceptual model are the choice of the real world details to be taken in or left out the model and the precise specification of the output parameters of the simulation. This paper describes the modeling technique and discusses its value in teaching and in the formulation of decision problems regarding operational processes.

Keywords

Discrete Event Simulation, Conceptual Modeling, Computer Independent Model, UML, Petri Nets.

1   Introduction

Model-driven systems development is gaining importance. The models at the three OMG MDA layers (Miller and Mukerji, 2003) (computation independent, platform independent and platform specific) matter in different development phases, each of which offers its own contribution to the reduction of risks and to the system design (Dick & Chard, 2003). The Computation Independent Model (CIM) shows the system in the environment in which it will operate, and thus helps in presenting exactly what the system is expected to do. Useful as an aid to understanding a

Please use the following format when citing this chapter:

Jan Pels, H., Goossenaerts, J., 2007, in IFIP International Federation for Information Processing, Volume 246, Advances in Production Management Systems, eds. Olhager, J., Persson, F., (Boston: Springer), pp. 305-312.

problem and for communication with the stakeholders, it is essential to mitigate the risks of addressing the wrong problem, or disregarding needs. These risks matter also in simulation studies (Law and Kelton, 2000, Chapter 5). By articulating the CIM a simulation project can be precisely scoped and the project results can be prepared for acceptance. The Platform Independent Model (PIM) describes the system in reference to a particular architectural style (e.g., agent based or client/server) but does not show details of platform use. Its structure may be quite different from the structure of a CIM of the same system. The Platform Specific Model (PSM) is produced from the PIM or the CIM by transformation. It specifies how the system makes use of the chosen platform and technologies. In our setting the executable model for the Arena tool is compared to a PSM model. It is obtained by a rather systematic mapping from the CIM model, taking into consideration the specifics of the simulation tool.

2   Research problem

The research problem of this paper is how to make a formal and precise discrete event process model, preferably on basis of proven modeling techniques. Operational processes are built out of a number of interacting parallel processes. Process algebras exist for formally specifying such systems, but such languages are not part of the industrial engineering curriculum. Also, these languages are not suitable for communicating with the stakeholders about the operational processes.

In our course on Simulation of Operational Processes, we require that students first make a Computer Independent Model, to define the problem and the proper level of detail, have this model approved by the teacher and after that implement it in Arena. The problem so far was that the available languages, ExSpect and CPN-Tools, resulted in models either too complex and with too many student specific design choices to be verified with reasonable effort, or models at a too generic level to expose the design choices that matter.

3   Approach

Petri nets (Jensen, 1992) are well defined and have been proven to be suitable for modeling the layout of processes as well as the synchronization between parallel processes. Their visual nature facilitates communication with stakeholders. By adding time and color they support the modeling of discrete time behavior. Several executable language systems are available to specify and run timed colored Petri nets [ExSpect (van Hee, 1994), CPN (Jensen, 1992)]. Our experience with ExSpect is that it heavily relies on a functional programming language for specifying fire conditions and state changes. Our Industrial Engineering students are not able to express themselves in such a language, and often stakeholders will not understand the resulting expressions. However, without the specification of fire conditions and state changes the Petri net model is incomplete and ambiguous, and therefore it cannot be checked for correctness (almost any graph can be good).

CPN tools provides a complete graphical language, but it appears that this language requires to specify additional places to implement state variables that are not really tokens. Also the functional programming style of specifying pre- and post-conditions and the tool-specific approach to the specification of the state variables is demanding for students with little programming skills. As a result the graphical models often loose visual resemblance with the operational process and their annotations become quite complex as an integral part of the model. Though the models are executable, their validity for a specific problem is difficult to confirm.

These reasons, and the fact that our students do understand UML static structures for the modeling of the domains in which operational processes affect system states have prompted us to use UML class and object diagrams to specify all necessary state variables in terms of such a diagram. Hence, a UML static structure diagram is used to model the state space of the operational process, while the transitions in the Petri Net model are used to specify state changes.

Our idea was to annotate each transition with a specification of its state changes in terms of logic expressions over the UML static structure. CPN-tools and UML are chosen because they are already part of the curriculum. Moreover, UML is a de facto standard in information modeling.

4   Design

In this section we explain how we build a complete discrete event model that is computation independent and unambiguously specifies a system and its behaviour. The following viewpoint specifications are used:

1. the process flow is specified as a Petri net constructed with CPN-tools,
2. the state space is specified using a UML static structure diagram,
3. the initial state of the system is specified as a marking of the Petri net with instances of the classes in the UML static structure diagram,
4. for each transition a pre and post condition is specified using predicate logic and an extension of UML OCL,
5. output variables and eventual additional functions are specified in terms of UML OCL.

4.1 Process Flow

The process structure is modeled using a timed, colored Petri net. Figure 1 shows the model for a simple queuing system, using the notation of CPN [Jensen, 1992]. The transition Arrive models the arrival of a client, who joins the queue modeled as place Wait. At the same time the next client is prepared in place NewClient. When a free server resource is available in place Free, the serving operation will start by putting the client-server pair in place Serve. The end of the operation is modeled as transition EndServe, which puts the client in place Served and returns the server in place Free.

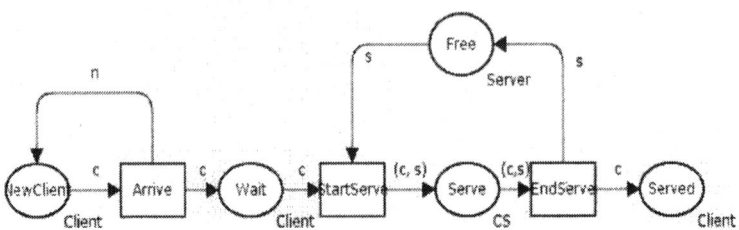

Fig. 1. Simple queuing system with client generator.

4.2 State Space

In a colored Petri net the tokens are objects with attribute values. Modeling UML static structure diagram is a very well suited tool to model the different types of orders and resources as object classes with specific attributes. Associations between object classes enable to relate attribute values of different objects in formal expressions. UML object constraint language (OCL) provides the necessary constructs to specify pre- and post-conditions in terms of attribute values (Warmer & Kleppe, 1999; OMG 2005). An advantage of this approach is that any expression can refer to any attribute value in the total state of the model. This makes an important difference with purely Petri net based languages like CPN-Tools that only supports references to attributes of tokens in the input places of the particular transition. This forces the modeler to model auxiliary places and maintain additional tokens and attribute values to make certain state variables accessible from a particular places.

Figure 2 shows the UML static structure diagram that specifies the state space for the Petri net in figure 1. A Petri net consists of Tokens and Places. The association between Token and Place expresses that every Token is always in one Place. Typical for Discrete Event Models are the object classes Clock and Random. A clock has an attribute time, which is supposed to increment implicitly. The class Random holds the concepts of random generator and random functions as used in stochastic simulations. These four classes are generic for every simulation model.

In the example, tokens can be either Clients or Servers, each having the attributes required to model the essential characteristics of the specific problem to be modeled. On the bottom row the objects are specified that populate the initial state of the system.

According to CPN-Tools every place has a type and consequently we label each place in the Petri net with an object class name from the UML static structure diagram.

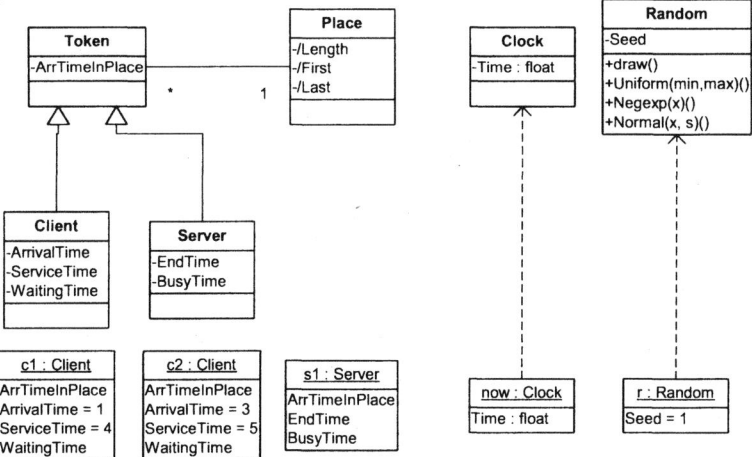

Fig. 2. A UML static structure diagram for simple waiting line simulation

The model shown above is very simple. In the problems we feed to our students complexities occur like multiple servers, multiple rows, clients that leave or switch rows under specific conditions, server failures, servers shared by different processes etc. All these situations can be elegantly modeled. However the size limits of this paper do no allow showing examples.

4.3 Transitions

From ExSpect we took the mechanism to specify transitions in terms of pre and post conditions. By default a transition fires when a token is available in each of its input places. The pre condition adds additional conditions these tokens and/or other objects must satisfy in order to allow the transition to fire: a transition fires if and only if there exists a set with exactly one token from each input place for which the pre condition holds. The clock object enables to make pre conditions time dependent.

The post condition must hold for the system state after the transition and thus enables to specify the change of system state because of the transition. The pre- and post condition are expressed in predicate logic using OCL to refer to specific attribute values,

The labels on the arcs are lent from CPN-Tools and are used as variable names in the condition specification. The labels of input arcs refer to the tokens selected for firing. The labels of output arcs refer to tokens put in the respective places and enable to specify changed attribute values. The following shows the specifications for transitions of the Petri net in figure 1 in combination with the state space specified in figure 2:

```
Init
      c1.Place = NewClient   ∧
```

```
        s1.Place = Free ∧
        s1.BusyTime = 0;

Arrive
Pre    c.ArrivalTime = now.Time
Post   Wait <- c    ∧
       NewClient <- n    ∧
       n.ArrivalTime := now.Time+r.Negexp(5)    ∧
       n.ServiceTime := r.Uniform(3,6);

StartServe
Prec = Wait.First
Post   Serve <- (s, c)  ∧
       s.EndTime := now.Time + c.ServiceTime      ∧
       c.WaitingTime := now.Time - c.ArrivalTime   ∧
       s.BusyTime := s.BusyTime + c.ServiceTime;

EndServe
Pre    now.Time = s.EndTime
Post   Served <- c   ∧
       Free <- s;
```

The block labeled with Init specifies the initial state of the system. In particular it specifies the places of the initial tokens. Initial tokens can be specified using UML as well as in the Init block. As an alternative, the places of the initial tokens can also be specified as markings in the Petri net. More often more than one of the Petri net, the UML or the transitions allow to specify some detail. In those cases the rule is that redundancy is to be avoided.

Note that the expressions are predicate logic and not programming statements. Proper interpretation of the post conditions requires some additional semantics because they must allow referring to both the old and the new state:
- Any variable that is not mentioned in the post condition remains unchanged,
- Where the <- operator (token is in output place) is used, all variables in the right operand refer to the old state, while the left operand is the place that holds the token resulting from the right operand,
- the := operator is used to specify that the left operand has, in the new state, the value resulting from the right operand, in which all variables refer to the old state,
- in all other cases variables refer to the new state.

In all small problems modeled so far we found these constructs able to express sufficient detail. For larger models, hierarchical Petri nets and place-connectors can be used.

4.4 Output Variables

Simulation models are to be used in experiments that yield output values. In order to specify the desired outputs we use functions, expressed in OCL. The expression below refers to the model of figures 1 and 2.

```
Functions
    NrServed = Served.Length;
    AverageWait = ∑(c ∈ Served: c.Waitingtime) / NrServed;
    Occupation = s1.BusyTime / now.Time * 100;
```

The expression AverageWait for the average waiting time of clients shows how expressions can reason over sets of objects. Note that objects are persistent: once created they remain in the state space. The place Served, for instance, holds at any point in time all clients served so far. In the expression for Occupation, s1 is the object name for the server, as defined in the UML model.

Functions can also be used as macros to simplify complex expressions in pre and post conditions.

5 Mapping to an Arena Executable Model

Once the Conceptual Model (CIM) has been created, and its validity (validation during early project phases) has been confirmed with the problem owners, it is important to carry the validity and relevance of the CIM to the executable model. This is achieved by the mapping approaches explained below, and finally confirmed by the verification.

The state-space and causal or flow logic expressed in the CIM must be mapped into a model that uses the ARENA building blocks. The "Petri-net token game" must be mapped on commands such as those associated with the use of resources (Seize, Delay, Release), and the life cycle of the token (e.g., Create, Dispose).

To document the derivation of the executable model from the conceptual model, it is convenient to use two tables:

- a table to map the classes and instances of the UML information model to Arena constructs
- a table to map the Petri net process model and the transition specifications to an Arena process model.

Furthermore each transition specification must be mapped to suitable specifications of the ARENA building blocks:

- for each transition:
    - identify corresponding module(s) in Arena
    - check post conditions,
- for each multi input transition:
    - check proper resources,
- for each pre condition
    - check corresponding priority rules or hold modules.

Students, who must mutually verify each other's models on consistency between the CIM and the Arena implementation, use this mapping.

6 Evaluation and Conclusions

Use of this approach in several courses has shown that students are able to construct and validate operational process models, from which teachers can easily recognize the problem and check correctness of the model.

So far the expressive power of this modelling approach appeared to be sufficient for the problems we give our students to solve. However this does not prove that the language is able to model every possible operational process situation. One apparent problem concerns the rules for reference between old and new situation in the post conditions. They may be not able to express certain complex transitions or they may become ambiguous in such situations.

Another problem is the use of OCL. Our students get only an elementary introduction, while the OCL definition is quite extensive. When using associations in expressions the result may be a set of objects. When several associations are followed in sequence, the result is a nest of sets. OCL follows the rule to flatten those nests, so that a simple set of objects results. However one must remain careful in evaluating such expressions.

Teaching discrete event simulation can benefit from the use of model-driven systems development techniques during the early project phases. The principles of model driven architecture can be applied to derive platform specific executable models from conceptual models that are precise and convenient to support communication with the problem owner. Combining Petri Net and UML static structure diagrams, one can build abstract, well defined and complete models of operational processes.

Further work will include the extension with hierarchy.

References

J. Dick, J. Chard, Requirements-driven and Model-driven Development: Combining the Benefits of Systems Engineering, Telelogic White Paper, www.telelogic.com, 2003.

K. Jensen, Coloured Petri Nets. Basic Concepts, Analysis Methods and Practical Use. EATCS Monographs on Theoretical Comp. Science, Springer-Verlag, Berlin, 1992.

A. M. Law and W. D. Kelton, Simulation, modeling and Analysis. Third edition. McGraw-Hill series, 2000.

J. Miller, J. Mukerji (eds.) MDA Guide Version 1.0.1, OMG, Object Management Group, 2003.

K.M. van Hee, Information Systems Engineering: A Formal Approach, Cambridge University Press, Cambridge, 1994.

J. Warmer,A. Kleppe, A., The Object Constraint Language: precise modeling with UML, Addison-Wesley, 1999.

Object Management Group (OMG), OCL 2.0 Specification. OMG document ptc/2005-06-06, June 2005.

Managing the After Sales Logistic Network– A Simulation Study of a Spare Parts Supply Chain

Fredrik Persson[1] and Nicola Saccani[2]

1 Linkoping University, Department of Management and Engineering,
Linköping, Sweden
fredrik.persson@ipe.liu.se

2 Università degli Sutdi di Brescia, Department of Mechanical and
Industrial Engineering, Via Branze 38, 2523 Brescia, Italy
nicola.saccani@ing.unibs.it

Abstract.
The after-sales services and in particular the spare parts business have acquired, in recent years, a strategic role for firms manufacturing durable or capital goods, as they represent a source of revenue, profit and a mean to achieve customer satisfaction and retention. Nonetheless, the huge variety and the characteristics of the demand of spare parts make the configuration and management of the spare parts inventory and distribution systems critical decision areas for managers. These decisions, in fact, may lead to very different cost and service performance by the system itself. The case study analyzed concern a world player of heavy equipment based in Sweden. Its spare parts distribution system is described, and the paper analyzes the configuration and allocation decisions concerning a second European warehouse and the transfer to that warehouse of a set of suppliers. A simulation model has been developed in order to support these choices. Discrete event simulation is well suited for studies where time-dependant relations are analyzed. Supply Chain Simulation applied to the case study provides useful insights on the decision choices and the cost structure related to the spare parts distribution system.

1 Introduction

Although durable goods for the consumer market as well as capital goods are designed to be reliable, *i.e.* to last for a long time without failures, it may happen that failures occur or that specific parts wear out with the usage. After-sales services and spare parts provision may be relevant sources of revenue, profit service

Please use the following format when citing this chapter:

Persson, F., Saccani, N., 2007, in IFIP International Federation for Information Processing, Volume 246, Advances in Production Management Systems, eds. Olhager, J., Persson, F., (Boston: Springer), pp. 313-320.

differentiation, and customer retention (Cohen *et al.* [1]; Yamashina [2]; and Goffin [3]).

The management of a spare parts inventory and distribution network presents some critical aspects. First of all, a company manufacturing a durable product, notwithstanding the degree of complexity of the product (*e.g.* from a dishwasher to a piece of complex machinery) has to manage a huge number of spare parts. All those products, in fact, are made of quite a high number of components or subsystems that can be subject to failure and substituted. Moreover, the rate of introduction of new products has indeed increased smoothly in the last decades: the product lifetime at the customer tends to be for most durable or capital goods much longer than the lifetime of the product offer on the market. Therefore, the manufacturer should manage the spare parts of its present product catalogue as well as for the old ones, thus reaching a very high number of parts. The second aspect is inherent to the demand for spare parts. It tends to be rather low (in relation with the amount of products sold) and irregular. Demand is often related to exceptional (failure) events, which happen stochastically in time and space. Of course, the knowledge of the reliability characteristics of the product and its parts and of the installed base characteristics may help in forecasting or interpreting spare parts demand [2], but this may work correctly at an aggregate level, while it is rather difficult to predict where and when the demand will take place.

Manufacturers, in configuring and managing their spare parts inventory and distribution systems, are then faced with the typical cost vs. service trade-off. They should act, in fact, in order to satisfy their customers by ensuring spare parts availability at a short notice, while on the other hand they should minimize the costs related to inventory holding and to the risk of obsolescence. This paper tackles the above issues through the case of a manufacturer of heavy equipment located in Sweden. In particular, the aim is to evaluate different choices about the location of spare parts inventory through a simulation model. The next section provides a review of spare parts classification and the relation with inventory decisions; the third section describes the case company, the specific problem analyzed and the simulation model designed. Section four provides the main simulation findings, while some conclusions and direction for future research are drawn in the last section.

2   Background

2.1   Spare parts distribution and inventory decisions

According to Huiskonen [4], when designing and managing any logistic system, and in particular a spare parts logistic system, there are four decision levels:
• The strategy/policies/processes level concerns the objectives to be pursued by one actor or the entire supply chain (e.g. in terms of service level or response time);
• The network structure defines the number of inventory echelons and locations in the supply chain;

• The coordination and control mechanisms include decisions about the inventory control principles, the incentive and performance measurement systems, and on the information support tools or systems to be used;

• The supply chain relationships, finally, consist of the degree of cooperation or reciprocal influence among supply chain actors that may impact on the achievement of the objectives through the implementation of the control and coordination activities.

Cohen *et al.* [5], focus on the interplay of spare parts allocation decisions. Decisions should be taken in order to optimize the cost-service trade-off related, as shown in figure 1, to the product and the geographic hierarchy. For example, a company may decide to replace a failed product with a standby end product stocked at the customer's premises. This solution provides the fastest response time, but it is much more costly then deciding to replace only the broken parts with spares stocked at the company central warehouse.

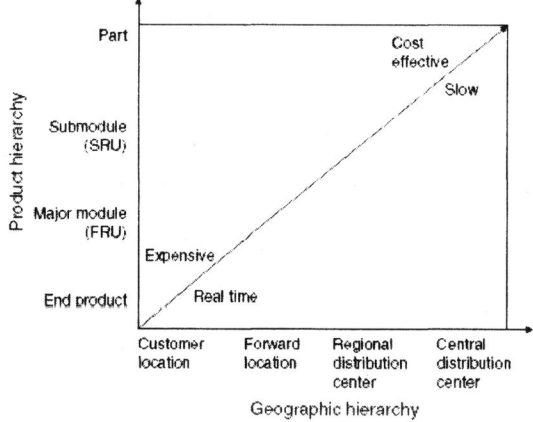

Fig. 1. The product and geography hierarchy for allocation decisions [5]

2.2   Spare parts classification

Some authors have treated the classification of spare parts as a way to provide guidelines on inventory management. Huiskonen [4] classifies spare parts according to four dimensions: *i.) criticality, ii.) specificity, iii.) demand pattern* and *iv.) value*. Resorting to a case example, he finds out five spare part groups, for which suggestions for inventory control policies are provided. Williams [6], and later Eaves and Kingsman [7], build their classification on the variability of spare parts demand. In particular, the lead time variance is decomposed into the contribution of: the number of orders during the lead time, the size of orders and the lead time duration. Eaves and Kingsman [7] identify 5 groups in reason of their demand pattern: *smooth, irregular, slow moving, mildly intermittent, highly intermittent.* Gajpal *et al.* [8] and Braglia *et al.* [9] use the AHP methodology for spare parts classification. Gajpal *et al.* [8] aim at assessing the spare part *criticality*, through a weighted measure of stock out implication, type of spare required (level of standardization) and of the lead time. Braglia *et al.* [9], instead, develop a multi attribute spare tree analysis. The classification of a spare part according to one criterion (e.g. *spare part plant criticality*) may lead to different decision diagrams, in which classification according to other criteria are requested. The other criteria proposed are *spare supply characteristics*, *inventory problems* and *usage rate*. Each classification is made through the AHP technique, and finally parts are grouped in four classes, to which inventory management guidelines are associated.

The reviewed classification methods present common elements such as the attention to criticality aspects and to the demand pattern of spare parts, although with different degrees of importance and different evaluation procedures. What emerges, also, is the fact that no classification considers directly the product and part lifecycle, that are suggested as relevant elements by Yamashima [2], and strongly influence the demand for spares.

3   The case study

3.1   Simulation Methodology

Supply Chain Simulation (SCS) is defined as the use of the simulation methodology, incorporating discrete event simulation, to analyse and solve problems found relevant to supply chain management.

Banks *et al.* [10] discuss what makes supply chain simulation different from other simulation applications. One major difference from e.g. simulation of manufacturing systems is that supply chain models contain information flows together with the flow of materials.

The used simulation methodology follows the steps described in Persson [11]. The first step (i) is the project planning or problem formulation where the outline of the study is determined. The next step (ii) is the conceptual modeling. The conceptual model describes the system under investigation. The conceptual model is validated as the next step (iii). The computer-based model is created as step (iv). This model must be verified (v) and validated (vi). Model verification aims at estimating if the simulation model is a valid representation of the conceptual model while model validation aims at estimating if the model is a valid representation of the system. The experimentation step (vii) consists of experimental runs with the simulation model. The results of these runs are then analyzed (viii) and the result of that analysis is the base for the recommended decision or implementation (ix). In this case study, simulation is used to calculate transportation costs in a complex setting incorporating a vast amount of data.

3.2   The Case Company

The case company is one of the world's leading manufacturers of heavy equipment. Production plants are located in Sweden, Western Europe, the North and South America, and in Asia. To obtain maximum profitability from the equipment and to respect scheduled activities, uptime is a critical performance for the company's customers. Therefore, great effort is devoted to the after-sales service, managed by a Service division headquartered Sweden. The logistics network of spare parts is depicted in figure 2. The company's suppliers are located in all different parts of the world, shipping spare parts to the two warehouses located in Europe, of which the main warehouse (WH 1) is located in Sweden.

Spare parts inventory levels and shipments to regional warehouses and dealers are controlled by reorder point systems. If the inventory level of any spare part in the two warehouses or at the regional warehouses are below the safety stock level, an express freight is utilized. That accounts for about 10% of all freights: otherwise ordinary transports are used.

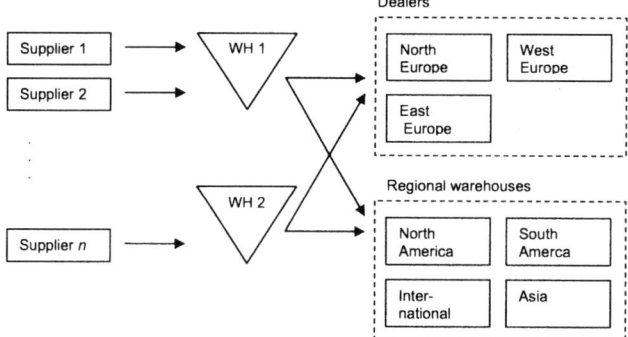

Fig. 2. Spare parts logistics network of the case company

3.3   Problem formulation

Today some of the suppliers of spare parts are dedicated to either the main warehouse (WH 1) located in Sweden or to warehouse WH 2. In the near future, the capacity needs to increase in WH 1 due to new forecasted higher sales volumes. One way to solve the problem would be to transfer some of the suppliers to WH 2, thereby reducing the number of suppliers storing their spare parts in Sweden and thus making storage capacity available. One restriction to the problem is that, if a supplier is moved, *all* spare parts supplied by that supplier must be moved.

The purpose of the case study is to investigate this transfer of suppliers from WH 1 to WH 2. This is done given the existing structure of the supply chain and taking all relevant costs into consideration. Simulation is chosen to investigate the supply chain cost structure due to two reasons: (i) The ability to handle stock outs and delays in transportations, and (ii) The reactive characteristics of the inventory control system that, among other things, determines if transportation should take place with ordinary transportation or with express.

The supply chain, as depicted in the previous section, is modeled in the simulation software Arena. The model contains 20 suppliers (out of over 800 suppliers in total). These 20 suppliers are chosen because they stand for 88,1 % of the total weight, 89,7 % of total purchasing cost, and 92,5 % of total number of order lines. All internal suppliers (suppliers owned by the same company) are not considered in this study. The model still covers hundreds of different parts and thousands of order lines per year.

4   Experiments and Results

The simulation model in Arena was used to calculate the transportation cost for the two different cases of localisation, WH 1 or WH 2. Transportation cost is calculated as the sum of all transports, taking the cost for weight, the weight of the parts and the order class into consideration. An order class is the type of transport that the part utilises. An express transport is more expensive than an ordinary transport.

The case company presented four different scenarios of future market growth to be analysed with the simulation model. The future growth is represented by percentage increase in the demand year by year in each market. The differences between Scenarios represent solely alternative changes in the European markets; thus, two Scenarios in the same year will have the same growth in sales in Latin America, North America, Asia and the International market, but the percentage of sales in the three European markets will be different. Each scenario contains sales data for the years 2008, 2010, 2012, and 2014 for both warehouses.

The first simulation used the demand for 2006 and the results of this experiment were used as a base case to compare to the rest of the simulations. Each experiment was run twice, one run considering the cost matrix of WH 1, the other time considering the cost matrix of WH 2.

For each future scenario the costs for each supplier was calculated in the simulation model. The difference between to place the supplier in WH 1 compared to place the supplier in WH 2 was calculated as the main result variable. If this difference is positive it means that the cost is lower when the supplier is moved from WH 1 to WH 2. The higher the positive value is in the calculations, the higher the cost savings when the supplier is moved to WH 2. In the cost calculations, the costs are split between inbound costs (transports to WH 1 and WH 2) and outbound costs (transports to regional warehouses and dealers in Europe). In some cases, both the inbound cost and the outbound cost are negative (indicating the spare part to stay at WH 1) and the conclusion to keep the part in WH 1 is unambiguous. In other cases, both costs are positive (indicating the part to be moved to WH 2) and here as well, the conclusion to move the product is clear. In the rest of the cases, the inbound and outbound costs differ and the cost levels decide the conclusion, see Table 1. This indicates that if a supplier is closer to a warehouse, it does not directly follow that the costs will decrease if it is moved to that warehouse. If the outbound costs become very high, they will surpass the benefits obtained in the inbound cost. Therefore, the location of a supplier with respect to the warehouse is not a decisive factor in finding the lowest total cost

Table 1. Cost relations

Inbound cost	Outbound cost	Total cost	Consequence
< 0	< 0	< 0	Stay in WH 1
< 0	> 0	?	?
> 0	< 0	?	?
> 0	> 0	> 0	Move to WH 2

Considering the results obtained in the first scenario, it is clear that costs can be saved if two of the suppliers (out of total 20) are moved to WH 2. Continuing with analyzing the other scenarios, the same two suppliers are suitable to be moved in every scenario. As the demand increases according to the sales forecasts there are two inferences: (i) for the first supplier (supplier 7) the cost savings decrease as the demand increases, and (ii) for the second supplier (supplier 13) the cost savings increase as the demand increases, see Table 2 for the costs in the base case.

Table 2. Total cost difference in base case

WH 1 - WH 2	Inbound*	Outbound*	Total cost*
supplier 1	-4 kr	-14 kr	-17 kr
supplier 2	-4 kr	-4 kr	-8 kr
supplier 3	-1 kr	-4 kr	-5 kr
supplier 4	-13 kr	-11 kr	-24 kr
supplier 5	-2 kr	-4 kr	-6 kr
supplier 6	-3 kr	-2 kr	-4 kr
supplier 7	**4 kr**	**-3 kr**	**2 kr**
supplier 8	2 kr	-7 kr	-5 kr
supplier 9	0 kr	-11 kr	-11 kr
supplier 10	1 kr	-1 kr	0 kr
supplier 11	-6 kr	-3 kr	-9 kr
supplier 12	-43 kr	-48 kr	-90 kr
supplier 13	**110 kr**	**-10 kr**	**100 kr**
supplier 14	3 kr	-14 kr	-11 kr
supplier 15	3 kr	-11 kr	-8 kr
supplier 16	-18 kr	-1 kr	-19 kr
supplier 17	-2 kr	-5 kr	-7 kr
supplier 18	-17 kr	-8 kr	-25 kr
supplier 19	7 kr	-10 kr	-3 kr
supplier 20	1 kr	-10 kr	-10 kr

* Numbers are indexed to hide real value

In the case of supplier 7, outbound costs to the international warehouse from WH2 and WH1 are very similar. As well, supplier 13 is much closer to WH2 than to WH1. Thus, the benefits from reduced inbound costs are far greater than the losses from the outbound costs.

On the other hand, the inbound costs savings are decreasing compared to the outbound costs increase with increasing demand. If the demand becomes high enough, this supplier should be moved back to WH1. Therefore, the benefits of moving it to WH2 might be rejected if the case company considers the dynamics over the years.

5   Conclusive remarks

As illustrated in the earlier sections of this paper, the spare parts business is critical for companies both at a strategic and at an operational level. Supply chain management is then a critical lever in order to obtain the highest customer satisfaction at a reasonable cost. Three relevant aspects have been touched by this paper: the configuration of the spare parts distribution system, the allocation decisions concerning suppliers and parts, and supply chain coordination mechanisms and inventory control mechanisms. Through the case of a heavy equipment manufacturer it was clear that these issues are also interesting to a company that is trying to cut costs in the supply chain. Classification of spare parts was touched in the reduction of suppliers to include in the simulation model. Focusing on weight, order lines and capital tied up provided a list of suppliers that are important to the company. The case study also provided useful information about cost structures in the supply chain for spare parts. Moving a supplier from a warehouse to another is not a straight forward decision. In this case it turns out that there needs to be costs savings both inbound and outbound or that the cost saving in one of the two needs to be large to overshadow the other.

Acknowledgements

The authors would like to thanks the students Pisut Changmai, Rajesh Maddineni, Emanuele Uberti, Iuliana David for their contribution in the development of the simulation model described in the paper.

References

1.   Cohen M.A., Agrawal N., Agrawal V. , (2006). Winning in the Aftermarket, *Harvard Business Review*, May, 84 (5), 129-138
2.   Yamashima H., (1989). The service parts control problem, *Engineering Costs and Production Economics*, 16, 195-208
3.   Goffin, K., (1999), Customer support - A cross-industry study of distribution channels and strategies, *International Journal of Physical Distribution and Logistics Management* 29(6), 374-397.
4.   Huiskonen J., (2001). Maintenance spare parts logistics: Special characteristics and strategic choices, *Int. J. Production Economics*, (71), 125-133
5.   Cohen M.A., Agraval N., Agraval V. , (2006). Achieving Breakthrough Service Delivery Through Dynamic Asset Deployment Strategies, *Interfaces,* 36 (3), 259-271
6.   Williams T.M., (1984). Stock Control with Sporadic and Slow-Moving Demand, *Journal of the Operational Research Society*, 35 (10), 939-498
7.   Eaves A., Kingsman B., (2004). Forecasting for the ordering and stock-holding of spare parts, *Journal of the Operational Research Society*, 55, 431-437
8.   Gajpal P.P., Ganesh L.S., Rajendran C., (1994). Criticality analysis of spare parts using the analytic hierarchy process, *Int. J. Production Economics*, 35, 293-297
9.   Braglia M., Grassi A., Montanari R., (2004). Multi-attribute classification method for spare parts inventory management, *Journal of Quality in Maintenance Engineering*, 10 (1), 55-65
10.   Banks, J., Buckley, S., Jain, S., Lendermann, P., and Manivannan, M., (2002). "Panel session: Opportunities for simulation in supply chain management", *Proceedings from the 2002 Winter Simulation Conference*, pp. 1652-1658.
11.   Persson, F. (2003). *Discrete Event Simulation of Supply Chains*, Doctoral Thesis, Department of Production Economics, Linköping Institute of Technology, Linköping, Sweden.

A Stochastic Single-vendor Single-buyer Model under a Consignment Agreement

Ou Tang[1], Simone Zanoni[2] and Lucio Zavanella[2]
[1] Department of Management and Engineering,
Linköping Institute of Technology,
SE-581 83 Linköping, Sweden
[2] Dipartimento di Ingegneria Meccanica e Industriale,
Università degli Studi di Brescia,
Via Branze, 38, 25123 Brescia, Italy

Abstract.
In the recent years, companies have begun to strengthen their supply agreements, such as sharing the management of inventories. This type of co-operation implies that the members of the supply chain share information and arrange a mutual agreement on their performance targets. The increased interest on supply chain topics has attracted researchers' attention to the problem of co-operation between the buyer and vendor, the two actors directly interacting in the supply mechanism. The present research investigates the way how a particular VMI policy, known as Consignment Stock (CS), may lead to a successful strategy for both buyer and vendor. The previous study [1] developed an analytical model of the CS policy, with reference to the centralised decision and deterministic settings. In order to fully explore the potentiality of CS policy, an extension of the model is proposed in this paper. The results indicate that the CS policy could be a strategic and profitable approach to improve supply chain performance in uncertain environments.

1   Introduction

Firms are no longer competitive as independent entities such as buyer and vendor, but rather as an integral part of supply chain. Thus the success of a firm depends on its managerial ability to integrate and coordinate the network of business relationships among the supply chain members. According to this scenario, Vendor Managed Inventory (VMI) represents an interesting approach to stock monitoring and control, progressively considered and introduced in both service and manufacturing industries. One VMI policy, known as Consignment Stock (CS), may suppress the vendor's inventory, as this actor will use buyer's warehouse to stock its

Please use the following format when citing this chapter:

Tang, O., Zanoni, S., Zavanella, L. 2007, in IFIP International Federation for Information Processing, Volume 246, Advances in Production Management Systems, eds. Olhager, J., Persson, F., (Boston: Springer), pp. 321-328.

finished products. Furthermore, the vendor guarantees that the quantity stored in the buyer's warehouse will be kept between a maximum and a minimum level, and thus further reduces the costs eventually induced by stock-out conditions. The buyer picks up the quantity of material needed to meet his production plans and, consequently, the material used is paid to the buyer.

The present study focuses on those situations where the deterministic conditions, necessary so as to apply Hill's model [2], do not prevail. The major effect determined by uncertain conditions is represented by stockout events (i.e., buyer's inability to promptly satisfy demand). This problem is generally approached by increasing reorder points and keeping safety stocks. Of course, an increased safety stock leads to the downstream movement of inventories, similarly to the CS approach [3]. In addition, in this study, we extend CS to a decentralised decision system in order to investigate the motivations and incentives for applying this policy.

A brief literature review is presented to summarise early studies in CS principle and relevant issues. The first study dealing with the integrated single-supplier single-customer problem is the paper by Goyal [4]. In [5] Banerjee considered the vendor manufacturing for stock at a finite rate and delivering the whole batch to the buyer as a single shipment. Goyal in [6] showed how lower-cost policies result from production batch splitting and multiple shipment delivery. Lu in [7] sets out the optimal production and equal-size shipment policy. Goyal in [8] demonstrated that lower cost policies may be obtained when shipments increase in size by a given ratio. Hill [9] derived the form of the optimal policy if shipment sizes may vary. More recently, Corbett [10] examined the impact of incentive conflicts and information asymmetry on performance in a two-player decentralized supply chain, which follows a continuous review (Q, R) policy. Corbett found out that, in the absence of a central planner with full information, no party may induce a joint optimal behaviour of all agents without sacrificing his own profits. In [11] a CS industrial case is presented while its analytical approach is treated in [1], together with some performance evaluation. Following in [12] a full analytical solution to CS is proposed. In [13] there is a summary of the previous research on the single-vendor single-buyer integrated production-inventory problem, in the deterministic cases.

According to the topics outlined, Section 2 presents the notation and system description. Section 3 focuses on the model formulation in a decentralised system and Section 4 discusses the results obtained by numerical examples. Finally, conclusions are drawn in Section 5.

2 System description

A single-vendor single-buyer supply chain is studied with a stochastic demand. The notations adopted are given below:

A_v: production setup cost, paid by the vendor, ϵ
A_b: order emission cost, paid by the buyer, ϵ
h_v: out-of-pocket inventory holding costs of vendor, ϵ/unit/year
h_b: out-of-pocket inventory holding costs of buyer, ϵ/unit/year
α: interest rate

D: demand rate, *units/year*
P: buyer's purchasing price of the product, €*/unit*
m: vendor's marginal cost of product, €*/unit*
Q: transportation batch size, *units*
n: number of transportations within one production batch
b: backorder cost, €*/unit*
TC_v total costs of vendor, €
TC_b total costs of buyer, €
σ_L: standard deviation of demand during lead time
k: safety stock factor

$f(\mu_0)$: probability density function of a standard normal distribution

and we also have

$$F(k) = \int_i f(\mu_0)d\mu_0$$

$$G(k) = \int_i (\mu_0 - k)f(\mu_0)d\mu_0$$

In this system, the vendor supplies products to a buyer facing a stochastic demand, which is assumed to be normally distributed. The lead time for transportation is constant and equal to L. The lead time for production at the vendor site is null (i.e., the vendor does not need safety stocks). The buyer uses a continuous review reorder point inventory control system. Thus, the buyer's decision variables are the order quantity Q and the reorder point s (further related to the safety factor k). Once given Q and s, the vendor will consequently determine the number of transportations n.

The present study distinguishes between the capital tied-up and the out-of-pocket inventory holding cost. Conventionally, the total inventory holding cost is assumed to be increasing when items move downstream in a supply chain. According to our industrial experience, this statement may sometime fail. In particularly, the out-of-pocket inventory holding cost does not have to follow the above assumption, e.g. when a specialised producer with limited storage space supplies a large manufacturer with low-cost bulk storage facilities [13].

3 Model formulation in a decentralised system

The early literature about CS [1,11,12] assumes that production and transportation decisions are centrally made. This assumption may not always be true (e.g., for multiple business entities with conflicting objectives within a supply chain). The present study investigates the motivation of using CS contract and evaluates the performance of a supply chain with a powerful buyer. Therefore, the buyer is assumed to impose his optimal decisions Q and k, and the vendor consequently may decide his production batch size ($n \cdot Q$). Finally, the expected on-hand inventory is considered equal to the expected net one (on-hand minus backorders), as a standard approximation in inventory modelling (e.g. [14] and [15]).

3.1 Without a consignment stock contract

The following formulae propose the vendor and buyer total costs:

$$TC_v = \frac{A_v}{n}\frac{D}{Q} + (\alpha m + h_v)\left(\frac{nQ}{2}\right) - (\alpha m + h_v)\left(\frac{Q}{2} + k\sigma_L\right) \tag{1}$$

$$TC_b = A_b\frac{D}{Q} + (\alpha P + h_b)\left(\frac{Q}{2} + k\sigma_L\right) + \frac{Db\sigma_L G(k)}{Q} \tag{2}$$

The buyer will minimize his cost by using optimal order quantity and safety factor. In case an optimal Q has been used, Equation 2 can be rewritten as

$$TC_b* = \sqrt{2D(A_b + b\sigma_L G(k))(\alpha P + h_b)} + (\alpha P + h_b)(k\sigma_L) \tag{3}$$

Formula (3) may be differentiated with respect to k and set equal to zero

$$\frac{\partial TC_b*}{\partial k} = -\frac{D(b\sigma_L F(k))(\alpha P + h_b)}{\sqrt{2D(A_b + b\sigma_L G(k))(\alpha P + h_b)}} + (\alpha P + h_b)(\sigma_L) = 0 \tag{4}$$

This ends with the optimisation conditions:

$$\frac{A_b / b\sigma_L + G(k)}{F(k)^2} = \frac{Db}{2\sigma_L(\alpha P + h_b)} \tag{5}$$

and

$$Q = \sqrt{\frac{2(A_b D + Db\sigma_L G(k))}{(\alpha P + h_b)}} = \frac{Db}{(\alpha P + h_b)} F(k) \tag{6}$$

Once the buyer's optimal safety factor and order quantity have been obtained, the optimal value of n may be found from the difference equation of TC_v, i.e. the optimal n should be the smallest integer satisfying

$$n(n+1) \geq \frac{2A_v D}{(\alpha m + h_v)}\frac{1}{Q^2} \tag{7}$$

3.2 A consignment contract

A CS contract implies that the vendor pays the capital tied up cost of the stocks delivered to the buyer but not yet sold. Thus, the total costs change as follows

$$TC_v' = \frac{A_v}{n}\frac{D}{Q} + (\alpha m + h_v)\left(\frac{nQ}{2}\right) - h_v\left(\frac{Q}{2} + k\sigma_L\right) \tag{8}$$

$$TC_b' = A_b\frac{D}{Q} + h_b\left(\frac{Q}{2} + k\sigma_L\right) + \frac{Db\sigma_L G(k)}{Q} \tag{9}$$

Using the same approach as in Section 3.1, the optimisation conditions for the buyer are:

$$\frac{A_b / b\sigma_L + G(k)}{F(k)^2} = \frac{Db}{2\sigma_L h_b} \quad \text{and} \quad Q = \frac{Db}{h_b} F(k) \tag{10}$$

The corresponding optimal n should follow:

$$n(n+1) \geq \frac{2A_{v}D}{(\alpha m + h_{v})Q^{2}} \cdot \frac{1}{}$$ (11)

3.3 Comparison and analysis

Proposition 1: $F(k)^{2}/G(k)$ is a decreasing function with respect to k, for $k > 0$. This can be easily observed in the function curve given in Fig.1 below.

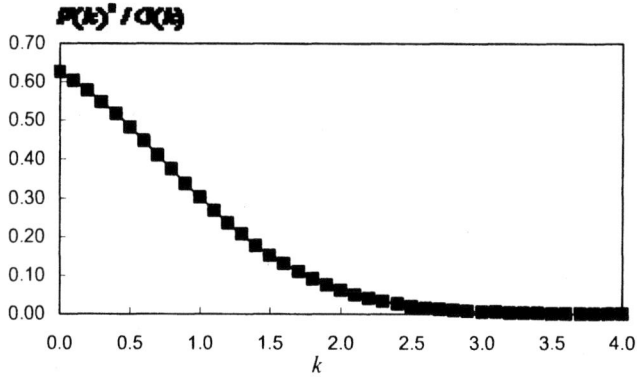

Fig. 1. $F(k)^{2}/G(k)$ as a function of k for a standard normal distribution

Proposition 2: the service level, compared to a policy without CS, is improved by CS adoption.

Both $F(k)$ and $F(k)^{2} \cdot G(k)^{-1}$ are decreasing functions of k. Thus, $(A+G(k)) F(k)^{-2}$ is an increasing function with respect to k (A is a positive constant). Since

$$\frac{Db}{2\sigma_{L}(\alpha P + h_{b})} < \frac{Db}{2\sigma_{L}h_{b}},$$ a larger optimal k value is found for the CS case, as

compared to the one without consignment. In addition, a higher safety factor will then lead to a higher service level.

Proposition 3: with a CS contract, the optimal order quantity shall follow the ratio

$$\frac{Q'^{*}}{Q^{*}} = \frac{\alpha P + h_{b}}{h_{b}} \frac{F(k'^{*}))}{F(k^{*}))}$$

The change in the order quantity is a balance between the inventory holding cost ratio and the changes of k value (correspondingly, the $F(k)$ value).

Proposition 4: The CS contract always reduces the buyer's cost.

Any Q and k values (including Q^{*} and k^{*} for TC_{b}) determine $TC_{b} > TC_{b}'$, as easy to see comparing Equations 2 and 9.

Proposition 5: Vendor's optimal cost can be reduced when

$$\frac{(\alpha m + h_v)}{h_v} < \frac{(Q'^* + 2k'^*\sigma_L)}{(Q^* + 2k^*\sigma_L)}$$

This result emerges from the following relationship:

$$TC_v'^* - TC_v^* = (\alpha m + h_v)\left(\frac{Q^*}{2} + k^*\sigma_L\right) - h_v\left(\frac{Q'^*}{2} + k'^*\sigma_L\right) < 0$$

4 Illustrative examples

The set of data offered in the present examples are slightly changed from the case of an Italian company producing braking systems for the automotive industry. Focus is given on brake pads (an assembly component in the company manufacturing cycle). The specialised vendor of the brake pads is close to the company (for a detailed description of the case study see [1]). Parameters values are given below:

$A_v = 400 \ €$ and $A_b = 70 \ €$

$h_v = 0.175 \ €/unit/year$ and $h_b = 0.15 \ €/unit/year$

$\alpha = 10\%$

$D = 50\ 000 \ units/year$

$P = 2.5 \ €/unit; \ m = 2 \ €/unit; \ b = 0.25 \ €/unit$

$\sigma_L = 300$ (Lead time L is about 1week)

Let us consider the policy without CS. In the first step, the safety factor is evaluated by Equation 3. $F(k)$ and $G(k)$ values can be easily obtained by standard normal distribution table or, alternatively, using Waissi and Rossin approximation [15]. An upper bound $\bar{k}$ can be used as an initial value for searching Equation 3 solution by examining

$$\frac{A_b / b\sigma_L}{F(\bar{k})^2} = \frac{Db}{2\sigma_L(\alpha P + h_b)} \tag{12}$$

Since $F(k)$ is a decreasing function of k, and $G(k) \cdot F(k)^{-2}$ is positive, the optimal k must be smaller than $\bar{k}$, as obtained by (12) (for the example, $\bar{k} = 1.11$ and $k = 1.09$). Inserting the results into Equation 6, optimal $Q = 4309$ and Equation 7 provides the optimal $n = 2$. The results are summarised in the left side of Table 1 below.

An additional case (Case 2) illustrates how the vendor's cost can be reduced by CS as well. Case 2 differs only in the marginal production cost, which is set to $m = 0.5$. Results are given in the right part of Table 1.

It can easily be seen that the service levels of the system are improved from 86.2% to 91.6%. The buyer's cost is significantly reduced, both in terms of ordering, inventory holding and stockout costs. Vendor's costs may increase or decrease, depending on the production batch size and the changes in inventory holding cost (base case and Case 2, respectively)

Table 1. Optimal decisions and costs

Base case				$m = 0.5$			
Without consignment		With consignment		Without consignment		With consignment	
k	1.09	k	1.38	k	1.09	k	1.38
Q	4309	Q	6983	Q	4309	Q	6983
n	2	n	2	n	3	n	2
SL^*	86.2%	SL^*	91.6%	SL^*	86.2%	SL^*	91.6%
TC_v	3006	TC_v	3369	TC_v	2443	TC_v	2320
TC_b	1866	TC_b	1108	TC_b	1866	TC_b	1108
$TC_v + TC_b$	4872	$TC_v + TC_b$	4477	$TC_v + TC_b$	4309	$TC_v + TC_b$	3428

$SL^* =$ service level $= 1 - F(k)$

5 Conclusions

This study presented a model to optimise the production and inventory decision in a single-vendor and single-buyer system with a stochastic demand. The optimisation models and the propositions of this paper can be applied to support a CS decision. The results show that, by using a CS principle and moving a part of the inventory holding cost to the vendor, the system performance can be improved, both in terms of total costs and service levels. Furthermore, a CS contract consistently reduces buyer's costs, but it may impact differently on the vendor's cost. Nevertheless, as already observed in the introduction, the vendor still perceives some advantages as a counterpart. For instance, he may manage his production plans more flexibly and the relationships between the vendor and the buyer can be improved. Moreover, a part of the channel costs reduction may be shared between the two actors and it can be used as an important element of the negotiation between the two actors of the chain. Finally, Sensitivity analysis and numerical experiments may allow an improved understanding of the circumstances determining the vendor benefits. Extending the model to positive and stochastic production lead time could be of great interest.

References

1. M. Braglia and L. Zavanella, Modelling an industrial strategy for inventory management in supply chains: the 'Consignment Stock' case, *Int. Jour. of Prod. Res.* **41**, 3793–3808 (2003).

2. R.M. Hill, The single-vendor single-buyer integrated production-inventory model with a generalised policy, *Europ. Jour. of Oper. Res.* **97**, pp.493–499 (1997).

3. L. Abdel-Malek; G. Valentini and L. Zavanella, Managing stocks in Supply Chains: modelling and issues, in Seuring S.; Goldbach M., *Cost Management in Supply Chains*, Heildeberg: Physica Verlag, 325-335 (2002).

4. S.K. Goyal, Determination of optimum production quantity for a two-stage production system, *Oper. Res. Quart.* **28**, 865–870 (1977).

5. A. Banerjee, A joint economic lot-size model for purchaser and vendor, *Dec. Science* **17**, 292-311 (1986).

6. S.K. Goyal, A joint economic lot size model for purchaser and vendor: A comment. *Dec. Science* **19**, 236 – 241 (1988).

7. L. Lu, A one-vendor multi-buyer integrated inventory model. *Europ. Jour. of Oper. Res.* **82**, 209–210 (1995).

8. S.K. Goyal and Y. P. Gupta, Integrated inventory models: the buyer-vendor coordination, *Europ. Jour. of Oper. Res.* **41**, 261–269 (1989).

9. R.M. Hill, The optimal production and shipment policy for a single-vendor single-buyer integrated production-inventory problem, *Int. Jour. of Prod. Res.* **37**, 2463 – 2475 (1999).

10. C.J. Corbett, Stochastic inventory systems in a supply chain with asymmetric information: cycle stocks, safety stocks, and consignment stock, *Oper. Res.* **49**(4), 487–500 (2001).

11. G. Valentini and L. Zavanella, The Consignment Stock of Inventories: Industrial Case and Performance Analysis, *Int. Jour. of Prod. Econ.*, **81-82**, 215-224 (2003).

12. S. Zanoni and R.W. Grubbstrom, A note on an industrial strategy for stock management in supply chains: modelling and performance evaluation, *Int. Jour. of Prod. Res.* **42**, 4421–4426 (2004).

13. R.M. Hill and M. Omar, Another look at the single-vendor single-buyer integrated production-inventory problem, *Int. Jour. of Prod. Res.* **44**(4), 791-800 (2006).

14. M. A. De Bodt and S. C. Graves, Continuous-review policies for a multi-echelon inventory problem with stochastic demand, *Manag. Science* **31**(10), 1286-1299 (1985).

15. E.A. Silver, D.F. Pyke and R. Peterson, Inventory Management and Production Planning and Scheduling, John Wiley and Sons, New York (1998).

Supply-chain Simulation Integrated Discrete-event Modeling with System-Dynamics Modeling

Shigeki Umeda

Musashi University, 1-26 Toyotama-kami Nerima Tokyo 176-8534 Japan,
shigeki@cc.musashi.ac.jp

Abstract

This paper describes a novel simulation framework that integrates discrete-event modeling with system-dynamics modeling. The former has strength in system performance evaluation; meanwhile, the later has an advantage of representing feedback mechanisms in complex systems. We are currently developing a hybrid-modeling framework, which combines discrete-event modeling with system dynamics modeling. The objectives of this framework are: (1) to simulate feedbacks of supply-chain activities in social system mechanisms, (2) to enable management simulation in long time terms, and finally (3) to clarify requirement specifications towards supply-chain management gaming. This paper summarizes this framework and represents application examples.

Keywords

Supply-chain management, Enterprise modeling, Production control, System-dynamics simulation, Discrete-event simulation.

1 Introduction

A supply-chain is a network of autonomous and semi-autonomous business units collectively responsible for procurement, manufacturing, distribution activities associated with one or more families of products.

Supply chain planning is, in a sense, restructuring a business system for supply chain members to collaborate with each other by exchanging information. Supply chain managers, in both planning phases and operational phases, face various kinds

Please use the following format when citing this chapter:

Umeda, S., 2007, in IFIP International Federation for Information Processing, Volume 246, Advances in Production Management Systems, eds. Olhager, J., Persson, F., (Boston: Springer), pp. 329-336.

of problems, such as capacity planning, production planning, inventory planning and others. Systematic approaches are needed to support planning and control of such supply chain systems.

Simulation is an effective tool for system performance evaluation. Many commercial simulation software products have been developed and used for manufacturing and logistics systems, when practitioners evaluate system performances in "what-if" scenarios. These software products belong to discrete-event types that represent target system behaviors as a set of events and activities. Meanwhile, system dynamics has been mainly used to predict social systems' behaviors such as urban transportation systems, national growth and its effect on environmental problems [1][2]. These are very well known as "Urban dynamics" and "Environment dynamics". Discrete-event simulation has strength in system performance evaluation; on the other hand, system dynamics simulation has an advantage to represent feedback mechanisms in target systems.

The authors are developing a new simulation-modeling framework based on a hybrid modeling approach, which combines discrete-event modeling with system dynamics modeling. The objectives of this framework are: (1) to simulate feedbacks of supply-chain activities in social system mechanisms, (2) to enable management simulation in long time terms, and finally (3) to clarify requirement specifications towards supply-chain management gaming. This paper describes its modeling framework and an application example.

2   Supply-chain simulation model

2.1   System views and Features-elements model

Any system modeling requires clarifying system view. This effort proposes modeling a supply chain system from different views relevant for its performance management. We introduced four views for supply chain systems that the stakeholders of implementation of supply-chain models may be interested in [3]. These views are "Organization", "Control", "Activity", and "Communication".

"Organization view" clarifies the roles of members that belong to a supply-chain system. Although various supply-chain members, the types of these members are countable. We defined six categories of chain members' organizations. These abstracted types are as follows.

- *Supplier*: A member gets and provides materials, parts, or products in the chain.
- *Source*: A member provides primal materials or parts in the chain. This member would be a start point of material-flows in the chain.
- *Storage*: A member stores materials, parts, or products.
- *Consumer*: A set of members that send purchase orders and acquire products.
- *Deliverer*: A member transports products, parts, and/or materials between chain members.
- *Planner*: A member that controls material-flows and information-flows in the chain.

"Control view" elucidates material management control policies. Two types of broad operational policies are introduced to control operations of members in a supply chain system: "Schedule-driven control" and "Stock-driven control". The former is a "*push*" control method, which is based on a central "Master Production Schedule" (MPS), meanwhile, the later is a "*pull*" control method, which is based on stock volume information in the down-stream supplier.

"Activity view" defines core activities that are classified into following seven groups: these are *Resources and facilities management, Planning, Manufacturing, Transportation, Storing, Material management, and Communication.*

"Communication view" clarifies the role of communication that activates chain member. The communication would be a driving force for sharing information and exchanging data among chain members. The collaboration among chain members is activated through information sharing and data exchanges. Supply chain systems usually own a special member (Planner), which plays a central role in communication among chain members. The planner produces processed data and information, which are needed to manage all over the chain. These major data items are used for planning and control in both production and inventory management activities. The communication data can be classified as below.

"Feature-elements model" is a set of activity models that represent business processes by using abstracted descriptions of the chain members. This is also a set of library of simulation models based on commercial simulation software. Each member's model is classified into two types by generic control methods: "Schedule-driven" control and "Stock-driven" control [4].

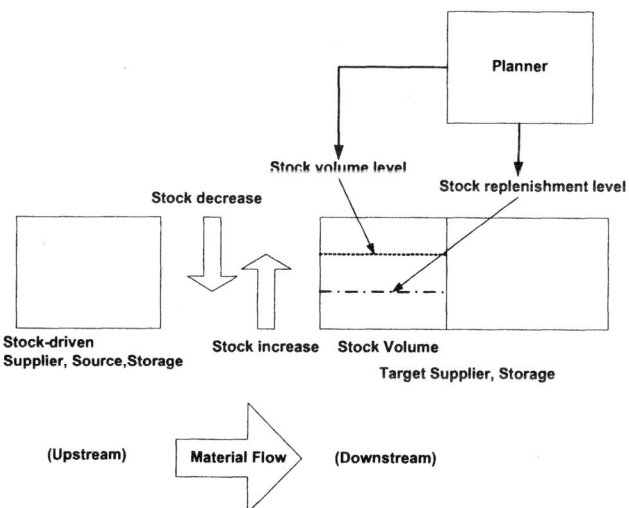

Figure.1 Stock-driven control

The schedule-driven control uses a production schedule, which is a so-called "Master Production Schedule" (MPS), generated by a supply chain planner. The chain planner periodically collects demand data from marketing channels in a constant cycle time, and it updates the MPS by using accumulated, demand prediction data. MPS is a schedule about when finished-goods are delivered to consumers. Meanwhile, the stock-driven control is based on autonomous operations of upstream suppliers using buffer inventory data of its down-streams suppliers (Figure.1). These models are expansions of our previous research and abstractions of material management policies in discrete manufacturing systems [5][6][7]. In the "Feature Elements models" layer, representations of various types of management controls in a chain are available by using combination of these control methods.

2.2 Scenario models

An activity is a sequence of events. Discrete-event models, accordingly, fit with operational activity in business and operational processes such as manufacturing, inspections, shipping, transportation, and their planning. Meanwhile, simulation will need another different modeling methodologies, if the simulation considers the feedback of the simulated activities from the external world. This is because input parameters to simulation are often considered as feedback data from the system behaviors.

Suppose that a supply-chain system realized a high performance and it shortened the consumers' purchase lead-time. The demand volume in a market, in this case, would increase, because the supply-chain could establish customers' satisfaction. The system would be busy for the increased demand, and consequently it would not be able to realize short lead-time operations. Similar scenarios would be appropriate to other supply-chain systems' activities such as quality improvement program in factories; manufacturing processes automation programs, and efficient transportations operations. System dynamics modeling is one of the typical simulation techniques that provide such feedback scenarios (Figure.2).

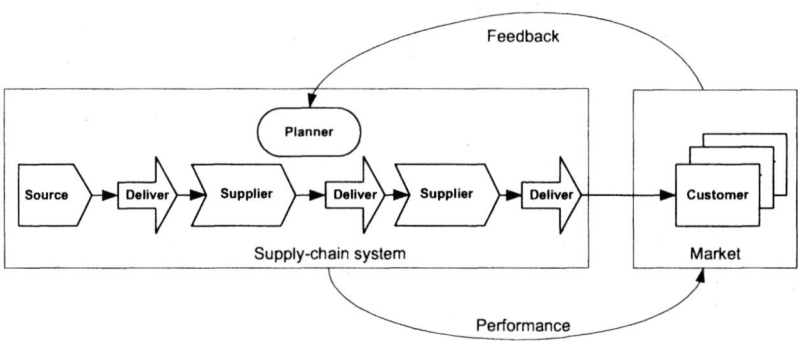

Figure. 2 A relationship between supply-chain performances and feedback from market

Figure.3 shows an outline of scenario of a simple dynamics model. Performance of product manufacturing depends on its manufacturing capacity and order volume from customers. The order volume increases, if customers' satisfaction goes up. Increased order volume would be another constraints on manufacturing and shipment process. This will be a cause of lead-time reduction and a downturn of customers' satisfaction. The average demand will be periodically going up and down.

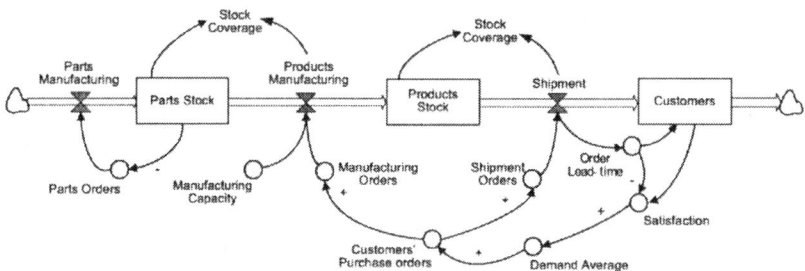

Figure. 3 Supply lead-time reductions and customers' satisfaction

Figure.4 is another scenario model using system dynamics notations. The system is a small chain, which poses two manufacturing resources (Product manufacturing and Parts manufacturing) and two manufacturing buffers (Parts inventory and Product inventory). The products and parts manufacturer work in Make-To-Stock operational strategy. "Desired inventory level" and "Desired WIP level" are major control parameters in this system. In this case, many factors form a complex causal relationship. Desired Production volume, for an example, produces positive effect on the "Desired WIP volume", and the "Desired WIP volume" is effectual on manufacturing cycle-time.

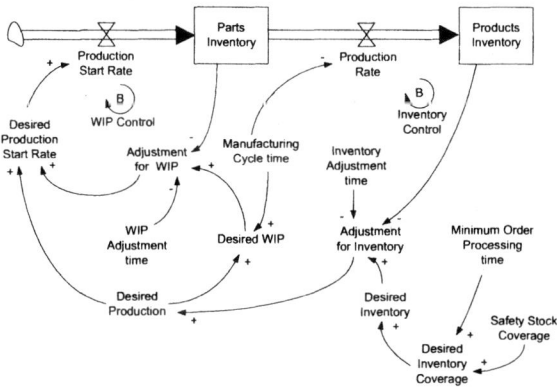

Figure. 4 Decision processes of parts & product inventories in Make-to-Stock supply-chain

3 Application to a manufacturing supply-chain

3.1 Target system

We discuss here an application of the proposed simulation to an actual manufacturing supply-chain. Final assembly factory must keep its DGR (Daily-Going-Rate) on the level from 100 to 250 products. Almost supplier provides part to final assembly in MTO (Make-To-Order) mode. The planner generates schedules on suppliers by using order data from customer. The planning cycle-time is one week. Only one supplier (S3) receives orders every other week and it delivers parts to the assembly factory two times in a week. This means that the order changes often occur in this operation. Particular two suppliers (S2 and S4), meanwhile, provide parts to the final assembly by MTS (Make-To-Stock) mode. A difficulty arises here, because these two are located in long distance from final assembly factory. The delivery lead-time between them will be one of key issues to determine both "Stock volume level" and "Stock replenish level".

System includes two 2nd-tier suppliers. One of them (S41) has a hard resource constraint, because it is a small-sized manufacturer. Its downstream supplier needs to pose much volume of parts inventory as its buffer.

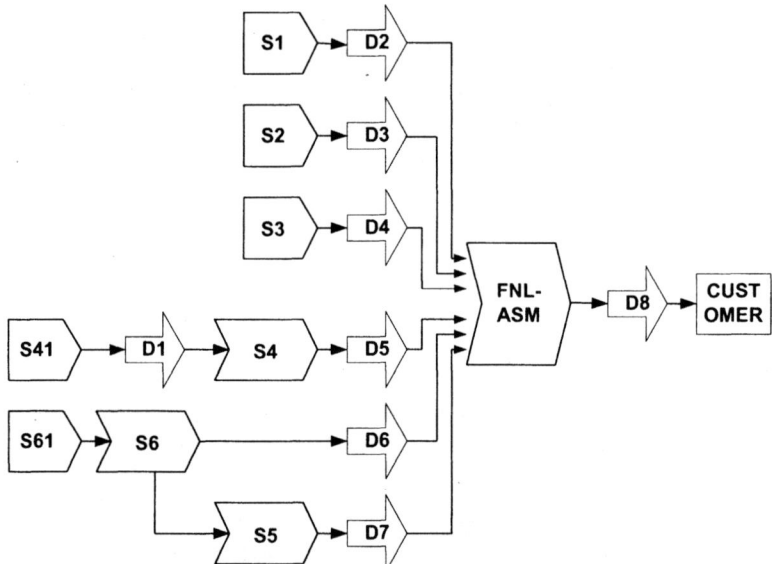

Figure. 5 Configuration of the target supply-chain system

3.2 Potential problems

A previous research discussed typical supply-chain management problems [3]. It would include the following items: "Rough-cut Supply-chain Capacity planning",

"Supply-chain Capacity Requirement planning", "Resource planning", "Lead-time planning", "Aggregate Production planning", "Operational production planning", "Supplier selection", "Outsource planning", and finally "Operational strategy selection". We analyzed the target supply-chain system, and found that the lead-time planning problem and operational strategy problem are critical issues in this case.

The term "Lead-time" has basically two meanings: a span of time required to perform a process (or series of operations), and the time between recognition of the need for an order and the receipt of goods. The second one is often used in a logistics context. Individual components of lead-time can include order preparation time, queuing time, processing time, move or transportation time, and receiving and inspection time. We use this term in this paper with its second meaning. This problem directly impacts the inventory planning problems through the Lead-time inventory, the inventory that is carried to cover demand during the lead-time. The examples of this class of problems are:

- When and what suppliers should produce, and associated due dates?
- When and how much volume of products or component parts should be transported?
- Which transportation channels should be used?
- Suppose that all of the factories in the chain use a common database for purchase ordering process, what impacts occur on total lead-time in the chain?

The operational strategy selection problem includes selecting the strategy to operate the supply chain. Suppose that the supply chain designer has solved the primary problems, has selected the best business partners as his/her suppliers and has decided the non-core processes to be outsourced, s/he still needs to decide how to control the flow of products through the supply chain. The problem examples are as follows:

- How to choose between PUSH, PULL, and Hybrid PUSH-PULL?
- How to choose the strategy such as STS, MTS, ATO, MTO, at each stage of the supply chain?

4 Conclusion

The system configuration discussed here has been partially scale-downed from the original configuration. We, however, believe that the configuration poses essential supply-chain management issues. We, here, designed system dynamics models representing interlocking mechanism that couples stock replenishment levels with market demands in stock-driven control operation, and tried a simulation. We used a method that occurs events with a constant interval time to represent implicitly parameters relations described in system dynamics models. We also chose to a method to decompose dynamics models into functions that integrate with discrete-event models.

We got a simulation result, however it is our future work to get more detail results that enable us to clarify systems' characteristics. We used a model that is partially simplified. The scale of the model is still more, vary large, and the mechanism is very complex. The efficient simulation methods and systematic modeling libraries will be needed to integrate discrete models with dynamics models in real system modeling.

5 References

[1] R.G. Coyle, System Dynamics Modeling – a Practical Approach, Chapman & Hall, London, 1996
[2] J.D. Sterman: Business Dynamics – System Thinking and Modeling for a Complex World, Irwin McGraw-Hill, 2000
[3] S. Umeda: SCMM: Supply Chain Maturity Model, Proc. of 2004 PROMAC conference, 2004
[4] S. Umeda and T. Lee: Design specifications of a generic supply chain simulator, Proc. of 2004 Winter Simulation Conference, pp.1158-1166, 2004
[5] S. Umeda: Planning and Implementation of Information system in supply chain system, Journal of Society for Project Management, Vol.5, No.4, pp.42-48, 2003
[6] A. Davis and J. O'donnell, Modeling complex problems: system dynamics and performance measurement, Management Accounting Today, May, pp.18-20, 1977
[7] S. Umeda and F. Zhang, Supply chain simulation: generic models and application examples, Production Planning & Control, Vol.17, No.2, pp.155–166, 2006

Methodology for the Analysis of Simulation-based Decision-making in the Manufacturing Area

Gert Zülch, Thilo Gamber and Patricia Stock
University of Karlsruhe, ifab-Institute of Human and Industrial
Engineering, Kaiserstrasse 12, D-76131 Karlsruhe, Germany
{zuelch, gamber, stock}@ifab.uni-karlsruhe.de,
WWW home page: http://www.ifab.uni-karlsruhe.de

Abstract

Growing market demands on enterprises and the resulting challenges for their organization have been discussed for many years now. The flexibility and mutability of an enterprise are thereby considered as a significant factor for success. As a reaction to this, many enterprises have realigned their enterprise along the value-added chain. The implementation of flat hierarchies and process-oriented work organization make up focus of current discussions about organizational structures in the manufacturing area. New departmental structures, however, often require that decisions are delegated to operative positions, thus increasing the decision-making and action leeway of the operative employees. However, there are currently no methods capable of prospectively examining the suitability of such decentralized decision-making systems. In this context, this paper presents a method with which the suitability of decision-making systems can be examined in a prospective and quantitative manner. In order to attain this goal, the simulation procedure *OSim-Ent* developed at the ifab-Institute of Human and Industrial Engineering of the University of Karlsruhe (Germany), was expanded through generic elements, with which relevant decision-making system elements can be modeled and examined through simulation.

Keywords

production planning and control, decision-making system, personnel-oriented simulation

Please use the following format when citing this chapter:

Zülch, G., Gamber, T., Stock, P., 2007, in IFIP International Federation for Information Processing, Volume 246, Advances in Production Management Systems, eds. Olhager, J., Persson, F., (Boston: Springer), pp. 337-344.

1   New Demands on Manufacturing Systems

Enterprises in the manufacturing area normally pursue various, partially conflicting goals. One can distinguish between customer-related goals, such as the demand for short delivery times and a high degree of due date adherence, and enterprise-related goals, such as the endeavour to attain a high and even resource utilisation as well as a low capital tied-up, which should be realised to the best degree possible with the help of a suitable manufacturing control (e.g. see [1, p. 19 ff.]). However, in the past few years the enterprises' situation has been aggravated by changes of the market as well as by a rapid progress in technology. Beside cost and quality, time has taken on an increasingly important role, forcing enterprises to become ever more dynamic and versatile [2, pp. 155]. Conventional organisational structures have often an issue with the handling of such an agile environment [2, p. 156]. In current discussions about organizational configuration approaches in the manufacturing area, one can notice two trends of decentralization:

- New approaches for the organization of manufacturing systems have achieved growing attention since the beginning of the eighties from research and industry. The trend is moving to self-organising concepts, often referred to as Intelligent Manufacturing Systems (IMS). The core of self-organization is that a system achieves a stable state without interference from outside and especially without a central control [3]. For example, the concepts of Holonic Manufacturing System, the Fractal Factory or the Bionic Manufacturing work out this paradigm for the context of production management [2, p. 158; 4, pp. 13].
- The introduction of flat hierarchies into operations organization is seen as a fundamental factor for success (cf. e.g. [5, p. 71; 6, pp. 12]). Flat hierarchies, however, require a decentralized decision-making system, i.e. the required decisions have to be delegated to operative positions thus significantly increasing the decision-making and action leeway of the employees.

Currently, there is no method allowing for a prospective examination of the suitability of such decentralized decision-making systems (cf. [7, pp. 44]), in particular while considering the interdependency of operative decisions and efficient manufacturing operations. The question, as to whether the delegation of operative decisions to employees in the lower levels of the hierarchy is actually advantageous in a certain manufacturing system and dependent upon the given order program, has, therefore, yet to be answered. The quality of the decision-making system (in terms of the decision-making processes and decision makers' abilities) is thus dependent upon the knowledge and experience of the organizer.

In this context, this paper presents a methodology for the examination of the suitability of decision-making systems in a prospective and quantitative manner. In order to attain this goal, the simulation procedure *OSim-Ent* (German acronym for Object Simulator for Operative Decisions; see for *OSim* e.g. [8, 9]; for *OSim-Ent* [7]) developed at the ifab-Institute of Human and Industrial Engineering of the University of Karlsruhe, was expanded through generic elements, with which relevant decision-making system elements in manufacturing systems can be represented and examined through simulation. Thereby, the paper focuses on personel-centred decision-making systems. Nevertheless, the generic architecture of *OSim* would also allow for the implementation of algorithmic and self-organizing decision-making systems (e.g. see [10]).

2 Decisions and Decision-making Systems in Manufacturing

In order to make decisions, goals must first be defined. One can thereby differentiate between strategic and operative goals. In the following, merely operative decisions in the manufacturing area are taken into consideration (for these kinds of decision cf. e.g. [11, p. 68]). Furthermore, one can differentiate between global and local decisions (cf. [12]). A distinction between internal (e.g. "short lead times") and external goals (e.g. "high delivery reliability") can also be made [1, p. 20]. Finally, it cannot be assumed that decisions of the operative personnel will harmonize with the goals of the entire system (differentiation between personal and organizational goals [13, p. 24]).

The result of a decision is highly dependent upon the quality of the information, meaning its availability, its timeliness and its genuineness: The better the information, the better the foundation for an efficient decision. Operative decisions, in particular decisions made by manufacturing employees, are influenced by both short-term, invariant information as well as by the dynamic manufacturing situations. In the face of these system dynamics, it seems promising to analyze decision-making systems prospectively through simulation in order to help create suitable configuration solutions.

Finally, the decision-making strategy must also be specified, which the decision-maker applies in order to select the best action alternative regarding the relevant goal system. The decision-making strategy joins components of the decision-making process to build up the decision.

Each decision is thus characterized by four elements: goal system, information system, decision-making strategy and decision maker, referred to in the following by the composite term decision field [7, pp. 82]. The entirety of the decision-making tasks and their respective decision fields makes up the decision-making system for the manufacturing system.

3 Concept for Modeling Decision-making Systems

Commercial simulation procedures possess only a limited ability to represent human decision-making, and in particular to vary the entire decision-making system. In order to make up for this deficit and to provide a simulation-based methodology for the planning of decision-making systems, the feasible decisions and decision-making strategies occurring in the manufacturing area, as well as the underlying goals and information, were first classified. A generic concept allowing decisions and decision-making systems to be modeled and then be evaluated with respect to their productions logistical and monetary goal criteria was subsequently derived.

The activities in the manufacturing system are modeled by so-called activity networks which are directed graphs with a logical sequence of activities for the winding-up of orders [9, p. 373]. A modeled manufacturing order program can thus be seen as a collection of activity networks. Activity networks can be released by external or internal events such as customer orders or internal requirement orders. In this way, it is also possible to model indirect activities [8, p. 373].

The spectrum of decisions arising in the manufacturing area is very broad. One can thereby differentiate between two types of decisions, which can be illustrated through varying representations of the decision nodes (cf. Fig.1):

- First, there are decisions which occur during the processing of an order (decision type a in Fig. 1), e.g. decisions about alternative work sequences, the order sequence or order lot sizes. In such decision-making tasks the various action alternatives are in competition with one another. Action alternatives can be illustrated by modeling the decision-making task as a decision node with alternative paths.

- Conversely, there are also decisions in a manufacturing system, which are independent upon a certain order (decision type b in Fig. 1), e.g. decisions about a short-term capacity increase through extended personnel working times. This type of decision cannot be modeled using activity nodes for alternative work sequences. Therefore, these decisions are modeled as decision nodes within a separate activity network, which arises either according to a fixed schedule or cyclically.

The decision nodes are assigned to one or more decision-making fields whose elements are specified as follows (cf. Fig. 1):

- The goal system presents assignment of relevant objectives from a catalogue of goals to the decision-making task. Employees from operative execution levels often act more in line with local goals that do management employees, who are primarily oriented towards global organizational goals ([14]; cf. also [13]).

- In addition to the differing goal systems of the various hierarchical levels, one must also assume a varying supply of information to the employees. Employees in the manufacturing system may also possess varying degrees of information [15], which may be dependent upon e.g. their functions or qualifications. Moreover, it can generally be assumed that the decision-maker does not have complete information. However, in the first step of the model creation, the last assumption was ignored and it was assumed that decisions are made based on complete information.

- In addition to the goals and the information available regarding the state of the manufacturing system, multi-criterion assessments are necessary in order to aggregate various goals e.g. through the application of additive or lexicographical preference functions.

- The decision-maker is either a foreman or an operative employee of the manufacturing system or a member of the centralized work planning department.

If a decision-making task is initiated during the simulation run, an assigned decision-making field is called upon. The decision-maker then makes a decision based on his decision-making strategy, which falls back on the goal system and the information available for the decision-maker (cf. [12]). In the simplest case, this can be carried out based on a multi-criterion assessment of the action alternatives. This means that the decision-maker evaluates the information about each action alternative with respect to his own goal system (e.g. using a lexicographic preference function) and then chooses the alternative promising the most benefit according to his goal system.

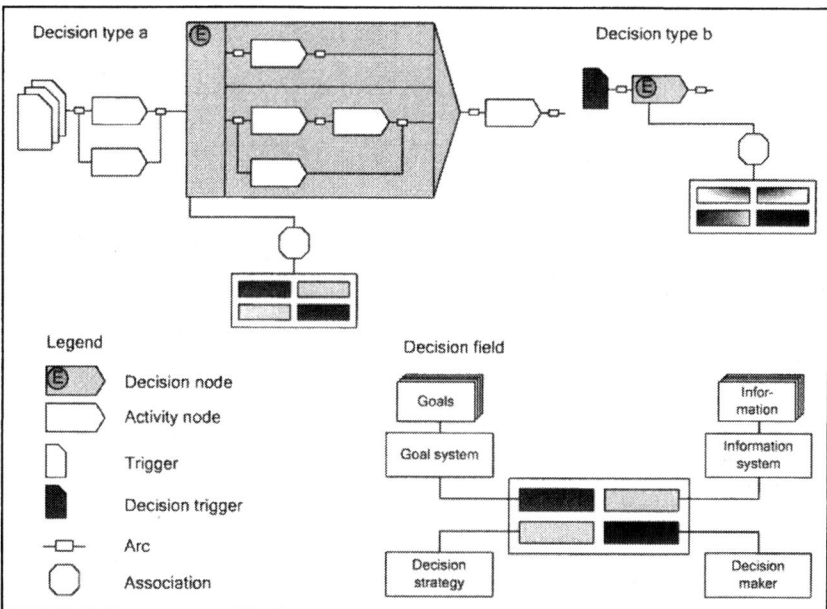

Fig. 1. Modeling decisions within the activity network (cf. [16, p. 102])

5   The Simulation Procedure *OSim-Ent*

The transfer of the modeling and assessment concept into the simulation procedure *OSim-Ent* (see Fig. 2) allows for a prospective and quantitative assessment of various decision-making systems. This enables the delegation of decisions to be planned more efficiently. Through simulation it is possible to assess decision-making systems and organizational structures (departmental and process organization) with respect to their efficiency and to derive configuration recommendations therefrom.

The modified simulation procedure *OSim-Ent* has been verified in test examinations. For this, several decision-making systems were modeled based on real manufacturing systems and then assessed:

- An ongoing simulation study investigates the manufacturing system of a mechanical parts manufacturer. In this parts manufacturing system, the modeled decisions concern the processing of jobs through a sequence of several single machines or a machining centre (i.e. decision type a; cf. Fig. 1). In the initial situation, the decision is made by chance whereas the probability of choice of each alternative is equal to 50 %. The performance of the manufacturing system and especially the lead time of the orders are poor. Therefore, several alternative decision-making systems are proposed and evaluated. The decision-making systems vary in the goal systems of the decision makers (e.g. different weighting factor for manufacturing costs and lead time) as well as in the decision-making strategy (e.g. additive or lexicographical preference function, maximin criterion). The first results are very promising. E.g. one decision-making system, which uses a lexicographical preference function, increased the goal achievement degree of the lead time by 4 % and the goal achievement degree of the manufacturing costs by 11 %.

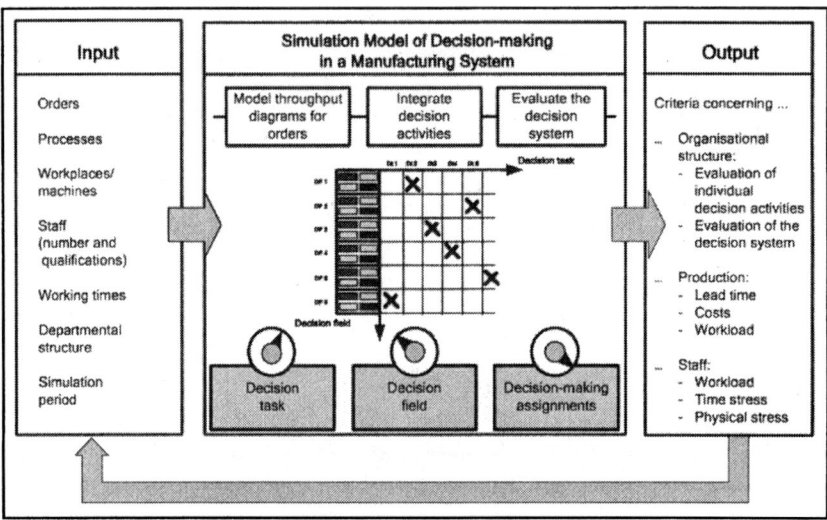

Fig. 2. The simulation procedure *OSim-Ent* (cf. [16, p. 105])

- Another project encompassed the improvement of disassembly and re-assembly processes for of electric devices. Therefore, a repair line, in which hammer drills in various weight classes are repaired, was chosen as the main object of investigation. The core problem was the broad variability of the temporal load to the system due to the heavy fluctuations in and the low predictability of the incoming orders. This had, on the one hand, a negative effect upon the service rate, while simultaneously creating disproportionately high idle-times. Therefore, the most important goal was to attain high service ability. In order to attain this goal, several strategies for the short-term employment of the staff were developed, e.g. "go home" or "help another employee". The foreman decided about the application of the strategies day-to-day, based upon information about the order waiting queues and the prognosis of the capacity requirements. In order to evaluate the strategies as well as the decision-making system, a simulation examination with *OSim-Ent* was conducted, using the decision type b for modeling (see Fig. 1). By this, very positive results could be attained, increasing the service degree by about 20-25 % and reducing the idle times by 44 % (please refer to [7, 12, 17] for details of this project).

6   Further Research Needs

The modeling and simulation of decision-making systems reveals as a great support for planners in the design of various decision-making systems. These can be assessed prospectively using the simulation procedure in order to identify improvement potentialities, and thus improve the organizational structure. The future aim is to build upon the insights gained in order to evaluate decisions made using elements of risk and uncertainty into the models.

The existing insights into the representation of human decision-making in manufacturing systems require further investigations. On the one hand, the quality of decision processes, with respect to their later effects, should be taken into account when modeling. Especially the efficiency of decision processes, i.e. their costs and duration, should be included into the decision process. On the other hand, the possibilities for creating an interface to algorithms and methods of artificial intelligence or for integrating learning decision-makers are also conceivable in order to improve the decision-making strategy, i.e. effectiveness of a decision.

7   Acknowledgments

The authors would like to thank the German Research Association (Deutsche Forschungsgemeinschaft - DFG) for its support in the research work referred to in this paper. These endeavours are currently supported within a project dealing with the development of a simulation-based method for the analysis of decision-making systems in the production area ("Entwicklung einer simulationsbasierten Methode zur Analyse von Entscheidungssystemen im Produktionsbereich").

8   References

1. H. Lödding, *Verfahren der Fertigungssteuerung* (Springer, Berlin et al., 2005).

2. J. Käschel and T. Teich, *Produktionswirtschaft. Band 1: Grundlagen, Produktionsplanung und –steuerung* (guc Verlag der Gesellschaft für Unternehmensrechnung und Controlling, Chemnitz, 2004).

3. H. Haken, *Synergetics* (Springer, Berlin et al., 1983).

4. A. Tharumarajah, From Fractals and Bionics to Holonics, in: Agent-based manufacturing: advances in the holonic approach, edited by S. M. Deen (Springer, Berlin et al., 2003), pp. 11-30.

5. S. Hauss and J. Brucke, Wettbewerbsfähig durch optimierte Organisationsstrukturen, *VDI-Zeitschrift,* 142(7/8), 71-73 (2000).

6. E. Frese and A. v. Werder, Organisation als strategischer Wettbewerbsfaktor, in: Organisationsstrategien zur Sicherung der Wettbewerbsfähigkeit, edited by E. Frese and W. Maly (Verlagsgruppe Handelsblatt, Düsseldorf, 1994), pp. 1-27.

7. J. Fischer, *Ein generisches Objektmodell zur Modellierung und Simulation operativer Entscheidungen in Produktionssystemen* (Shaker Verlag, Aachen, 2004).

8. U. Jonsson, *Ein integriertes Objektmodell zur durchlaufplanorientierten Simulation von Produktionssystemen* (Shaker Verlag, Aachen, 2000).

9. G. Zülch, J. Fischer, and U. Jonsson, An integrated object model for activity network based simulation, in: WSC'00 Proceedings of the 2000 Winter Simulation Conference, Volume 1, edited by J. A. Joines, R. R. Barton, K. Kang and P. A. Fishwick (2000), pp. 371-380.

10. G. Zülch and P. Stock, Self-organizing manufacturing control based on the ant colony approach, in: Modeling and Implementing the Integrated Enterprise – Proceedings of the International Conference on Advances in Production Management Systems, edited by the IFIP WG 5.7 (CD-ROM, 2005).

11. A. Hax and G. Bitran, Hierarchical planning systems - a production application, in: Disaggregation: Problems in Manufacturing and Service Organizations, edited by L. P. Ritzman et al. (Martinus Nijhoff Publishing, Boston et al., 1979), pp. 63-93.

12. G. Zülch, Modelling and simulation of human decision-making in manufacturing system, in: Proceedings of the 2006 Winter Simulations Conference, edited by L. F. Perrone et al. (The Institute of Electrical and Electronics Engineers et al., Piscataway, NJ, 2006), pp. 947-953.

13. A. Klinger, *Referenzmodelle für die Abbildung von Personalsteuerung in der Simulation* (Society for Computer Simulation International, San Diego et al., 1999).

14. G. Reinhart and W. E. Lulay, Koordination dezentraler Produktionsstrukturen durch begleitende Simulation, *Zeitschrift für wirtschaftlichen Fabrikbetrieb* 93(1-2), 35-38 (1998).

15. J. Krüger, *Entscheidungstheorie-basierte Simulation der Handlungsorganisation im Fertigungsbereich* (Shaker Verlag, Aachen, 1999).

16. G. Zülch, T. Gamber and P. Stock, Analyse von Entscheidungssystemen im Fertigungsbereich mittels Simulation, in: Simulation und Visualisierung 2007, edited by T. Schulze, B. Preim and H. Schumann (SCS Publishing House, Erlangen, 2007), pp. 97-109.

17. G. Zülch and J. Fischer, Increasing the Flexibility of Working Times and Personnel Control in an Industrial Repair Work System (May 30, 2007); http://www.simserv.com/white_papers.php.

PART V

Improving Operations

Lean Practices for Product and Process Improvement: Involvement and Knowledge Capture

Jannis J. Angelis[1] and Bruno Fernandes[2]

1 Warwick Business School, University of Warwick, Coventry CV4 7AL,
UK, jannis.angelis@wbs.ac.uk

2 Centro Universitario Positivo, Curitiba 81280 Brazil, bruno@unicenp.br

Abstract

Innovation is key source of a company's competitiveness in the knowledge economy, and continuous improvement is a key element of such corporate pursuit. Lean production is a globally competitive standard for product assembly of discreet parts. Successful Lean application is conditioned by an evolutionary problem-solving ability of the rank and file. Such ability is in itself contingent on employee involvement in improvement programs and the implementation of appropriate practices. But the challenge of operating innovative Lean systems lacks statistically valid guidance. This empirical study is based on 294 worker responses from twelve manufacturing sites in four Brazilian industry sectors. It identifies particular practices that impact employee participation in change or improvement activities and their performance outcomes.

Keywords

Lean Operations, continuous improvement, worker involvement

1   Introduction

As described by Lewis Carroll in *Through the Looking Glass*, today's businesses can be described as operating in a Red Queen economy, where it takes all the running you can do, to keep in the same place. Innovation is key source of competitiveness in the knowledge economy [1]. Continuous improvement (CI) is a core element of such corporate pursuit [2, 3] and can be summarized as a company-wide focus to improve process performance [4]; using gradual step by step improvement [5, 6]) and organizational activities with the involvement of all people in the company [4, 7], while creating a learning and growing environment [8]. Lillrank and Kano [9] refer

Please use the following format when citing this chapter:

Angelis, J. J., Fernandes, B., 2007, in IFIP International Federation for Information Processing, Volume 246, Advances in Production Management Systems, eds. Olhager, J., Persson, F., (Boston: Springer), pp. 347-354.

to CI, or kaizen as the "principle of improvement". CI programs were initially developed in organizations with product-focused processes or repetitive processes [10]. Special teams were organized to work on improvement tasks, which were separate from their typical organizational tasks. As such, through their commitment and involvement employees become a source of sustainable competitive advantage. CI activities are a core philosophy in Lean operations as a means of improving product quality and reducing waste throughout the operations. [11]. Lean, if applied correctly, results in the ability of an organization to learn. Lean operates with balanced, synchronised material flow and minimum use of 'wasteful' contingencies of material, people and machinery. While there is literature on workforce effects of Lean [12-15], this study focuses on involvement. Lauder [16] links the 'high performance work organisation' with Lean, whereby a 'degree of power is devolved to teams to engage in a constant process of innovation and improvement'. Womack *et al.* [17] advocate that involved workers are necessary for the expanded roles effective CI program require, and that the worker involvement is enhanced by job enrichment such as improvement projects, self-inspecting tasks, and conducting routine maintenance. Contrary, Delbridge and Turnbull [18] and Bruno and Jordan [19] argue that Lean implementation may diminish worker commitment through fast-paced, high intensity operations with close monitoring, de-skilling and little job autonomy.

The Brazilian industrial complex has been built up over the last century through a combination of private, state-owned and multinational firms, and through joint ventures between them with an industrialisation pattern similar to Mexico, Argentina and Chile [20-23], Hungary and Bulgaria [24], India and China [25]. Moreover, Brazil has been considered a training "lab" for uncertainty and unfavourable conditions, where quality and productivity issues are being considered from the perspective of a globalised economy as a way for reducing costs and cycle times as well as improving sales and profits [26, 27]. Lean has been implemented across Brazil with little unions or worker resistance [28, 29]. While efforts have been made to implement a local Lean version, a Brazilian system has not emerged [30].

2   Method

Conti and Gill [31] developed the initial hypotheses by examining expected outcomes for a variety of Lean practices. The independent variables were twenty-one lean work practices and thirteen control variables recorded on five-point Likert scales. Several authors [32-34] argue that product and process innovations need to be separate in studies of CI, since they not necessarily share determinants. Hence, the dependant variables consist of worker suggestions to improve existing products and processes. Avoiding an object approach (i.e. innovation count) also reduces the favour of radical innovations over incremental ones and product over process innovations [35]. The independent lean implementation variable was measured using ten key elements: set-up reduction, Inventory and waste reduction, kanbans, supplier partnerships, continuous improvement program, mixed-model production, total quality management, Foolproof or design-for-assembly, total preventive maintenance, and standard operating procedures. Scale reliability, measured by the Cronbach alpha, was .816. Their levels were estimated on a five-point scale in the management questionnaire, using categories of Powell [36]. Plant tours and interviews helped to verify management responses. The sampling plan of Cook and Campbell [37] was used to recruit sites differing in work practices. Sample space is the population of Brazilian sites with *60* or more assemblers. All 12 sites are in four US SICs: three in 35 (machinery), three in 36 (appliances and electronics), three in 37 (motor vehicles) and three in 38 (instruments), similar to the distribution reported by Fullerton *et al.* [38]. Sites are a mix of union and non-union workplaces. All assemblers received instructions and were given questionnaires in stamped envelopes for anonymous posting. 294 questionnaire responses were obtained out of 840, generating a response rate of 35%.

3   Analysis of work practices and improvement

Continuous improvement through worker participation is a core lean principle. The study results reveal that there is a significant correlation between lean implementation and product and process suggestions (r=520, p<.001), as well as their implementation (respectively r=582, p<.001 and r=418, p<.001). Analysis of Variance was used to check the means and 95% confidence intervals for the five levels of affective commitment responses. Multiple regression, using redefined variables, identified relationships significant at .05 or less. The work practice hypotheses were tested using stepwise regression, with product and process suggestions as response variables. The former model F=7.281, p<.001, adjusted R square .041, and the latter is F=5.911, p<.001, adjusted R .063. There is no evidence of collinearity, with VIF values well below the usual cut-off of 10.

Description, beta coefficient and significance level is given for each work practice with an association as measured by the standardized coefficients significant at .05 or less. As for control variables, there were no significant relationships for age or years of employment at the site, or for perceived job security. It appears that demographic and life-style factors do not materially affect the study results.

Lack of tools The relationship between process suggestions and the lack of proper tools is significant and positive. (Beta=.132, p=.026). The lack of appropriate tools indicates inadequate technical support, and can also lead to quality problems, and raises managerial competence issues .

Work pace and intensity The significant and positive relationship between pace and intensity and worker suggestions on process improvements (Beta=.162, p=.005) fits the lean notion of employing resource removal as a change catalyst [39]. But a high pace and intensity may also be perceived as 'unfair' and hence erode worker commitment.

Flow interruptions The relationship between process suggestions and flow interruptions is significant and positive. (Beta=.129, p=.031). Flow interruptions have mixed worker effects. The broken repetition may relieve task monotony, but may also interrupt the steady rhythm of job task that many workers value [17]. Nonetheless, flow interruptions are beneficial in that they may provide workers with more time to think about existing processes, or that workers' dislike for interruptions galvanize them into making improvement suggestions.

Team work The relationship between process suggestions and the utilisation of team work is significant and positive. (Beta=.117, p=.041). Support from peers or supervisors is not significant. Team work allows for worker job task expansion and supports peer support for time and quality standards, and it also indicates management confidence in workers' ability to multi-task. Nonetheless, given some workers preference for working alone, it may be prudent to offer alternative choices.

Feeling of being blamed for defects There is a negative association between product suggestions and the feeling of being blamed for defects. (Beta=-.171, p=.003). Lean pinpoints specific defect locations which may make individual workers feel they are being blamed. Moreover, blame feelings persist long after actual defect episodes, perhaps due to lingering apprehension about future defects. For successful employee involvement, workers must be given the opportunity and responsibility for organizational change and improvement, but they must also be motivated to avail themselves of this opportunity and responsibility.

Working longer hours than desired There is a positive association between product suggestions and working longer hours than desired. (Beta=.153, p=.008). It may indicate the undesired overtime intrusion into private life that hampers worker commitment, and in turn involvement in improvement schemes.

Martinez-Ros [40] found that product and process innovations are interdependent. Neglecting process innovations can weaken firm capacity to develop new products, and undermine the innovation process entirely. The research results indicate the prevalence of three conditions. First, working conditions perceived to be harsh or difficult appear to motive workers to make (process) suggestions, as seen by the significant variables of lack of appropriate tools, high pace and intensity, and flow interruptions. The study results shows that worker commitment is significant for suggestions made on products (Beta=.171, p=.003) but insignificant for process suggestions. This make sense, since a committed worker may have an interest in improving the product, while a non-committed worker may primarily be motivated to improve his or her immediate working conditions. Second, supportive work practices in the form of mutually beneficial human resource practices and industrial relations, such as team work, a flexible work schedule, and the absence of a blame culture strengthen worker improvement involvement. Third, poorly designed or implemented processes may make them easy improvement targets for the workers – an improvement form of low hanging fruit.

As Midgley [41] points out, there is limited advantage in developing worker commitment and involvement if there then is no commitment on the part of the management to provide the environment in which the workers' involvement can be applied. The relationship between workers making improvement suggestions on existing products or processes and worker participation in formal improvement schemes is not significant. Hence, formal participation in improvement exercises is not necessarily a good indication of worker involvement. Similarly, on a firm level, there appears to be weak links between a formal innovation strategy and actual worker involvement on the shop floor. This gap indicates poor management by-in to improvement through worker involvement [41].

Ergonomics The relationship between ergonomics and implemented process suggestions (Beta=-.218, p<.001) and product suggestions (Beta=-.163, p=.006) is significant and negative. Positioning hard to handle items shows lack of technical support. Indirectly poor ergonomics restricts access to physically demanding jobs, and in turn fail to capture all potential innovators [42].

Comments on change The relationship between implemented process suggestions and comments on change is significant and positive. (Beta=.169, p=.003). Similarly, the relationship between implemented product suggestions and comments on change is significant and positive. (Beta=.132, p=.020). Such capture of suggestions indicates good management practice.

Pace control The relationship between implemented process suggestions and pace control is significant and positive. (Beta=.161, p=.004).

Blame for defects There is a negative association between implemented product suggestions and the feeling of being blamed for defects. (Beta=-.158, p=.006).

Job rotation The relationship between implemented product suggestions and job rotation is significant and positive. (Beta=.163, p=.005). Task expansion, as well as greater co-worker interaction, may lead to better suggestions through improved shared understanding.

Change autonomy The relationship between implemented product suggestions and worker change autonomy is significant and positive. (Beta=.131, p=.021). Danford [43] notes that job autonomy, rather than team working, can have a positive impact on workers' sense of trust, commitment and satisfaction. But to reduce the likelihood of errors induced due to human error probability, enhancement of worker autonomy must at the same time limited discretion [44]. Complexity can be minimised through product and assembly design, while variability can be minimised through poka-yoke systems and non-discretionary tasks.

The results indicate that managers should pursue employee involvement rather than intensification approach [45]. Berggren [13] calls this a team rather than JIT-driven lean approach, achieving gains through high employee commitment rather than through cost reduction and work intensification.

There is a strong and positive correlation between productivity and quality (r=.947, p<0.001) and delivery (r=.993 p<0.001), and between delivery and quality (.996 p<0.001). Following the Sandcone model, operational advantage is based on high product quality. There is a positive and significant association between product improvement suggestions and product quality (r=.187, p=.013) and there is a positive and significant association between process improvement suggestions and speed of delivery (r=.210, p=.005). But there are no other significant relationships between product and process improvement suggestions and realised improvements in quality, productivity or delivery. Perhaps unsurprising, speedy introduction of new products appears to have a negative effect on improvement suggestions from the workforce (r=-.176, p=.020 and r=-.201, p=.008 for product and process suggestions respectively). Similarly, the relationship to implemented process improvements is significant and negative (r=-.270, P<.001). There is a significant and positive relationship between the implementation of suggestions on process improvements and manufacturing unit cost (r=.152, p=.044), ability to change product mix (r=.207, p=.006) and speed of delivery (r=.350, p<.001). Appelbaum *et al.* [46] and Pil and MacDuffie [47] state that high involvement practices employ workers in improvements activities to improve quality and not to achieve cost reductions. The results show that worker suggestions on process improvements mainly have an impact on product quality ($r=.157$, $p=.008$), which indicates that the involvement aspect on Lean may fit into the same category. This shows both the importance of tracking implementation of suggestions and the use of appropriate product design and process design other than on the shop floor. For instance, product improvements can be pursued through dedicated design teams while workers on the shop floor focus their suggestions to error proofing activities. This has the benefit of reducing the probability of human error and reducing discretion while at the same time retaining a degree of job autonomy and capturing employee skills and knowledge.

4   Conclusion

The results show that there is a significant correlation between lean implementation and product and process suggestions and their implementation, even when controlling for firm size or age, unionized workforce and compensation systems. The statistical analysis of particular work practices and their relationship with product and process suggestions made by individual workers or teams reveals that process suggestions are driven by a combination of difficult working conditions that the workers seek to improve and team-based work. However, for suggestions on product improvements significant practices are worker favourable industrial relations and human resource practices. In terms of implementation of suggestions, both product and process suggestions are significantly and positively correlated with management capturing ideas voiced by the workers, worker discretion in pace and task, and job rotation. To control for human error probability and ensure product quality and consistency, a degree of job autonomy may be needed but adverse effects of job discretion on product quality need to be built out through poka-yoke fool-proofing designs. The results also indicate that the main direct business benefit is in enhanced product quality through process, rather than product, improvements. This suggests that management should pursue worker involvement on continuous process improvements, and employ designated design teams for product improvements. For the unfavourable practices, actions should minimise their negative effects. First, if possible, overtime should be voluntary, aided by cross-training of workers to expand the pool of volunteers. Second, task time standards should be set with pace and intensity set at 'normal' levels as defined by industrial practice. Third, process designs should emphasise eliminating ergonomic difficulties, providing adequate tools and minimising flow interruptions. Finally, supervisory training and disciplinary policies must emphasise "blame free" defect investigations.

5   References

1 G. Hulta, G., Hurleyb, R. and Knight, R., Innovativeness, *Industrial Marketing Management*, 33, 429–438 (2004).
2 R. Cooper, From Experience: the invisible success factors in product innovation, *Journal of Product Innovation Management*", 16, 115-133 (1999).

3 C. Wu and C. Chen, An integrated structural model towards successful continuous improvement activity, *Technovation*, 20, 1-11 (2005).

4. M. Imai, *Kaizen: the Key to Japan's Competitive Success* (Random House, NY, 1986).

5. C. Berling, Continuous improvement as seen from groups and improvement agents, *Total Quality Management*, 11(4-6), 484-489 (2000).

6. A. Brunet and S. New, Kaizen in Japan, *International Journal of Operations and Production Management*, 23 (12), 1426-1446 (2003).

7. J. Bessant and S. Caffyn, High involvement innovation through continuous improvement, *International Journal of Technology Management*, 14(1), 7–28 (1997)

8. R. Delbridge and H. Barton, Organizing for continuous improvement, *International Journal of Operations and Production Management*, 22(6), 680–692 (2002).

9. P. Lillrank and N. Kano, *Continuous Improvement*, Center for Japanese Studies, University of Michigan, Ann Arbor, MI. (1989).

10. N. Bhuiyan and A. Baghel, An overview of continuous improvement, *Management Decision*, 43(5), 761-771 (2005).

11. C. Forza, Work organization in lean production and traditional plants, *International Journal of Operations and Production Management*, 16(2), 42-62 (1996).

12 S. Kamata, *Japan in the Passing Lane* (Allen and Unwin, London,1983).

13. C. Berggren, Lean production, *Work, Employment and Society*, 7, 163-188 (1993).

14. R. Conti, J. J. Angelis, C. Cooper, B. Faragher and C .Gill, Lean production implementation and worker job stress, *International Journal of Operations and Production Management*, 26(9), 1013-1038 (2006).

15. J J. Angelis, R. Conti, C. Cooper and C. Gill, Building a high commitment lean production culture, *Working Paper,* Institute for Manufacturing, University of Cambridge (2006).

16. H. Lauder, Innovation, skill diffusion, and social exclusion in P. Brown, A. Green and H. Lauder (eds.) *High Skills* (Oxford University Press, Oxford,2001), pp. 161-203.

17. J. Womack, D. Jones and D. Roos, D. *The machine that changed the world* (New York: Rawson Associates, 1990).

18. R. Delbridge and P. Turnbull, Human resource maximisation, in P. Blyton and P. Turnbull (eds), *Reassessing Human Resource Management* (Sage, Newbury Park,1992).

19. R. Bruno and L. Jordan, Lean production and the discourse of dissent, *Working USA*, 6(1), 108-134 (2002).

20. M. Bertin, A view of quality trends in South America in the twenty-first century, *The TQM Magazine*, 11(6), 409-413 (1999).

21. A. Fleury, The changing pattern of operations management in developing countries, *International Journal of Operations and Production Management*, 19(5/6), 552-564 (1999).

22. B. Flores, F. Burgos and A. Marcias, Manufacturing practices in Mexico, in D. Whybark and G. Vastag (Eds), *Global Manufacturing Practices*, (Elsevier, Amsterdam, 1993).

23. A. Kovacevic, C. Lopez and C. Whybark, Manufacturing practices in Chile, in D. Whybark and G.Vastag (Eds), *Global Manufacturing Practices* (Elsevier, Amsterdam, 1993).

24. A. Chikan and K. Demeter, Manufacturing strategies in Hungarian industry, International Journal of Operations and Production Management, 15(11), 5-19 (1995).

25. A. Fleury and M. Fleury, Competitive strategies and core competencies, Integrated Manufacturing Systems, 14(1), 16-25 (2003).

26. L. Carpinetti, F. Santos and M. Gonçalves, Human resources and total quality management, *The TQM Magazine*, 10(2), 109–114 (1998).

27. C. Lemos, Innovation and industrial policies for small firms in Brazil, *4th International Conference on Technology Policy and Innovation*, Curitiba, 28-31 August (2000).

28. J. Humphrey, Japanese production management and labour relations in Brazil, *Journal of Development Studies*, 30(1), 92-114 (1993).

29. T. Wallace, Innovation and hybridization, *International Journal of Operations and Production Management*, 24(8), 801-819 (2004).

30. S. Pires, New productive systems in the auto industry, *GERPISA 11th colloquium*, Palais de Luxembourg, Paris, June 7-9 (2001).

31. R. Conti and C. Gill, Hypothesis creation and modelling in studies, *International Journal of Employment Studies,* 6(1), 149-173 (1998).

32. N. Becheikh, R. Landry and N. Amara, Lessons from innovation empirical studies in the manufacturing sector, *Technovation*, 10, 1–21 (2005).

33. J. Linder, S. Jarvenpaa, and T. Davenport, Toward an innovation sourcing strategy, *MIT Sloan Management Review*, summer, 43–49 (2003).

34. J. Michie, and M. Sheehan, Labour market deregulation, 'flexibility' and innovation, *Cambridge Journal of Economics*, 27(1), 123–143 (2003).

35. M. Flor and M. Oltra, Identification of innovating firms through technological innovation indicators, *Research Policy*, 33, 323–336 (2004).

36. T. Powell, Total quality management as competitive advantage, *Strategic Management Journal*, 16, 15-37 (1995).

37. C. Cook and R. Campbell, *Quasi-Experimentation* (Houghton and Mifflin, Boston, 1979).

38. R. Fullerton, C. McWatters and C. Fawson, An examination of the relationships between JIT and financial performance, *Journal of Operations Management*, 21, 383-404(2003).

39. D. Buchanan, Cellular manufacturing and the role of teams, in J. Storey (ed), *New Wave Manufacturing Strategy* (Paul Chapman, London, 1994).

40. E. Martinez-Ros, Explaining the decisions to carry out product and process innovations, *The Journal of High Technology Management Research*, 10(2), 223–242 (1999).

41. D. Midgley, How can Australia improve?, *Enterprising Nation*, Canberra, AGPS (1995).

42. P. Adler, B. Goldoftas and D. Levine, Ergonomics, employee involvement, and the Toyota production System, *Industrial & Labor Relations Review*, 50(3), 416-435 (1997).

43 A. Danford, Workers, unions and the high performance workplace, Work, Employment and Society, 17(3), 569-573 (2003).

44. R. Conti and M. Warner, Technology, culture and craft, *New Technology, Work and Employment*, 12(2), 123-135 (1997).

45. H. Shaiken, S. Lopez and I. Mankita, Two routes to team production: Saturn and Chrysler compared, Industrial Relations, 36, 17–46 (1997).

46. E. Appelbaum, T. Bailey, P. Berg and A. Kalleberg, Manufacturing Advantage (Cornell University Press, Ithaca, 2000)

47. F. Pil and J. MacDuffie, The adoption of high-involvement work practices, Industrial *Relations* 35(3), 423-455(1996).

An Integrated Methodology of Manufacturing Business Improvement Strategies

S Berkhauer-Smith[1] and R Bhatti[1]

1 The School of Engineering, The University of Greenwich, Pembroke Building, Central Avenue, Chatham Maritime, Chatham, Kent. ME4 4TB. U.K. bsl2@gre.ac.uk. 00 44 (0)1634 883473.

Abstract

Business Environments need to react more efficiently and quickly to demand in today's global markets. Previous research by the authors entailed reviewing the current understanding of the different types of generic business improvements that are applicable to the production environment through a literature review. It highlighted that widely used methodologies have similarly links and differentiating characteristics, thus producing many types of implementation strategies. The research resulted in studying the inter-relationships between these Manufacturing Business Improvement Concepts including Cultural Issues surrounding process improvement initiatives, so they can be unified into an integrated Methodology creating a unique strategy that can be correctly tailored to a chosen environment. This paper outlines a methodology that involves ten stages of change including the planning, creating, data collection, analysis and strategic implementation to apply. The approach flows through the change process systematically highlighting how to achieve the best outcome and feedback into the system is also visible.

Keywords

Process Improvement, Improving Operations, Integrated Strategy Application, Improvement Strategy Methodology, Manufacturing Operations.

1 Introduction

Research was initiated [2] by understanding currently practised Manufacturing Business Improvement Strategies centred around Lean Manufacture. This work studied the associations of these methods and the results highlighted that there are many techniques being applied and investigated at many different levels. From this

Please use the following format when citing this chapter:

Berkhauer-Smith, S., Bhatti, R., 2007, in IFIP International Federation for Information Processing, Volume 246, Advances in Production Management Systems, eds. Olhager, J., Persson, F., (Boston: Springer), pp. 355-362.

there was a requirement to understand the cultural aspects surrounding these Manufacturing Business Improvements to eliminate the theory of "85% of change projects fail" [6]. These writers also stated that;

.....''That many such initiatives failed because they had concentrated on the purpose and ignores the people who made them work." [6]

This further research was undertaken and the results confirmed the need for an integrated methodology of both technical and cultural strategies that highlighted a "specific development path to follow" [3]. This research work was taken further and a comprehensive methodology was created to meet these requirements found from the previous work of the authors [4]. The hypothesis from the literature review created two main objectives to be tested within the methodology. This methodology satisfies the following two main objectives;

1) An Integrated Business Improvement system that analysed many improvement concepts.
2) Implementing this theoretical design through analysing and evolving the cultural aspects.

The main significant advantaged gained from the methodology is having a specific development path to follow.

2   Methodology

A methodology was created and consists of two parts; the Design Phase, which creates the project, collecting the information and analysing it. This information will be used to create the second part; the Implementation Phase. This element shall understand the strategy and the implementation of the theoretical design. The methodology will be used in case studies to secure a higher success rate of implementation and the continuation of improvement.   The diagram, **Fig. 1** highlights the stages of change; the planning, creating, data collection, analysis and strategic implementation to apply.   The diagram also shows the process flow and how at some point the stages work in parallel to achieve the best outcome.   The feedback into the system is also visible; it highlights how at every stage it is very important to feedback the information gained. This will ensure the project definition remains true to the project brief.

The stages created in **Fig. 1** have been developed further to create a number of guidelines to enable greater use in cases to test the hypothesis. Each stage is designed so that there are a number of guidelines to follow and then within these guidelines there are suggestions on how to achieve these requirements.   The methodology is colour coded to indicate different parts of the methodology; there is a distinction between those aspects that are planning, data collection and implementation, benchmarking therefore making statements and documenting and finalising those areas that require feedback loops.

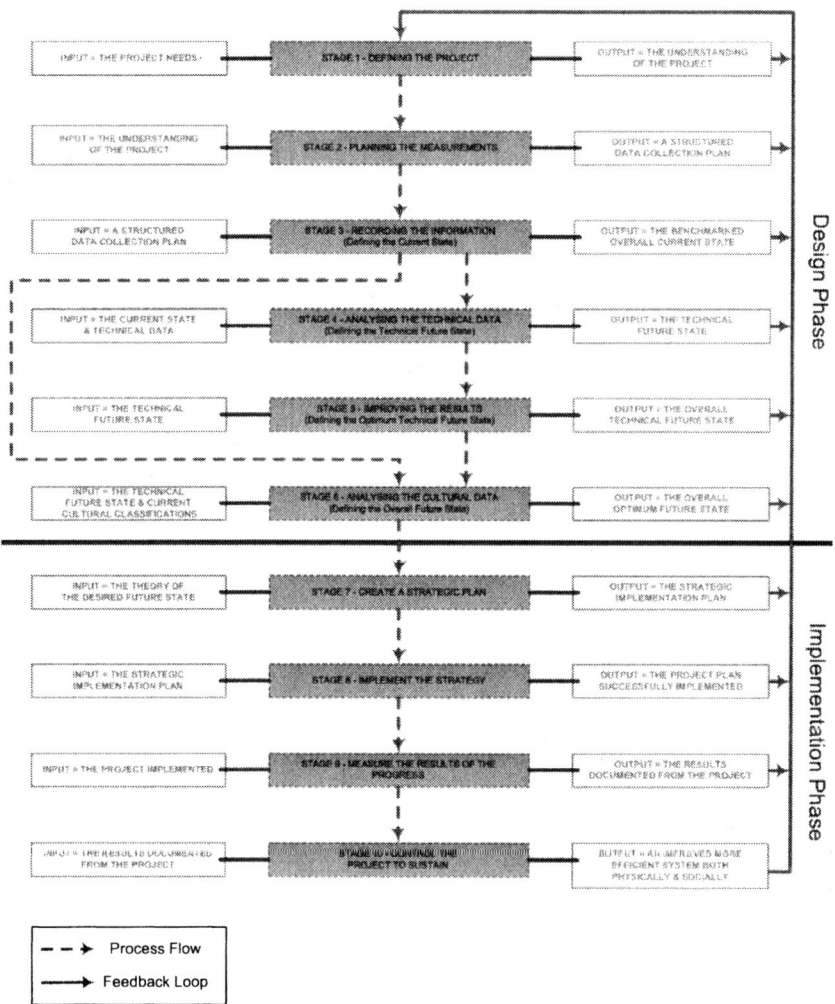

Fig. 1. The Process Flow of the Integrated Business Improvement Methodology.

2.1   Design Phase

This phase was created to guide the methodology through the different stages of the design of the project.   It was also derived to highlight the importance of the integration of the current Business Improvement ideas and how different scenarios require different aspects of these concepts to be used.   So it is not necessary to assume that one technique is applicable to all and so the art of this methodology is to have the ability to select what is suitable.

The First Stage of the methodology involves taking the project needs and defining them. Objectives as to why the project is needed are clarified and this will in turn define the strategy importance of the project within the business. It is also necessary at this stage to understand the project strategy; this would entail investigating the options and path ways of the project, to appreciate the future market share, as well as the desired outcome of the improvement process. Other areas that are required at this stage are the scope of the project, as well as the project due date and the deliverables. The outcome from this stage is the knowledge of the current flow of the business project. From this the measurements (Stage Two) that are required for the project can then be planned. A feedback loop into the system is essential to ensure that the project remains true to the brief. It also ensures that the stakeholders are happy with the monitored progress.

The Second Stage of the improvement process is to plan the measurements that are required. As the project is defined and the scope of the project analysed, a plan of the data to be measured should be calculated. It should also involve how these measurements will be collected, by whom and in what format, so that the results can easily be analysed.

This stage is the most significant part of the methodology and this highlights the originality of the research. An important factor of this stage is the integration of the Operational Business Improvements used in many businesses, mainly manufacturing. All of the key business improvement concepts are evaluated for their suitability to the project. This methodology incorporates a checklist facility that enables the user to select the specifics required, through a weighted method, therefore directing the pathway of the development. It is expected that in most cases not all techniques will be applicable; however it is thought that it will be necessary at some point in the transformation process.

By using this method of analysis the user will gain a better understanding of the project requirements as well as the improvement techniques. The integrated system will uniquely prioritise the amalgamated changes suggested, therefore creating a direction of physical change. From this stage a detailed data collection plan can be created. The research hypothesis is tested to understand how to fully integrate these concepts to absorb the better qualities of each technique, to create a fully improved system specific to the project. The Third Stage requires all of the information planned to be collected through the structured data collection plan that was derived within Stage Two of the methodology. The outcome from this Third Stage will be the overall current state of the project.

It is here in the improvement process that the methodology becomes divided; the cultural issues concerned with improvements do not exist without the technical changes forming. In this approach the technical advances will be documented theoretically and then analysed. Thus, allowing these technical aspects to be involved when deriving the cultural future state, this also allows greater change of successful implementation. So after technical issues have been assessed, these ideas influence the cultural changes required and so both features have then fused together to create a unified methodology, that will be introduced into the old system together.

The Fourth Stage uses only the technical data for analyse, from this benchmarked current state the information is analysed by using the Improvement Strategies. This transforms the model into the Technical Future State. In turn these technical changes should be documented for investigation, by highlighting the significant technical changes. With this benchmarked technical future state vision, the methodology creates questions to be answered about the reliability and achievability of the results produced from the technical theory. The approach that the methodology uses to accomplish this is through computer simulation modelling; this then allows the results of the system to be tested on a model before making these changes in the real system. Once more these results shall produce further questions that may have to be considered by feeding back into the system through the feedback loops.

This model can then be used to simulate other aspects that are not studied in the previous theoretical techniques, thus improving and justifying the results of the change process to make more realistic and achievable. Again it will be necessary to investigate the project to ensure that it is making the initial requirements that were set and that all people involved are satisfied with the progress the project is taking. From this stage, Stage Five, it is felt that the Overall Technical Future State should be defined.

Stage Six is the last part of the Design Phase and involves using mainly the unprocessed culture data recorded in Stage Three. This information is directly affected on the outcomes created from Stages Four and Five, therefore all of the data collected, processed and analysed up to this point in the methodology shall be used. This Sixth stage in the methodology analyses the Current State Culture, by taking the current cultural classifications and evaluating how change and business improvements affects the business and the success of the project. This covers a number of different aspects through investigating the thoughts and opinions of those involved and how they think these changes will affect the business. The methodology explores the different ideas and the possible multiple approaches instead of pursuing a single approach. This stage understands and investigates the use of employee relations as well as empowerment and company ethics. This different approach shall again test the hypothesis to establish how to actually take the technical processed theoretical data and use that information to also consider and investigate the cultural aspect. It is this feature that is very often forgotten when transforming a system through a change process. The Sixth Stage is also very significant as it highlights the originality of the amalgamation of both the technical and cultural transformations, which are two very different aspects, but highly important factors of change.

At this point in the methodology the ideal Future State Culture shall be documented. It is also important to remember the significant differences between the 'Creating Future State' stages. These ideas are then ready to be implemented as the Optimum Future State System into the transforming current situation.

2.2   Implementation Phase

The Implementation Phase of the methodology is mostly about Management Techniques. The hypothesis of this research is not only to understand how to integrate the many business improvement techniques available to transform a system, but to test the ability to successful implement this integrated approach. This stage shall create a strategic plan of implementation from the Design Phase, by integrating a number of strategic management policies by similar scientific weighted values as that in Stage Two.

Stage Seven being the first module of the Implementation Phase starts by understanding the project definition defined within Stage One. An important factor of the methodology is that those affected by the changes are involved at every stage of the plan. Internal stakeholders have an input into the creation of the Strategic Plan by discussing the success of the Design Phase, whether the Design Phase has meet the needs of the company and whether further adjustments can be made. The next step is to assess the financial situation of the change and the relationships of the customers and the suppliers are also investigated to ensure successful implementation of the methodology. Goals and targets are set to guarantee that all aspects of the design are met, creating loops as inspection process to verify that the implementation is in line with the design. Most importantly time constraints are added to these objectives to ensure the sustainability.

Once the Strategic Plan has been documented it is analysed on topics such as costs, profits, durability, work breakdown structures and organisational work breakdown structures. The change agents are involved in the plan, the transparency and reliability of the project is also considered to ensure that the improvements are generic for other improving areas, creating a strategic plan.

Stage Eight of the methodology is to "Implement the Strategy". This module concentrates more on the monitoring of the cultural environment, by unfreezing the current culture this highlights the need for change and how this will improve the working environment

Stage Nine is seen as a revision section, it is felt that this stage will analyse the results from the implementation, all previous documented stages in the methodology are referred to and compared to the final documentation gathered at this stage. Defects can be assessed and improved upon for future projects.

The final stage, Stage Ten is considered to be one of the most important as this determines the success of the project. This stage is about controlling the project to sustain the improvements made. The first area to focus upon is understanding how to control the physical system, ensuring that the designed physical system remains within it limits and constraints. However, the system is also required to be reactive to the customer demand and the method needs to be cautiously balanced. Variances in this system need to be understood and so the actual to required measurements of the improved system need to be also be monitored and adjusted accordingly.

Not only does the physical system need to be controlled but cultural system needs to also carefully be observed. This is carried out by creating objectives to control the cultural environment, variance again will have to be measured and reduced. Measuring variance is only useful if the information is fed back into the system, by feeding this back into the system, the improvements controlled and measured will be sustained. The entire project needs to be periodically restarted as a form of continuous improvement and to emphasize sustainability and control.

3   Case Studies

The methodology has been initiated within industrial environments', the flexibility of the methodology to differing but linked businesses has been tested and the outcomes are being finalised. The methodology has been applied within a supplying manufacturing environment and has been used to improve a process level production line. This case study has been used to test the methodology; the results of this preliminary implementation are limited due to the progression through the Design Methodology. This has created the project definition and an initial understanding of the current state of the production system. The data required has also been clarified, thus allowing a clear perception of what improvement techniques are applicable to the specific system. Data has been collected and the results show that there are issues related to inventory, delay periods and product transportation. It was decided to collect the data over a long period of time to ensure better accuracy of the system, through variation on product range, variety, volume and annual cyclic periods. It is often found that with Business Improvement techniques the data collection timeframe is a substantial proportion of the whole overall methodology implementation. This case study is at a point where all design and planning of the improved system is complete. These products have then been followed through into the customer and have been tested at the initiating part of the customer's production process. These case studies have enabled the hypothesis within the methodology to be successfully tested individually, within two differing manufacturing environments. This process of improvement (through the supplier and the customer) has also allowed the methodology to be tested throughout the product supply chain. With these results it has in turn allowed the methodology to be reanalysed within the supplier environment, thus understanding the effects of the supplier-buyer relationships and justifying how the hypothesis has again been tested throughout a supply chain.

4   Conclusions

This Design Methodology has incorporated and fulfilled a niche in the findings that were emphasised in the literature survey [3],[4]. It has integrated a number of Manufacturing Business Improvements within the Design Phase and highlighted how to apply these to a development project through weighted values of importance. The methodology has integrated both factors of change technical and cultural aspects within both phases [7]. It has also guided how to implement these findings through integrated management polices. This unified approach has ensured that a number of strategies that are not currently synchronised and assessed for project suitability can be. The literature review also highlighted that the failure rate of improvement initiative programs is quite high [6] due to lack of planning of the cultural aspects and because technical issues are easier to implement [1]. This methodology will automatically reduce factors associated to cost and time; however it also portrays as a confident view for success. Having this path of transformation alleviates cultural issues and resistance to change, it gives the project team assurance that the right changes are being made in the most efficient manner, therefore allowing a smoother acceptance to a change initiative program [13]. This time can be better spent on training and culture programs to ensure greater implementation success. All of these factors will aid to reducing the lead-time of a traditional change improvement program making the manufacturing environment a more competitive industry.

5   References

1. BALOGUN, J., & JENKINS, M., 2003. Re-conceiving Change Management: A Knowledge-based Perspective. *European Management Journal*, 21, p247-257.
2. BERKHAUER-SMITH, S., BHATTI, R., & SPEDDING, T.A., 2007. The Associations of Manufacturing Business Improvement Strategies. *International Journal of Operations and Production Management.* (Submitted)
3. BERKHAUER-SMITH, S., & BHATTI, R., September 2007. Integrating both Technical And Cultural aspects of Process Improvement. *5th International Conference on Manufacturing Research (ICMR 2007).*
4. BERKHAUER-SMITH, S., & SPEDDING, T.A., June 2007. Creating Integrated Improvement in the Manufacturing Industry. *5th Anzam Operations Management Symposium.*
5. CHIOU, E. F., HONG, J.C., HUANG, T. L., SUN, F. Y., WANG, L.J., & YANG, S.D., 1995. Impact of employee benefits on work motivation and productivity. *The International Journal of Career Management,* 7, p10-14
6. CHU, K.F., 2003. An organizational culture and the empowerment for change in SMEs in the Hong Kong manufacturing industry. *Journal of Materials Processing Technology,* 139, p505-509.
7. LIMAN MANSAR, S., & REIJERS, H.A., 2004. Best practices in business process redesign: an overview and qualitative evaluation of successful redesign heuristics. *The International Journal of Management Science*, 33, p283-306.
8. LINK, P., & MARXK, C., 2003. Integration of risk- and chance management in the co-operation process. *The International Journal of Production Economics*, 90, p71-78.

Cooperation of Lean Enterprises – Techniques used for Lean Supply Chain.

Marek Eisler, Remigiusz Horbal and Tomasz Koch

Wroclaw University of Technology,
adress: ul. Lukasiewicza 5
50-371 Wroclaw, Poland
email: marek.eisler@pwr.wroc.pl, remigiusz.horbal@pwr.wroc.pl,
tomasz.koch@pwr.wroc.pl

Abstract.
The paper presents the problems with integration of the companies within the supply chain. Usually the separate actions are undertaken by the companies to implement lean tools for production systems and external logistics processes. This situation leads to minor results or moving the costs between production and logistics processes instead of reduction. The purpose of the paper is to present the new version of Value Stream Mapping method, focused on synchronised reorganisation of company production system, external logistics processes between the company and its suppliers as well as suppliers' production processes. This paper shows the techniques currently used to support cooperation between enterprises and will demonstrate their incompleteness and how they can be improved.

Keywords
Supply Chain Management, Lean Management

1 Introduction

Supply chain management is a discipline that finds enormous interest in last two decades. It becomes now a forefront of business reorganisation. In the previous years supply chain was considered only as material transfer channel. Now, as Lancioni points out, it is and will be the main source of competitiveness [1]. Nowadays, there is no place for fighting only between brands, factories or stores. According to Lambert and Cooper [2] "individual businesses no longer compete as solely autonomous entities, but rather as supply chains." The need of competitiveness is

Please use the following format when citing this chapter:

Eisler, M., Horbal, R., Koch, T., 2007, in IFIP International Federation for Information Processing, Volume 246, Advances in Production Management Systems, eds. Olhager, J., Persson, F., (Boston: Springer), pp. 363-370.

now of paramount importance. Therefore, companies need to perform no longer as individuals, they must think about cooperation with other players in supply chain. The source of competitiveness of supply chain was also pointed out by numerous authors ([1], [2], [3]).

Zdzislaw Arlet, managing director of Fiat Auto Poland, claims that suppliers nowadays must be treated as business partners rather than just as suppliers [4]. This way of thinking is a result of policy presented by many OEM manufacturers. They currently need the superior quality as well as on time deliveries according to Frank Guyot, Head of Purchasing Office for Central Europe of PSA Peugeot-Citroën [5]

At the end of last century western manufacturing companies became very interested in the way Japanese produce their goods, especially cars. There was and still exist huge interest in those techniques. Numerous manufacturing companies become aware of the waste that exists on their shop floors and in the offices, especially after publication of The Machine that Changed the World [6] and Lean Thinking by Womack and Jones [8]. Companies have been implementing those techniques and nowadays can demonstrate significant achievements. Authors of the articles believe that the same situation takes place now in Central and Eastern Europe. Many companies try to implement the lean techniques already known well from literature.

In spite of the fact that managers around the globe are aware of problems that exist in their plants and even though they have some achievements with solving those problems, supply chains are not transparent to everybody involved in process of product creation. Many problems can be found not inside the isolated facilities but rather within the relations that occur between cooperating companies. Managers responsible for supply chains need to be equipped with tools that will help them to resolve such kinds of problems. Valuable method in this matter is Extended Value Stream Mapping eVSM proposed by Womack and Jones in 2002 [8]. However eVSM lets to recognize improvement areas within the supply chain, it needs to be complemented by the elements relevant to transportation network reorganization. The main purpose of this paper is to present how the existing method of eVSM might be improved.

2   Value Stream Mapping and Extended Value Stream Mapping

In 2002 Jones and Womack [8] proposed a method that allows managers of any supply chain to analyze and find solutions for any value stream that is comprised of manufacturing plants as well distribution centres, warehouses and stores. The method assumes that value stream as well supply chain need to be seen from market demand perspective. The method is based on experiences of Toyota that are described in general as Toyota Production System (TPS) (see [9]). The considered method was called Extended Value Stream Mapping (eVSM) and is an extension to Value Stream Mapping (VSM), a method described by Rother and Shook in 1998 [10].

2.1   Value Stream Mapping

The VSM method allows managers to perceive their companies from the final customer perspective. VSM helps the practitioners to understand how their plants work at present (the Current State Map) and to plan the improvements in approaching 9-12 months (the Future State Map). The VSM analysis is usually performed for plant level from raw materials to finished goods. It allows identifying the status of manufacturing system in any plant and to plan improvements with use of lean techniques such as level pull system, one piece flow cells, SMED, TPM and others. Usually the VSM analysis takes a few days. The result is a Future State Map drawn by managers and engineers depicting precisely what tools should be used in what areas of the plant.

2.2   Extended Value Stream Map (eVSM)

VSM method is not limited only to a single manufacturing plant. Jones and Womack [8], Rother and Shook [10] suggest starting VSM process on a plant level and then extend the analysis for supply chain level. Such analysis should encompass in the beginning only selected, manageable part of the whole supply chain. It is obvious that if eVSM would be used to analyze OEM and all of its cooperating companies, the map would be very complex and therefore difficult for analysis. That is why eVSM teams usually start to draw maps only for limited fragment of the supply chain. After recognizing problems and implementing the solutions for the chosen fragment the team might repeat eVSM analysis for other suppliers [8, 9].

The eVSM method functions in following way: the mapping team members draw a Current State Map, including both material and information flows, for selected branch of supply chain, then using the lean tools and methods they design the Future State Map. The output of this process, beside a Future State Map, is an implementation plan including the set of projects that must be put in action [8].

2.3   Limitations of Extended Value Stream Mapping

The eVSM method is a very supportive tool to begin supply chain improvement by implementation of such techniques as pull system, just-in-time deliveries and others. The main problem that may occur is lack of willingness of managers of cooperating companies to share the knowledge about their plants and warehouses. Another problem that may occur is a fear of managers of particular plants that they will not benefit from the whole improvement process [1, 10]. Witkowski in "Logistics of Japanese firms"[10] claims that Western companies (Europe, USA) are not willing to cooperate with their business partners to find better and cheaper solutions for their problems. They usually focus on unit price and quality level accepted by costumers. It is hardly to find fair rules that would allow benefiting both supplier and costumer from the outputs of improvements made together within supply chain. It is observed that common effort of suppliers and consumers is mostly made to improve quality, but the efforts to reduce the waste are rather limited to the separated actions within the plants.

Another problem that may occur while using eVSM method is existence of other costumers and suppliers for individual plants that are not involved in an analyzed branch of a supply chain. The main issue of the eVSM method is to reduce inventories that are waste from costumer's point of view. When users of eVSM try to reduce excessive inventories it is often related to increased frequency of deliveries. Higher frequency of deliveries results in smaller transportation batches, however it might lead to higher transportation cost. This will be the case if the delivery frequency is increased without redesigning of transportation routes [12, 13]. Jones and Womack suggest taking advantages of optimizing transportation with "milk runs". As Baudin claims "a supplier milk run is a scheduled pickup of parts from multiple suppliers in matching quantities (...)" [12, p.131]. Milk run idea is depicted in the fig 1.

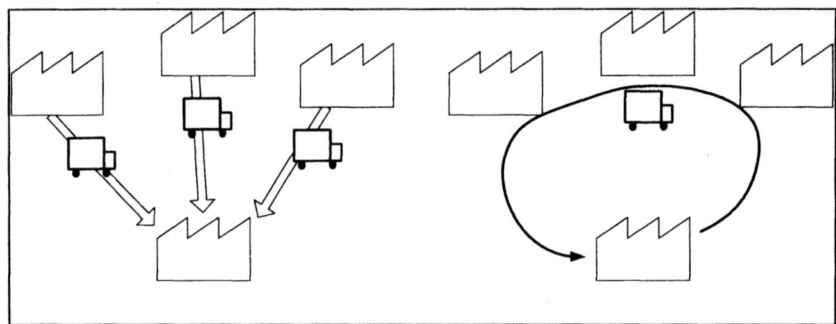

Fig 1. The concept of milk run versus direct shipments.

eVSM method does not support transportation route analysis and is focused only on the selected value stream. These both aspects are crucial to design milk run deliveries. The next section of this paper presents how to complement eVSM method with transportation route analysis.

3   Incorporation of transportation route design within Extended Value Stream Mapping.

In the proposed approach the new elements were added to eVSM to put more attention to visualization of transportation links. These elements allow designing more precisely the milk run deliveries in order to lower the inventory levels. There are several approaches of milk run design. One of them is to add few suppliers that provide small amount of parts to one or two suppliers providing large amount of parts for the costumer. This approach leads to better vehicle utilization as well as to higher frequency of deliveries. To design well operating milk runs, the following issues must be taken into consideration:

- Cost of vehicles being currently in use and its utilization level to calculate existing transportation cost.
- Number of suppliers located in the proximity and delivering to the one, common customer to design potential milk-run route.
- Distances among suppliers and customer to calculate the transportation costs for designed milk-run route.

To visualize those issues on the map the three new map elements are proposed:

The truck cost and utilization

This entity is shown in the fig.3 and represents the information about the transportation cost of the vehicles currently in use. It provides as well information about actual utilization level of vehicle transportation space. For example the symbol in the fig.2 shows utilization at the level of 25%; the transportation costs is 0,75 € per 1 km. These kind of data let to have a closer look for actually incurred costs of transportation. Including this data on the map lets to estimate roughly if designing milk runs will be economically justified.

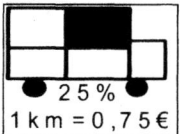

Fig. 2. Proposed symbol of transportation vehicle being currently in use.

The milk run area

The second proposed element allows gathering data related to geographical location of the suppliers considered to be served by designed milk run transportation process. The symbol that is presented in the fig 3 summarizes data about number of existing suppliers that deliver to the customer as well as the size of the area. For example in fig 3 there are 3 cooperating companies that should be taken into consideration as points for a milk run. All these companies are located in the radius of 50 km from the centre of the area.

Fig. 3. Proposed symbol that would summarize data about number of existing plants and warehouses that supply particular customer and the distance between them.

The milk run route

For the Future State Map the milk run route symbol is proposed (see in fig.5). During designing of the future state map the chosen milk run areas will be transformed into milk run routes. The proposed milk run route symbol will let to estimate the transportation costs for the designed milk run. It provides data about number of points in a particular milk run loop (for example in the fig 5 there are 6 points in the loop) and about overall length of the loop (150 km in the fig.4).

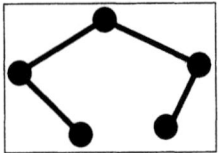

Fig. 4. Proposed symbol for designed milk run route.

4　Example of usage of proposed method.

eVSM method was used for a case study conducted with a Polish producer of industrial valves. To simplify the analysis only two cooperating companies were taken into consideration:

- the manufacturfer of industrial valves – 'OEM',
- one of the suppliers of OEM– 'Supplier A'.

Both companies were analysed and the Current State Map was drawn (see in fig 5).

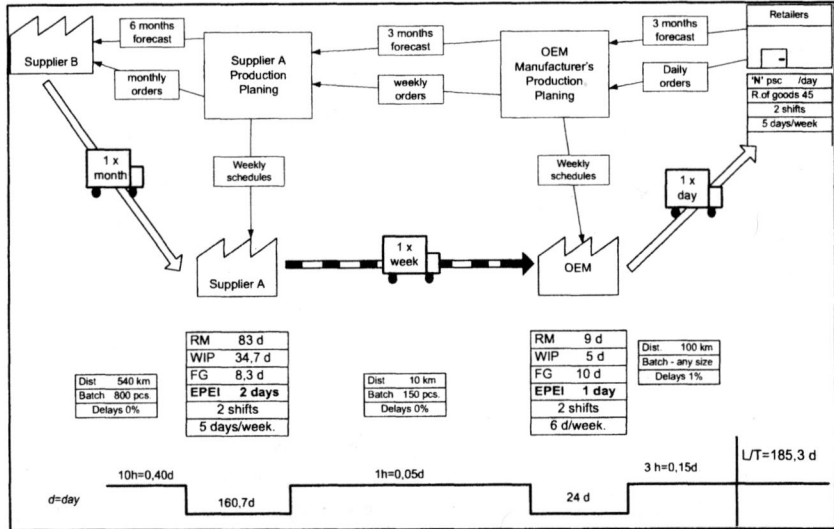

Fig. 5. The Extended Value Stream Map for analyzed case (Current State).

During the analysis it was decided that material flow from supplier to the customer will be reorganised and based in future on replenishment pull system, often called supermarket pull system or kanban system. Basing on the collected data and guidelines proposed by Harris and Wilson [14] the target inventory levels for both cooperating companies were calculated. Authors inferred that the levels of raw material (RM) inventory in customer company (OEM) mostly depends on frequency of deliveries from the supplier (Supplier A). This conclusion comes from analyzing of equations proposed by Harris and Wilson [14] to calculate target inventory level for supermarket pull system. The maximum inventory consists of three parts:

- cycle stock,
- buffer stock,

● safety stock.

These three types of stocks might be calculated as follows:

● cycle stock = (average daily consumption [pcs.]) x (replenishment time [days])
● buffer inventory = (maximum daily consumption [pcs.] - average daily consumption [pcs.]) x (replenishment time [days])
● safety stock = (cycle stock + buffer stock) x (safety factor [%])

As pointed out above, inventory level strongly depends on frequency of deliveries. Practitioners conscious about this issue may be confused while using eVSM method because increasing deliveries frequency comes together with significant increase of transportation costs. To keep those costs constant implementation of milk run is essential, but with eVSM method there is not enough data to consider implementation of such solution. Therefore the second Current State Map was prepared (see in fig 6). For the second Current Stat Map proposed new elements were used. With these elements the additional data necessary to design new milk run routes are available on the map.

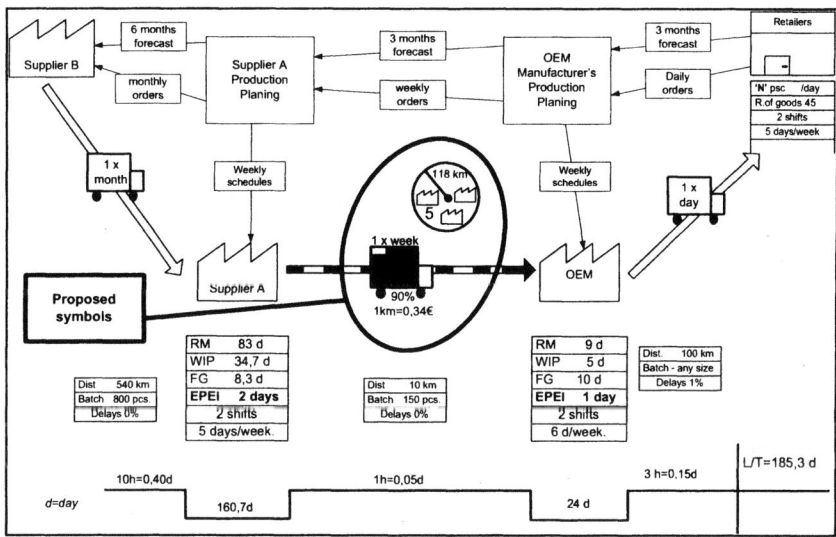

Fig. 6. The Extended Value Stream Map for analyzed case (Current State) with proposed symbols.

The Current State Map complemented with new elements led to the decision about designing daily milk-run encompassing Supplier 1 and 4 other suppliers. The length of the route is 300 km. Implementing it will let to increase the delivery frequency from Supplier A to OEM from one week to one day, which in turn will lead to raw material stock reduction (on OEM's side) almost by 80%.

5   Conclusion.

Complementing the eVSM method with transportation route data proposed in this paper would allow managers to make easily decisions about designing milk run routes and make right strategic decisions quicker, without time consuming investigation of all the possibilities. Thanks to such solutions supply chain will become more agile and more competitive while inventory levels will be decreased with higher frequency of deliveries. The described case-study proved that the milk run areas identified during current state analysis may be transformed in the real milk run routes later in the phase of designing new solutions for the supply chain.

6   References.

1. Lacioni, R. A., *New Developments in Supply Chain Management for the Millennium,* Industrial Marketing Management, 29, 1–6, 2000.
2. Lambert D., Cooper M., *Issues in Supply Chain Management*, Industrial Marketing Management 29, 65–83, 1999.
3. Porter M., *Competitive advantage: Creating and sustaining superior performance*, New York, NY: Free press.
4. Arlet Z., *Our way to success*, Conference proceedings of 3rd Annual Conference of Automotive Industry in Poland, Poznan, May 2007.
5. *Statements at the Autoevent May 23th-25th,* 2007, Poznan, Poland
6. J.P. Womack, D.T. Jones, *The Machine That Changed The World. The Story of Lean Production*, Harperperennial, New York, November 1991.
7. J.P. Womack, D.T. Jones, *Lean Thinking: Banish Waste and Create Wealth in Your Corporation,* Simon & Shuster, New York 1996.
8. J.P. Womack, D.T. Jones, *Seeing the Whole: Mapping the Extended Value Stream*, Lean Enterprise Institute, March 2002.
9. Ohno T., *Toyota Production System: Beyond Large-Scale Production*, Productivity Press, New York, 1988.
10. Rother M., Shook J., *Learning to see.,* The Lean Enterprise Institute, 1998.
11. Witkowski J., *Logistyka firm japońskich*, Wydawnictwo Akademii Ekonomicznej im. Oskara Langego we Wrocławiu, Wydanie 2, Wroclaw 1999.
12. Budin M., *Lean Logistics. The nuts and bolts of delivering materials and goods.*
13. Cooper R., Slagmulder R., *Supply Chain Development for the Lean Enterprise: Interorganizational Cost Management.*, Productivity Press, June 1999.
14. Harris R., Harris C., Wilson E., *Making Materials Flow: A Lean Material-Handling Guide for Operations, Production-Control, and Engineering Professionals.*, The Lean Enterprise Institute, September 2003.

Lean Maturity, Lean Sustainability

Frances Jørgensen, Rikke Matthiesen, Jacob Nielsen and John Johansen
Center for Industrial Production Aalborg University Fibigerstrade 16 9220
Aalborg E., Denmark
corresponding author: frances@production.aau.dk

Abstract

Although lean is rapidly growing in popularity, its implementation is far from
problem free and companies may experience difficulties sustaining long term
success. In this paper, it is suggested that sustainable lean requires attention to both
performance improvement and capability development. A framework for describing
levels of lean capability is presented, based on a brief review of the literature and
experiences from 12 Danish companies currently implementing lean. Although still
in its emerging phase, the framework contributes to both theory and practice by
describing developmental stages that support lean capability development and
consequently, lean sustainability.

Keywords

Lean Capability Model, Lean Sustainability, Lean Manufacturing

1 Introduction

During recent years lean as a process management philosophy has rapidly gained
popularity in the manufacturing sector, and in some countries, in both private and
public administrative and service organizations as well. With its focus on just-in-
time production, elimination of waste, and continuous improvement, many
companies are reporting impressive performance gains [1]. According to Womack *et
al.* [2], lean also provides opportunities for a positive and fulfilling working
environment for employees, due the employees' involvement in and ownership of
problem-solving and improvement activities, more diversified work functions
requiring varied skills and abilities, and increased cross-functional and inter-
organizational functions.

A number of empirical studies have shown, however, that implementation of
lean may be anything but a positive experience for employees [3] and studies by
Landsbergies *et al.* [4] suggest that the increased intensity and higher degree of

Please use the following format when citing this chapter:

Jørgensen, F., Mathiessen, R., Nielsen, J., Johansen, J. , 2007, in IFIP International Federation for Information
Processing, Volume 246, Advances in Production Management Systems, eds. Olhager, J., Persson, F., (Boston:
Springer), pp. 371-378.

standardization of the work functions associated with lean may have detrimental effects on employees, both psychologically and physically. These negative consequences on the workforce may be aggravated by a skewed focus on the implementation of lean tools and methods that sacrifice the development of a lean culture [5]. In the most extreme cases, companies may become almost obsessed with cost and/or staff reduction that the work becomes "more intense, stressful and hazardous" [6], thereby trading potential long term, sustainable improvement for short term gains, sometimes even leading to what is referred to as organizational anorexia [7]. More common are situations in which companies have difficulty maintaining momentum for lean after often dramatic performance improvements are realized following the initial implementation of lean. In other words, companies have difficulties sustaining lean over time.

One explanation for the difficulties companies encounter in sustaining lean may be attributed to a lack of focus on the developmental progression of lean capabilities amongst the members of the organization. By focusing on developing lean capabilities, members of the organization should then become progressively better at doing lean while at the same time, creating a learning environment that supports a lean culture. The objective of this paper is to present a framework describing the development of lean capabilities towards a state of maturity conducive to long term sustainability of lean. The framework is derived from the current lean literature and descriptions of lean experiences in 12 large Danish organizations currently involved in a lean implementation process. In the next sections, the sustainability as it relates to lean is introduced; thereafter, the development of the concept of lean is discussed as the foundation for creating a lean capability framework.

2 Lean Sustainability through Lean Capability Development

When the term *sustainability* is used with respect to lean manufacturing, most think of how lean may support ecological preservation by reduction of waste of raw materials and energy supplies [8] or even the economic stability for an organization that provides opportunities for future growth and prosperity. Sustainability in this context generally refers to maintaining a balance between the exploitation of resources to fulfill the needs of today while ensuring protection of resources for survival in the future. The sustainability of lean itself is however also worth considering, both in terms of how to maintain momentum once initial pilots and "blitzes" are complete and on how members of the organization can actually develop their lean implementation capabilities. The former issue can be viewed as a relatively horizontal progression over time, as employees use lean tools for cost reduction in their daily work processes.

While the effects of these changes may increase as more people are involved and as lean is introduced to new areas, the activities and the capabilities of those performing them may remain relatively stable. On the other hand, as employees learn from their own and others' experiences, what may be viewed as a vertical progression would be expected to occur as well. Although both vertical and horizontal progression is implied in lean implementation models, there appears to be more focus on sustaining lean momentum in terms of spreading the concept to more organizational units than on ensuring lean development.

The issue of sustainability through vertical as well as horizontal development is central, albeit often implicit in many improvement initiatives. Researchers at the Center for Innovation Management (CENTRIM) at Brighton University addressed these issues specifically when developing what is referred to as the Continuous Improvement (CI) Maturity Model [9], which illustrates the gradual but steady development of CI through five stages of maturity through adoption of certain sets of behaviors that together build organizational capabilities. At the first level of the model, behavioral activities such as idea generation and problem-solving are implemented *ad hoc*, or what is often called "putting out fires". By creating systems, procedures, and processes in the organization, the CI development becomes more strategically oriented and structured until, at stage five maturity, a learning organization built on CI and organizational learning emerges.

There are distinct similarities between CI and lean—and in fact CI is a critical component in sustainable lean—and thus the model, with modifications to incorporate the specific characteristics of lean, should therefore be applicable to understanding how lean sustainability can be achieved through focus on capability development. Further, the development of lean as a concept parallels this maturation process, as described in the following section.

3   Development of the concept of lean

The origin of Lean can be traced back to Toyota in the 1950's, but did not become recognized until the beginning of the 1980's [10-14] and only first gained serious attention with the publication of The Machine that Changed the World [2]. Since its introduction, the concept of lean has developed considerably, beginning with an almost exclusive focus on shop floor workers in the automobile industry. In the 1990's, the lean concept was extended not only to other industries, but also to much more than simply a "Toyota" and "Tools" orientation to a more comprehensive philosophy encompassing both strategic and organizational components. Hines *et al.* [15] suggests that this developmental progression of the lean concept can be described as occurring in four stages, similar to the four stage classification McGill and Slocum [16] use to portray the a company's progression towards becoming a learning organization. From a prescriptive focus on shop floor practices, tools and techniques applied in production cells and lines for higher efficiency in the 1980'es to a widened focus on quality and Lean as a set of management practices in the early 90's, still mainly applied within the automotive industry.

This widened focus is also evident in a review of Lean constructs based on Lean practises by Shah and Ward [17]. Here practices such as focused factory production, process capability measurements and cellular manufacturing are introduced to such constructs in 1990. According to Hines *et al.*'s [16] review the scope of Lean conceptualizations increased to value streams and focus shifted to flow creation from mid 90's. Customer value as constituted by cost, quality and delivery was inherited from the automotive supplier sector. From 2000 onwards the Lean concept has involved a greater degree of contingency and the scope has increased to value systems with concepts such as demand chain management. Further, in their review of 12 lean production models, Paez *et al.* [18], allude to a developmental progression of the lean concept from being primarily technically-oriented to one that also more strongly emphasizes the human elements of lean during the past two decades. A movement from singular practices such as team work, multi-tasking, and autonomy over to human resource practices, involvement, and empowerment that contribute to establishment of a learning environment is seen in their review.

Although the central focus of the research conducted by Hines *et al.* [16] and to some extend that of Paez *et al.* [18] is on describing the developmental progression of the lean concept itself, they may also provide some insights into the type of developmental progression lean must undergo within a given company in order to steadily progress from the isolated use of tools to a more integrated lean philosophy that also encompasses an organizational learning perspective. Ballé [19] only refers to leaders, but also addresses the importance of progressing from simple tools and methods for lean to be successful over time when he describes the continuous learning that must occur for leaders to meet three phases of challenges: an early aha experience that inspires leaders to hunt for low hanging fruits and which results in a preference for quick action (1), a shift to a more rigorous, systematic and structure approach to problem solving and investigations to meet more complex challenges (2) and finally the realization that the leaders cannot solve all problems themselves, but must involve other employees (3).

How precisely this progression occurs, what is needed to allow the progression, and how the leaders are to know when they have reached one level of lean development and must become equipped to meet the next phase is however not directly addressed by the author. What is needed then is a framework that allows a company to assess its current status of lean development as well as identify areas in need of improvement, based on concrete descriptions of behaviors, mechanisms, processes, etc. necessary for successful, enduring lean. In the next section, the subject of assessment with respect to lean and the development of lean capability is discussed.

4   Assessing Lean

Assessment tools are critical to successful lean implementation—or in the successful implementation of any world class manufacturing principles for that matter. Assessment tools have many functions, most important perhaps as a "roadmap" that illustrates the company's current status among its most important performance parameters. A good assessment is also invaluable in identifying opportunities for improvement and the parameters in which action plans should be designed. For an assessment tool to fulfil these functions, it must accurately reflect the nature and complexity of what is being assessed. Based on the previous discussion on the importance of both "horizontal" and "vertical" lean development, this means that a lean assessment tool must address two perspectives or dimensions, each of which encompass a number of variables. Specifically, it is proposed that a lean assessment tool must include the following: (1) A technical perspective, which reflects performance, methods, and tools in relationship to the given company's strategic "scope", as described by Hines *et al* [16]; and (2) an organizational perspective, which reflects management, organizational and human capabilities, culture, and learning.

In addition to being able to evaluate variables related to each of these perspectives, a lean assessment tool should be able to measure the relative balance between the two elements and the possible synergy created by focusing attention on both perspectives simultaneously. The majority of available assessment tools, however, address primarily or exclusively the technical perspective [17] and only a select few refer aspects of the progressive lean development emphasized here (i.e. elements associated with the organizational perspective). Of those tools that do include mention of the organizational perspective of lean [20-21], even fewer consider the balance between the two perspectives and the potential synergy between them [22-23]. Finally, there do not appear to be any lean assessment tools that incorporate both perspectives while still emphasizing the processes necessary for ensuring developmental progression of lean in the organization.

In the following section of the paper, the methods used to gather data relevant to the construction of a model which emphasizes lean capability development and can be used to assess both the technical and organizational aspects of lean are briefly presented.

5   Methods

The study described in this paper is taken from a larger research project aimed at implementing lean with emphasis on creating a positive psychological environment for employees and involves data collection through workshops, seminars, interviews and observations conducted with and in 12 large Danish organizations currently implementing lean and representing the industrial, administrative, and service sectors. The collected data are used in this paper to develop the foundation of a framework for describing and assessing lean implementation that focuses both on performance improvement through lean (i.e. a technical perspective) and the lean capability development (i.e. an organizational perspective) and the balance of these proposed here as necessary for ensuring lean sustainability.

The four authors of the paper have been involved in the data collection following an action research methodology, with active participation in discussions of the companies' experiences. Data have been verified continuously through active dialogue with the participants in the study.

The lean capability model, which is presented in the following section, is derived from descriptions of both actual experiences with lean in the companies and what key persons implementing lean define as an ideal state for lean that would support lean capability development and long term sustainability. The data were classified into the maturity levels according to group consensus and are considered consistent with the lean literature. Further, the developmental phases (i.e. maturity levels) included in the model can be theoretically and practically compared to those found in the CI Maturity Model [9] which has been subject to numerous empirical applications [24].

6   Lean Capability Model

Experiences from these companies suggest a number of common trends, especially in the beginning or immature phases of the lean implementation. Much time is spent in trying to build a shared understanding of the lean philosophy and how it impacts the individual and the organization. Initially, training is limited to project leaders ("experts") and focuses primarily on skill training once extended to the general workforce (i.e. shop floor workers or those who will be responsible for working with lean). Indications of lean maturity first become apparent as technological and organizational mechanisms become aligned with the strategic objectives to be fulfilled through lean. At this point, HR functions become critical. Characteristics of the five identified levels of maturity are described below:

1. **Sporadic production optimization:** This level is characterized by occasional rather random efforts at optimization in various organizational unites, but these activities are not planned or implemented on the basis of an overall strategy or a specific manufacturing philosophy. The optimization projects are typically led by experts with little to no general employee involvement. Organizational mechanisms and systems are not integrated with lean philosophy and/or lean objectives.

2. **Basic lean understanding and implementation:** Lean has now been chosen as the manufacturing philosophy that will serve as the basis for production control and optimization. The experts and general workforce have received basic training and pilot projects have been initiated in isolated unites within the organization for the purpose of experimenting with the individual lean tools and methods. Isolated mechanisms developed to support lean (e.g. reward and suggestion schemes, training)

3. **Strategic lean interventions:** The implementation of lean is now a part of the organization's strategy and projects and activities are planned on the basis of established goals and objectives. Knowledge of and practical experience with lean tools and methods as well as a lean philosophy are widely acknowledged and recognized at all levels of the organization, although initiatives are still primarily implemented according to an established plan. Satisfactory performance improvements are achieved. Specific HR systems (i.e. selection, compensation, training functions) are aligned with lean objectives to support lean goals.

4. **Proactive lean culture:** Lean activities occur continuously from all areas of the organization. To think and act lean has become a part of the daily work, and CI is more of a habit than a specific task, although efforts have not yet been made to extend these efforts outside of the organization's own boundaries. The practical understanding of lean tools and methods is quite high and these are used actively by all members of the organization to develop and implement performance improvements. All HR functions are

aligned with lean objectives for the purpose of supporting long term sustainability. Focus on, e.g. career development via lean and extended developmental activities (e.g. external education).

5. **Lean in the EME:** The lean strategy is no longer just an internal strategy and its impact is visible in activities throughout the EME (Extended Manufacturing Enterprise) level. Lean activities are planned, implemented, and monitored across the EME's boundaries. Knowledge sharing and knowledge transfer are important components of the act ivies across the EME and organizational structures support inter-organizational network building.

7   Discussion and Conclusion

Many authors state that lean implementation efforts often fail to provide companies with the long term benefits promised in the literature. Generally there is agreement that successful sustainable lean involves more than the use of tools and methods and efforts should be made to support development of a lean culture, but there is little in the literature to serve as a "roadmap" for companies wishing to support this development. On the basis of a brief literature review and experiences in companies currently implementing lean, this paper presents an emerging framework describing five stages of lean capability development that are consistent with long term sustainability of lean. One of the important findings in this study was that Human Resource (HR) functions may play a critical role in supporting what is referred to here as "vertical" development of lean. Specifically, training and development targeted at learning and knowledge sharing, compensation and reward schemes, and focus on lean as a means towards career development may facilitate establishment of a lean culture that is sustainable. The importance of organizational learning is also emphasized by Emiliani [25], who proposes that a lean production philosophy provides excellent opportunities to couple personal growth and learning needs with organizational performance objectives. The paper contributes to both theory and practice by describing the stages of lean capability development that are necessary for sustained success with lean and suggesting a framework for assessing a company's current level of lean capability maturity.

References

1. J. P. Womack and D.T. Jones, *Lean Thinking: Banish Waste and Create Wealth in Your Corporation*, Simon and Schuster: New York, NY (1996)
2. James P. Womack, Daniel T. Jones, and Daniel Roos, *The Machine that Changed the World*, Harper-Collins: New York (1990).
3. B. Harrison, *Lean and Mean: The Changing Landscape of Corporate Power in the Age of Flexibility*, The Guildford Press: New York (1994).
4. P.Landsbergis, J. Cahill J, and P. Schnall, The Impact of Lean Production and Related New Systems of Work Organization on Worker Health., *Jn. of Occupational Health Psych.*, **4**(2), 108-30 (1999).

5. J.K. Liker, *The Toyota Way - 14 Management Principles from the World's greatest Manufacturer*, McGrawhill: New York, NY (2004).
6. N. Kinnie, S. Hutchninson, J. Purcell, C. Rees, H. Scarbrough, and M. Terr, *The People Management Implications of Leaner Working, Institute of Personnel Management*, London (1996).
7. R.J. Radnor and R. Boaden, Developing an understanding of corporate anorexia, *Journal of Operations & Production Management*, **24**(4), 424-440 (2004).
8. T. H. Johnson, Sustainability and lean operations, *Cost Management*, **20** (2), 40-46 (2006).
9. J. Bessant and S. Caffyn, High involvement innovation", *Int'l. Jn. of Technology Mgmt.* **14**(1), 7-28 (1997).
10. S. Shingo, *Study of the Toyota Production Systems, Japan management Association*, Tokyo (1981).
11. S. Shinko, *Non-Stock Production: The Shinko System for Continuous Improvement*, Productivity Press: Cambridge, MA (1988).
12. R. Schonberger, *World Class Manufacturing – The Lessons of Simplicity Applied,* The Free Press: New York, NY (1986).
13. Y. Monden, *The Toyota Production System,* Productivity Press: Portland (1983).
14. T Ohno., *The Toyota Production System: Beyond Large-Scale Production*, Productivity Press: Portland, OR (1988).
15. Peter Hines, Matthias Holwe, and Nick Rich, Learning to evolve – A review of contemporary lean thinking, *Int'l. Jn. of Operations & Production Mgmt.* **24**(10), 994- 1011 (2004).
16. M.E. McGill and J.W. Slocum, Unlearning the organisation, *Organisational Dynamics*, **22**(2), 67-79 (1993).
17. Rachna Shah and Peter T. Ward, Lean manufacturing: context, practice bundles, and performance, *Jn. of Operations Mgmt.* **21**, 129-149 (2003).
18. O. Paez, J. Dewees, A. Genaidy, S. Tuncel, W. Karwowski, and J. Zurada, The Lean Manufacturing Enterprise: An Emerging Sociotechnological System Integration, *Human Factors and Ergonomics in Manufacturing*, **14**(3), 285–306 (2004).
19. M. Ballé, Lean attitude - Lean applications often fail to deliver the expected benefites but could the missing link for successful implementations be attitude?, *Manufacturing Engineer*, **84**(2), 14-1 (2005).
20. Christer Karlsson and Pär Åhlström, Assessing changes towards lean production, *Int'l. Jn. of Operations and Production Mgmt.*, **16**(2), 24-41 (1996)
21. LESAT, LEAN ENTERPRISE SELF-ASSESSMENT TOOL, MIT and University of Warwick, www.lean.mit.edu, Version 1.0 (2001).
22. T. L. Doolen, M. E. Hacker, A Review of Lean Assessment in Organizations: An Exploratory Study of Lean Practices by Electronics Manufacturers, *Jn. of Mfg. Systems*, **24**(1), 55-67 (2005).
23. R. Sawhney, and S. Chason, Human Behavior Based Exploratory Model for Successful Implementation of Lean Enterprise in Industry, *Performance Improvement Quarterly*, **18**(2), 76-96 (2005)
24. Sarah Caffyn, Development of a continuous improvement self-assessment tool, *International Journal of Operations & Production Management*, **19**(11), 1138-1153 (1999).
25. M. L. Emiliani, Lean behaviors, *Management Decision*, **36**(9), 615-631 (1998).

Understanding the Interdependences Among Performance Indicators in the Domain of Industrial Services

Ingo Lange[1], Oliver Schneider[1], Matthias Schnetzler[1] and Lee Jones[2]
1   ETH Zurich, Center for Enterprise Sciences (BWI), 8092 Zurich,
Switzerland, ilange@ethz.ch, oschneider@ethz.ch, mschnetzler@ethz.ch
WWW home page: http://www.lim.ethz.ch
2   Ventana Systems UK Ltd., CH43 5RD Oxton Merseyside,
United Kingdom, lee.jones@ventanasystems.co.uk,
WWW home page: http://www.ventanasystems.co.uk

Abstract

Within the context of the EU-Project InCoCo-`S`, one of the key aims is to standardize integrative industrial service processes in order to facilitate transparency on service operation performance and the resulting customer benefit. Therefore the Service Performance Measurement System (SPMS) has been developed in order to quantify both the efficiency and effectiveness of industrial service operation activities and to support the measurement of customers' benefit through industrial service activities. But performance indicators are only a measurable expression of the underlying system performance, a system which is ordinarily complex in nature. It follows, therefore, that it would be beneficial to understand the interdependences between performance indicators in order to better utilize them in evaluating the options for improvements in system performance and the monitoring of an often complex system. Based on a comprehensive literature review and making best use of the tools and expertise available to the InCoCo-S consortium, a process to develop an understanding of the interdependences between performance indicators was created and executed. The results provide both the service provider and the manufacturing customer with an insight into those performance indicators to be targeted for improvement actions and those better suited to monitoring.

Keywords

Performance indicators, service performance measurement system, industrial service operations, interdependences among performance indicators, interdependency matrix

Please use the following format when citing this chapter:

Lange, I., Schneider, O., Schnetzler, M., Jones, L., 2007, in IFIP International Federation for Information Processing, Volume 246, Advances in Production Management Systems, eds. Olhager, J., Persson, F., (Boston: Springer), pp. 379-386.

1   Introduction

Increasing customer requirements, ever faster changing environments and increased competition are often mentioned as important, general business trends. The packaging machine industry is one example where these trends, which pressure companies into a continuous improvement cycle, can be observed. One challenge is to improve the level of satisfaction and quality of co-operation at the service interface by integrating service providers in the customers processes [1].

Based on intensive literature research and the results of a survey, the assumption was validated that there exists no standard performance measurement system addressing the special requirements for industrial services so far, but there is an unsatisfied business need for that [2]. After analyzing 1,352 papers published in 546 different journals, Andy Neely [3] stated in 2005, that an academic professionalism in the field of performance measurement has not yet occurred. Only since the late 1990s, an increasing shift in the performance measurement literature to more theoretical and methodological pieces occurred, which is an indicator for academic professionalism [3]. Basically, a performance measurement system`s (PMS) purpose is to measure process activities which are linked together and characterized by complexity and high interdependency. Based on the researched literature there are various PMS available, but only a few authors have given consideration to the relations between performance measures. However, in literature this area has not been adapted to the special needs of service operations performance. Due to this fact, the objective of the paper is to show the interrelations in a qualitative way. For this purpose, potential methods have been analyzed. The paper presents and makes use of Vester`s paper machine [4], which is basically an influence matrix for identifying and evaluating a system's critical variables. Based on the results, it will be possible to cluster important performance drivers and to support companies in the selection of relevant PIs, especially in the packaging business.

1.1   Methodology

The research methodology incorporates the principles of action research, which consist of involvement of industrial partners through workshops and a structured research process. Based on literature review (desk research) and workshops, relevant PIs have been identified and the Service Performance Measurement System (SPMS) was developed. The structure of the system and the PIs were validated in industrial settings by means of a proof of concept.

The System Dynamics (SD) software tool, Vensim®, has been used in the development of correlation between previously isolated PIs. Vensim has powerful causal analysis and sensitivity tools including Causal Tracing™ using causes trees. The Causal Tracing™ capabilities of the simulation software have been used to create causal trees for qualitative analysis of the connectivity between service processes and the hierarchy of the PIs. For the investigation of the interdependences among the PIs of the SPMS, potential methods have been reviewed by literature research. Ensuring practical relevance, the selected conceptual tool (Vester`s paper machine) for showing influences among the considered PIs has been put into practice by using industry partners experiences in the domain of industrial service business.

1.2   Service Performance Measurement System (SPMS)

Industrial service operations are usually a result of the interaction between customers (OEM) and the service provider, including the service staff, production data exchange, service equipment, service environment and facilities. Figure 1 illustrates an overview of the SPMS structure and presents the two dimensions Service Perception and Service Encounter Interface.

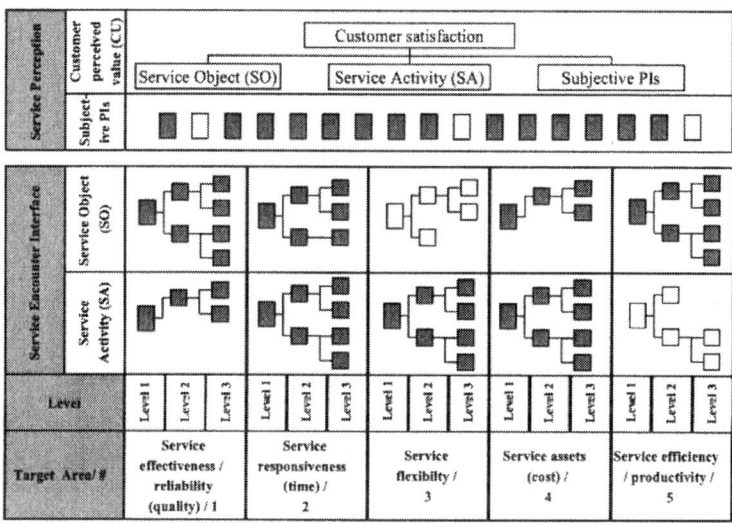

Fig. 1. The structure of the SPMS

The basic idea of the Service Perception is to have a dimension quantifying the gap between actual service operation performance based on objective measures and the perceived service operation performance from the customers' perspective. The dimension Service Encounter Interface provides a comprehensive set of approx. 120 PIs structured by different target areas to firstly measure the service providers' internal service operation performance and secondly to measure the performance of the processes at the interface with the customer which directly affect the performance of the service operation. In reference to the SCOR model [5], the differentiation in overall goals (target areas) helps in selecting specific PIs according to the companies strategic goals. As an enhancement of existing PMS the Service Encounter Interface is divided into the sections Service Activity (SA) and Service Object (SO): **Service activity (SA)** relates to all process steps to be taken to fulfil the specified level of service; e.g., alignment of required production capacities and planned maintenance service or productivity consulting. The SA view is measuring how efficient and reliable the service offerings like maintenance, modernization, and trainings are offered to the customer [6]. **Service object (SO)** refers basically to the element (process and/or object) in the customers' manufacturing supply chain that is being serviced. In reference to a physical asset this implies the resulting condition of the service object (e.g. machine, software) required to guarantee a specified level of service object performance; e.g. increased availability of a packaging machine [6].

Following this structure, each target area is providing a hierarchical tree of generic PIs on level 1-3. In addition, the performance indicators facilitating the measurement of industrial service operations are linked to reference service processes, the InCoCo-S Reference Model (IRM). In the paper, the focus is on the 38 Level 2 PIs which are representing the service operation performance in accordance to Level 2 service operation processes in the IRM.

2   Research background

In this section the focus is on the selection of available methods and tools which might be useful for analyzing interdependences between performance measures in the briefly introduced SPMS. The etymological meaning of interdependence (lat.) is "mutual dependence". Interdependence contains in contrast to dependence a retrospective among all involved objects. With a brief introduction into the world of systems the theoretical background of interdependences is outlined.

2.1   Systems Thinking and the Interdependency Matrix (IM)

Ulrich [7] defines a system as a whole consisting of elements. In this context "whole" denotes that the system is clearly distinguishable from surrounding things and even the inside is heterogeneous. Vester adds to this very generic definition another essential attribute of systems: Beside the fact that systems consist of distinguishable parts, it is important to mention that these parts are linked to each other in a certain structure [4]. The nature of these relations is, however, not further explained. In reality it could be flows of material, information, or energy as well as cause-effect relations. The extension of Vester also fits in the understanding of systems as sets of elements which are linked to each other. A better knowledge and understanding of interdependences is not only desired in the field of performance measurement. In respect of a more effective analysis of complex systems, a consideration of interdependences is necessary. The common way of analyzing a system focuses on the structure and the isolated system elements. By this means the gained information contains nothing about how the system elements interact with each other.

As a result of the literature research, the Interdependency Matrix (IM) was selected as a very simple method for analyzing interdependences among performance measures and is therefore a suitable method for the presented purpose. The IM is based on the so-called paper computer according to Frederic Vester [4], Ninck [8] and Ulrich [7], and helps identifying and evaluating a system's critical variables. The matrix allows the calculation of three approximate measures (called "influence indices") for the extent to which any variable: (a) influences other variables; (b) is itself influenced by them; and (c) is a critical leverage point for intervening into the system. Therefore the IM is useful to depict the interdependences among the PIs of the SPMS. The two-dimensional matrix contains all performance indicators arranged vertically and horizontally, with the matrix entries indicating the strength of influence of the PI on the vertical axis on those PIs arranged along the horizontal. In the IM, a value of "1" or "2" means that the PI in the corresponding row has a medium or strong influence on the PI of the corresponding column, while a value of "0" indicates that there is no significant influence in general. For the application of the method to the investigation of interdependences among PIs, the IM is defined as follows:

- IM nxn is a n x n-Matrix with n = number of PIs
- for each element $m_{ij} \in$ IM: m_{ij} is the weighted influence of PI_i on PI_j with $i, j = 1...n$ and $m \in \{0, 1, 2\}$

The objective of the IM is to assess the characteristic of each PI in order to better understand its role and use in the framework of the SPMS. In order to facilitate this, the following figures are calculated [8]:

- **Active sum of PI_i: $AS_i = \sum m_{ij}$ for $j = 1...n$**
  AS_i indicates the degree of influence of a certain PI_i on other PIs. The higher the AS, the higher the influence on other PIs.
- **Passive sum of PI_j: $PS_j = \sum m_{ij}$ for $i = 1...n$**
  PS_j indicates the degree of how a certain PI_j is influenced by other PIs. The higher the PS, the higher the PI is influenced by other PIs.
- **Product of PI_i: $P_i = AS_i \cdot PS_i$**
  P_i indicates the intensity of cross-linking of a certain PI_i. The higher P, the more a certain PI is cross-linked with other PIs.
- **Quotient of PI_i: $Q_i = AS_i / PS_i$**
  Q_i indicates the intensity of activity of a certain PI_i. A low Q (Q < 1) means that a certain PI is more influenced by other PIs than influencing other PIs.

Influence of ↓ on →	PI1	PI2	PI3	PIn	AS	P
PI1		1	1	0	2	6
PI2	1		1	1	3	15
PI3	0	2		2	4	8
PIn	2	2	0		4	12
PS	3	5	2	3		
Q	0.7	0.6	2	0.6		

Fig. 2. Interdependency Matrix (IM) (example)

Using the figures AS, PS, P, and Q, the results can be assessed and interpreted as follows [7,8]:

- **Active PIs with Q > 1** are influencing other PIs to a high degree and are hardly influenced by other PIs.
- **Passive PIs with Q < 1** are influenced by other PIs to a high degree and hardly influencing other PIs.
- **Cross-linked PIs with P = high** (e.g., P > 0.5 · Max (Pi) with i = 1…n) are highly connected to other PIs and are involved into many cause-effect relationships (interdependences). Their influence is high and they are influenced.
- **Isolated PIs with P = low** (e.g., P < 0.5 · Max (Pi) with i = 1…n) are hardly influencing other PIs and are not highly influenced.

The PIs can be positioned according to P and Q in an Interdependency Portfolio.

2.2   Research approach

The hierarchies of PIs developed in the SPMS were transferred to the Vensim software, in order to be able to exploit the causal tracing capabilities of the package. These were further combined with the IRM in order to present a complete causal picture of the relationship between processes, their performance indicators at level 3, and the hierarchy of PIs up to level 1. To assist in the assessment of interdependence between Level 2 PIs, individual causal trees for the 38 Level 2 PIs were created, and the influences between these pairs of Level 2 PIs have been recorded with the Interdependency Matrices for different service clusters.

3   Results

This section details the results of the interdependency exercise for the packaging service area in the project. An analysis of the Interdependency Matrices in MS-Excel is presented and the section concludes with an overall interpretation of the results.

Of the 38 Level 2 Performance Indicators, 33 (87%) were considered for the packaging service. Of these 33, 17 were active, 16 passive and 9 (27%) were considered active and isolated with PIs 18 *(Resource adaptability to modifications of the service object/ operation process)* and 21*(Service object production/ operating flexibility)* being the most notable (all within the target area service flexibility).

Based on the P and Q values, in Figure 3 the PIs are positioned in the Interdependency Portfolio (IP). The positioning of the PIs in the IP can be used to determine appropriate use of the PIs [7,8]. According to [9] the results are interpreted to develop improvement strategies:

- **Active and isolated:** *Intervention, Controlling*
  Interventions here can have a huge impact on a few other PIs. Therefore, they are levers that can be used to influence a system in a targeted way. See

e.g. *PI 19 (Resource flexibility in service operations)* in the IP.

- **Active and cross-linked:** *Accelerators, selective interventions*
  Selective interventions here are crucial and may be used as accelerators of trends. There may be feedback loops that intensify the impact. See *PI 4 (Service interaction reliability).*
- **Passive and isolated:** *Stabilizers, Monitoring*
  Since such PIs show influences with delays, they are stabilizers and should be used for long term monitoring. See **PI 32** *(Service interaction costs).*
- **Passive and cross-linked:** *Indicators, Monitoring*
  Such PIs should be primarily used as indicators for monitoring the status of a system on a mid or long term basis. See **PI 8** *(Service operating output quantity).*

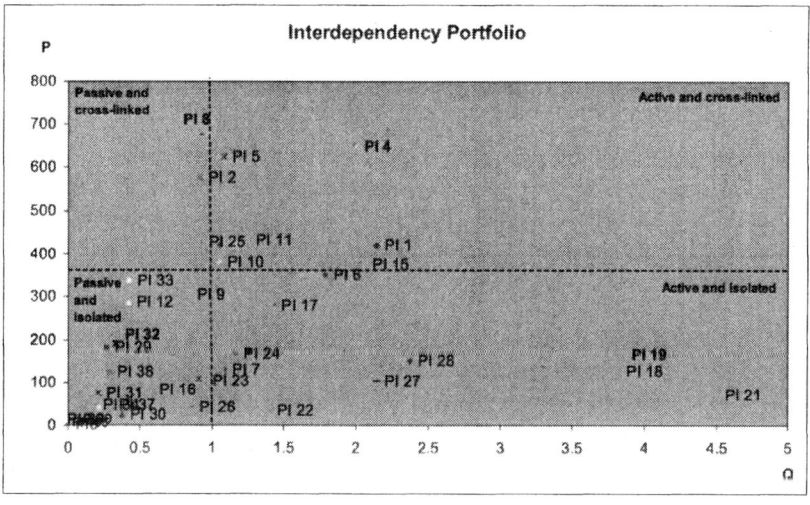

Fig. 3. Interdependency Portfolio for the Packaging Service

The IM which served as the basis for creating the IP contains information about all entries determined to be a strong influence (value=2). For example, *service interaction reliability* (PI 4) has a strong influence on *service adapt cycle time* (PI 22), *service build cycle time* (PI 23), *service interaction cycle time* (PI 24) and *service operate cycle time* (PI 25). The higher the interaction reliability in terms of availability and quality of shared information with customers and other service providers (3rd party suppliers), the fewer communication loops are necessary to ensure a timely delivery of material and man power. Another interesting observation is the strong interrelation between service adaptability/flexibility and the ability to *launch new services* (PI1) and to perform existing services to a higher level of performance (PI7). This is due to the fact that, e.g. better *Resource flexibility in service operations* (PI 19) comprising higher education level, number of languages spoken and investments in personnel trainings result in advanced service operations performance. This connection expresses the importance of staff trainings for the ability to offer new services (e.g. taking over of the complete packaging process) and the efficient processing of regular services. In addition, PI 19 has a medium impact on almost every Level 2 PI across the different target areas (cost and reliability) highlighting the power of PI 19 (active and isolated) to influence the system in a targeted way.

4   Conclusion and Outlook

The methods used to elicit the interdependences between PIs in the Service Performance Measurement System (SPMS) have served to better understanding of the appropriate use of a certain PI and its role in a cross-linked system, as well as the development of service operation improvement strategies for service providers (where and how to intervene in order to improve process performance). The results enable the service provider and the client manufacturer to evaluate the potential impact of improvements in an indicator on other areas of their business and illustrate areas where great care should be taken to fully understand the cause-and-effect and feedback structures present in their business. In future research activities the result will have an impact on the development of a methodology for the individual assignment (selection) of PIs to various services.

5   References

1. Lange I, Schneider O, Larsson A, Minkus A (2006): Logistics controlling concept for benchmarking service delivery performance. In: Proceedings of the 11th World Congress for TQM; December 4-6, 2006, Wellington, New Zealand

2. Osadsky, P.; Fischer, T.; Garg, A.; Lange, I.; Schaedlich, B.; Schneider, O.; Montorio, M.; Nitu, B.; Puetz, F. (2006): Innovation, Coordination and Collaboration in Service Driven Manufacturing Supply Chains (Brochure to EU-funded project InCoCo-'S', www.incoco.net)

3. Neely, A., (2005): The evolution of performance measurement research. In: International Journal of Operations & Production Management, 12 / 2005, S. 1264-1277.

4. Vester, F., (1999): Neuland des Denkens: Vom technokratischen zum kybernetischen Zeitalter, 11. Aufl., München, 1999

5. Supply Chain Council (2006): Supply chain operations reference model SCOR Version 8.0. Pittsburgh: Supply Chain Council (http://www.supply-chain.org)

6. Lange, I.; Schnetzler, M.; Schneider, O.; Osadsky, P., (2007): Design of a Performance Measurement System. In: Proceedings of the 2nd International Conference on Changeable, Agile, Reconfigurable and Virtual Production (CARV), 22.-24. July 2007, Toronto

7. Ulrich, H.; Probst, J.B., (1988): Anleitung zum ganzheitlichen Denken und Handeln: Ein Brevier für Führungskräfte, Bern, Haupt, 1991

8. Ninck, A. et al., (2004): Systemik: vernetztes Denken in komplexen Situationen, 4. Aufl., Verlag Industrielle Organisation, Zürich, 2004

9. Schnetzler, M.; Sennheiser, A., (2003): Identification of performance improvement strategies in production networks. In: Cunningham, P. et al.: Building the knowledge economy - issues, applications, case studies. Amsterdam: IOS Press / Omsa, 2003, pp. 306-312

From Toyota Production System to Lean Retailing. Lessons from Seven-Eleven Japan

Shinji Naruo[1] and Sorin George Toma[2]

1 University of Bucharest, Faculty of Business and Administration,
University from Bucharest, 36-46 Bd. M. Kogalniceanu, Bucharest
050107, shinjinaruo@yahoo.co.jp,
WWW home page: http://www.unibuc.ro

2 Academy of Economic Studies, Faculty of Commerce
13-15 M. Eminescu, Bucharest 010511, Romania
tomagsorin@yahoo.com,
WWW home page: http://www.ase.ro

Abstract

In order to face their global competitors, one solution consists for manufacturing companies around the world to implement the Toyota Production System (TPS). Lean manufacturing evolved from TPS. The purpose of the paper is to demonstrate that lean principles and concepts can be successfully applied in a company from the retail industry, such as Seven-Eleven Japan. Lean retailing is now a reality that has forced manufacturers to build standard products on-demand using build-to-order techniques. The retail markets are characterized by a strong competition, short product life cycles, long product development lead times, and highly volatile demand. Today's powerful retailers insist on low prices and refuse to carry inventory. The main objectives of the paper were achieved by using an extensive review of the literature and a case study.

Keywords

Toyota Production System (TPS), lean, manufacturing, retailing, Seven-Eleven Japan

1 Introduction

The automobile was introduced more than one hundred years ago. The automotive age began in Europe at the end of the 19th century when C. Benz started the first commercial production of motor vehicles in 1886, and when twenty-three

Please use the following format when citing this chapter:

Naruo, S., Toma, S. G., 2007, in IFIP International Federation for Information Processing, Volume 246, Advances in Production Management Systems, eds. Olhager, J., Persson, F., (Boston: Springer), pp. 387-395.

automobiles participated in a race from Versailles to Bordeaux and back to Paris in 1895 [2].

At the beginning of the 20th century, it was clear for everyone that the automobile represented modernity and progress. The Ford Motor Company was founded in Detroit in 1903. When H. Ford launched the famous Model T in 1908, he understood that a big business was born. All his plants and branches around the world assembled and sold the Model T, following the assembly line method developed by Ford at Detroit. Based on mass production and economies of scale, the Fordian model implemented a "volume" strategy, accompanied by an advertised brand name. Ford Motor became the epitome of the modern multinational company and the term "Fordism" symbolized the new corporate America. At the end of 1920's, Ford was overtaken by General Motors and then by Chrysler. Created to break away from preceding models, the Sloanist model applied a "volume and diversity" strategy.

2 From Fordism to toyotism

Later, adapting Fordism selectively, the Japanese developed flexible production systems. By using quality circles, team output, or pull systems, they transformed Fordism into Toyotism. The Toyota Production System (TPS) evolved from the Ford manufacturing system. The primary purpose of TPS is to obtain profit through cost reduction, or productivity improvement [6]. Cost reduction and productivity improvement are attained through the elimination of different kinds of waste, such as excessive inventory. The basis of the TPS is the absolute elimination of waste [7].

Today, TPS represents not only tools, methods and techniques, but also a way of doing business. The two pillars of the so-called "Toyota Way" are continuous improvement and respect for people. The principles that constitute the Toyota Way are related to a long term business philosophy. Based on his 20 years of studying the Toyota Company, Liker stated that 14 main principles constitute the so-called "Toyota Way" [1]. These principles represent the foundation of TPS and are divided into the following 4 categories (the 4 "P" model): philosophy (long-term thinking), process (eliminate waste), people and partners (respect, challenge, and grow them), and problem solving (continuous improvement and learning).

Among other elements, the kanban system is a key of TPS. This system is supported by smoothing of production, standardization of jobs, autonomation, and improvement activities [6].

In 2003, Toyota overtook Ford Motor to become the second largest automaker of the world. In the first quarter of 2007, Toyota unseated GM as the world's largest automaker.

Businesses today have struggled to successfully imitate Toyota. That is why other Japanese companies are trying to implement TPS in their plants (box 1).

Table 1. The principles of Toyota Way

Section	Principle
I. Long-Term Philosophy	1. Base your management decisions on a long-term philosophy, even at the expense of short-term financial goals
II. The Right Process Will Produce the Right Results	2. Create continuous process flow to bring problems to the surface.
	3. Use "pull" systems to avoid overproduction.
	4. Level out the workload (heijunka).
	5. Build a culture of stopping to fix the problem, to get quality right the first time.
	6. Standardized tasks are the foundation for continuous improvement and employee empowerment.
	7. Use visual control so no problems are hidden.
	8. Use only reliable, thoroughly tested technology that serves your people and processes.
III. Add Value to the Organization by Developing Your People and Partners	9. Grow leaders who thoroughly understand the work, live the philosophy, and teach it to others.
	10. Develop exceptional people and teams who follow your company's philosophy.
	11. Respect your extended network of partners and suppliers by challenging them and helping them improve.
IV. Continuously Solving Root Problems Drives Organizational Learning	12. Go and see for yourself to thoroughly understand the situation (genchi genbutsu).
	13. Make decisions slowly by consensus, thoroughly considering all options implement decisions rapidly.
	14. Become a learning organization through relentless reflection (hansei) and continuous improvement (kaizen).

Box 1- Introduction of TPS at Fujitsu Component Limited

In May 2001, Fujitsu Component Limited, one of the companies of the Japanese Fujitsu Group, was selected to put TPS into practice [9]. In order to improve its performances, outside TPS consultants were hired. In approximately 15 months, major improvements occurred as follows:

- Three-fold increase in productivity;
- 50 % reduction in inventories;
- 50 % reduction of work space;
- 48 % reduction in manufacturing lead time.

The successfully introduction of TPS in Fujitsu Component Limited was enabled by:

- A revolution in employees consciousness;
- A sense of crisis felt by all employees and management;
- Producing what is needed in only the required quantity when it is needed (the pull production system);
- Searching solutions for the complete elimination of waste.

In sum, the introduction of TPS comprised two main stages:

1. Instruction activities and elimination of waste activities;
2. Full-scale TPS principles implementation.

3 TPS and lean manufacturing

Lean means a systematic approach to identifying and eliminating "muda" (waste) through continuous improvement. It seems that the famous Henry Ford had been using parts of lean as early as the 1920's. But, the TPS was the fundamental basis for the "lean" movement that has dominated manufacturing environments for the last fifteen years (box 2).

Box 2- Applying lean manufacturing at Toyota plant at Georgetown (USA)

In 2006, the 7.5 million square foot Toyota factory located in Georgetown produced more than a half million vehicles [4]. The Georgetown plant is the biggest Toyota Motor's manufacturing facility in North America. Using pull system, andon, kanban and other TPS tools, Toyota Motor Manufacturing Kentucky (TMMK) is designed to produce vehicles in a continuous flow. As S. St. Angelo, the president of TMMK, stated, "there were a great difference between GM's mass production and the TPS" and "more car manufacturers are going towards what is called lean manufacturing, which is just another flavour of TPS". Its entire culture was built around the lean philosophy, based on waste elimination and on the continuous improvement of its manufacturing and business processes. Producing best quality cars, with fewer worker-hours, lower inventory and fewer defects, than any other competitor in the world, this lean factory is using technology on a large scale. In this respect, Toyota formed at the end of the 1990's a partnership with ABB Automation Technology, a Swiss- based maker of manufacturing robots. Due to the painting robots, Toyota saves about 29 USD per vehicle, and 2.1 hours off the time needed to produce a vehicle.

Step by step, the lean movement has penetrated other industries, like retailing. Lean retailing is now a reality that has forced manufacturers to build standard products on-demand using build-to-order techniques. The retail markets are characterized by a strong competition, short product life cycles, long product development lead times, and highly volatile demand.

Today's powerful retailers insist on low prices and refuse to carry inventory. That is why lean retailers are now expecting manufacturers to provide rapid and frequent replenishment of retail goods based on real-time sales. In other words, manufacturers are now facing lean retailing pressures. Lean retailing itself contributes to the transformation of manufacturing.

Lean gives Seven-Eleven Japan the clear vision to see what markets desire. Also, lean retailing provides competitiveness for this largest convenience store chain in Japan.

4 Expanding lean in the retailing industry. Study case: Seven Eleven Japan

We can compare the lean management structure between manufacture and retail industry. For example, the kanban system used in TPS imposes on the later process to send requests to the earlier one. It means that demand represents a trigger in the manufacturing industry. This concept is similar to Demand Chain Management (DCM) in retail industry.

If TPS has some key concepts like JIT, Kanban, or Autonomation, there are no established lean management concepts in retail, but there are other similar concepts like Supply Chain Management (SCM), DCM, franchise system, or in store merchandising. These concepts are not still organized as lean concepts in retail. However, when we have studied Seven-Eleven Japan (SEJ), we found similar corporate philosophy between Toyota and SEJ. They both trust human intelligence and they don't take automation easily. In case of Toyota, it is Autonomation, and in case of SEJ, it is individual store management. Both companies are excellent companies in their field. They have achieved high business performance, created unique corporate philosophy and global operation in the world.

Table 2. Comparison between manufacturing and retail regarding lean management

	Lean management		
	Manufacturing	Retailing	
	TPS (Toyota Production System)	General model and concept	Seven-Eleven key concept
1	JIT (Just In Time)	SCM (Supply Chain Management)	CDC, NDF
2	Kanban system	DCM (Demand Chain Management)	Store initiative ordering
3	Production smoothing	SCM (Supply Chain Management)	Team merchandising
4	Shortening setup time	SCM (Supply Chain Management)	Customer focus
5	Shortened Lead Time	Order-delivery	Dominant strategy
6	Standardization of Operations	Franchise system	Store initiative ordering
7	Autonomation	In store merchandising	Individual store managemen
8	Kaizen (Improvement Activities)	In store team meeting	Tanpinkanri

SEJ was opened in 1974 in Tokyo based upon international licensing agreement with Southland Corporation, USA (currently Seven-Eleven, Inc, USA). Now, SEJ operates 11,525 stores (as of December 2006) in Japan. The total store sales was 2,498,754 millions yen ($21,275 mil) and the revenue from operations was 492,831 millions yen ($4,176 mil) in Japan. SEJ achieves this huge profit with only 4,804 employees because of its franchise system.

SEJ is a franchiser of Seven-Eleven convenience stores in Japan. Each store has about 100-150 square meters sales floor and 2500 Stock Keeping Unit (SKU). Daily customer transaction in each store is about 1,000 and the average purchase amount is $5-6 per customer. However, SEJ keeps growing on sales and in the number of stores in past 30 years. They believe that "Reviewing and changing day-to-day business from the standpoint of the customer is the basis for breakthroughs". It is "TANPINKANRI" (item by item control), the basis of retailing. SEJ is facing three main lean challenges.

Lean challenge no. 1: Total integration in the process from ordering to delivery
Ordering in each store with store's responsibility is the key. Store employee orders each item in the store. The order goes to Data center. Data center analyzes the order data, and then generates order data to the vendor, picking data/shipping instruction to distribution center, and inspection data to each store. Based upon this integrated

information flow, delivery process is well managed among store, vendor and distribution center. After store inspection, inspected data goes to accounting center. Accounting center verifies the data, and then sends account payable data to vendors. The whole process looks like a TPS line. Data, paper, product and human resources work together efficiently without waste (Fig. 1). The delivery route and schedule are also carefully designed in each store group. In order to improve the delivery network, SEJ conducts the strategic process.

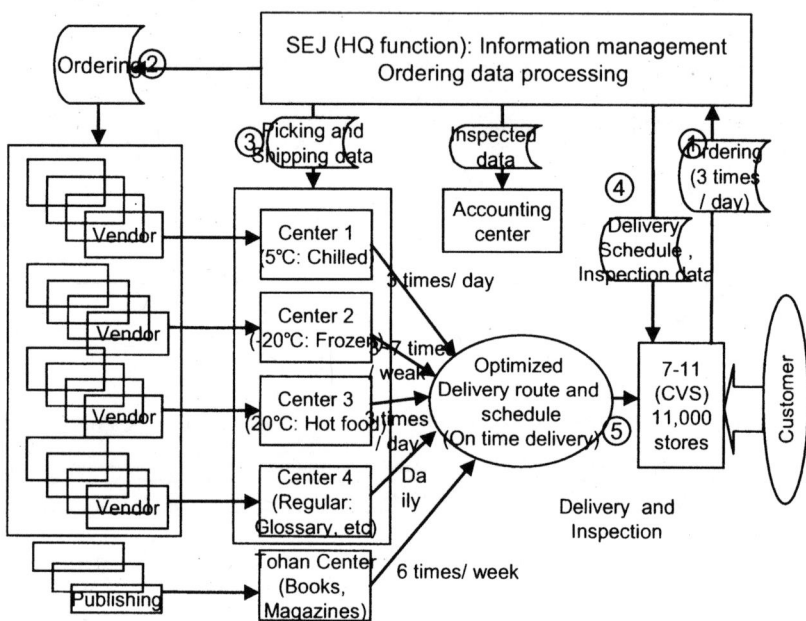

Fig. 1. Total integration system from ordering to delivery

Lean challenge no. 2: Increasing delivery frequency with reducing number of tracks
In 1974, there were 70 times deliveries to stores per day per store in SEJ. Now, it is only 9 times deliveries per day per store (Fig. 2). This enabled drastic reduction of the distribution costs. Furthermore, it reduced inspection workload in store side. On the other hand, it enabled the store operators to concentrate on sales and customers in the store. SEJ had succeeded in increasing delivery frequency by product category. The SEJ success relies on the designing of a combined distribution center (CDC). SEJ designed the CDC concept by category and by temperature. CDC is operated by a vendor's representative, or by an independent third party (Table 3).

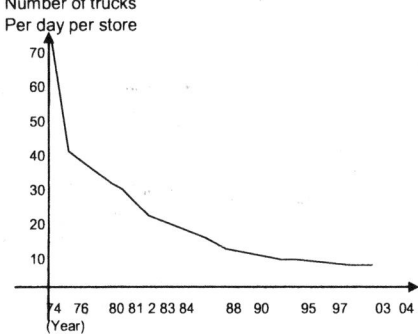

Fig. 2. Reduce number of trucks with increasing service level.

Table 3. CDC for integration distribution system

Combined Distribution Center (CDC)		Delivery frequency	Number of Centers
5 management	Chilled food Center	3 times a day	67 Centers
20 management	Hot food Center	3 times a day	69 centers
20 management	Frozen center	3-7 per week	47 centers
Ambient temperature	Process food center Liquor center General goods Center Sundries and Confectionery Center	Daily delivery	145 centers (Including 18 consolidated ambient temperature centers)

(Data from SEJcorporate outline 2005)

Lean challenge no. 3: Increase sales combined with decrease inventory.

It used to be wrong common sense in retail. For example, one was "much merchandise in store enables large sales". Many retailers believed in the common sense, but SEJ doubted. SEJ narrowed the number of SKU, and then reduced inventory level by deleting dead items. The results were a financial cost reduction and a store image change. Due to a lower inventory level, stores had the chance to introduce new items, and that attracted customers. The Japanese Seven-Eleven stores are a third of those in US, and that is why they are putting significant limits on inventory space at the store. During its existence, SEJ has proved the relationship between inventory and sales (Fig. 3).

This slim body approach of SEJ is similar to TPS inventory reduction. TPS reduces inventory, and then increases productivity and quality. TPS enables a better working environment, high performance and quality management. SEJ's store has achieved the same thing. Dead item eats space, and this is waste. By reducing SKU, space is creating, and this enable customers to walk and find easy the merchandises in the stores.

SEJ's corporate philosophy is "Corresponde to changes and execute basic principles thoroughly". Top management declares "Tomorrow is another day". In order to be agile in the market, SEJ must keep lean. In this respect, SEJ has implemented three strategies for keeping an agile constitution.

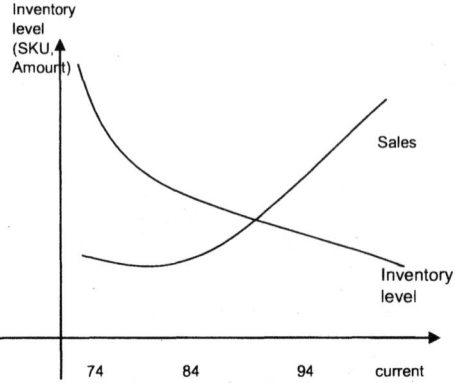

Fig. 3. Relation between inventory and sales

The first strategy is the the franchise system. The majority of the stores is franchised store. SEJ owns small number of stores only (about 5 %). The second strategy is partnership with distribution centers. CDC dedicatedly serves 7-11 stores. However, the CDC is managed by vendors or a third party. They have achieved a mutual prosperity of the business (Fig. 4). The third strategy is outsourcing. SEJ has to manage huge networks and data processing, and that is why, it has outsourced the system development and operation to system integration companies.

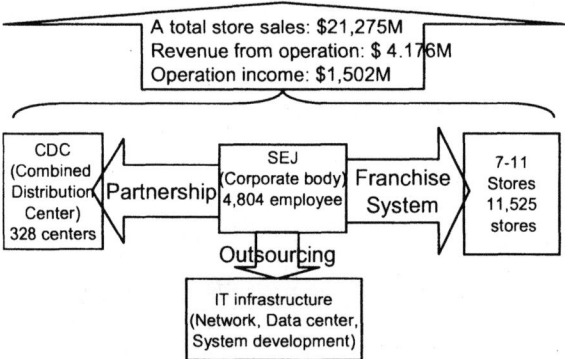

Fig. 4. Lean management in SEJ

SEJ is deeply focusing on the customer. SEJ is an excellent world retail company. SEJ doesn't intend to be a lean retail company. However, there is some unique lean management practice in its business operations.

5 Conclusions

In our paper, we aim to report on the introduction of lean to retailing industry. In order to demonstrate that lean principles and concepts are successfully applied in a company from the retail industry, we presented the case of Seven-Eleven Japan. Without being a lean company, Seven-Eleven Japan have obtained high business performances.

The clear conclusion that emerges from our paper is that lean companies are able to be more responsive to market trends, deliver products faster, and provide products less expensively than their non-lean competitors.

By putting significant limits on inventory space at the store, Seven-Eleven Japan needs rapid replenishment of items. In order to deliver products in time and with minimal amount of lead times, the retailing chain has created its own distribution system, which it has integrated with all the stores. Seven-Eleven distribution centers carry no inventory.   Seven-Eleven Japan has proved to be an excellent retail company.

References

[1] Liker J. K.- *The Toyota way. 14 management principles from the world's greatest manufacturer*, McGraw-Hill, New York, 2004

[2] Loeb A. L.- Birth of the Kettering doctrine: fordism, sloanism and the discovery of tetraethyl lead, *Business and Economic History*, vol. 24, no. 1, Fall 1995

[3] Fujimoto T.- *The evolution of a manufacturing system at Toyota*, Oxford University Press, New York, 1999

[4] Fujimoto H.- Supply Chain for the Synchronization of Production and Distribution, *Osaka Keidai Ronshu*, vol. 53, no.3, September 2002

[5] Harris A.- Made in the USA, *Manufacturing Engineering*, February 2007, http://www.itc.mh.ca/downloads (last accessed April 2007)

[6] Monden Y.- *Toyota Production System: an integrated approach to just-in-time*, EMP, Institute of Industrial Engineers, Norcross, 1998

[7] Ohno T.- *Toyota Production System: beyond large-scale production*, Productivity Press, Oregon, 1988

[8] Sakai Y., Sugano T., Maeda T.- Introduction of Toyota Production System to promote innovative manufacturing, *Fujitsu Science Technology Japan*, no. 43, p. 14-22, January 2007, http://www.fujitsu.com (last accessed March 2007)

[9] Shimizu K.- *Le toyotisme*, Editions la Decouverte, Paris, 1999

[10] Weil D.- Lean Retailing and Supply Chain Restructuring: Implications for Private and Public Governance, *Paper prepared for "Observing Trade: Revealing International Trade Networks"*, Princeton Institute for International and Regional Studies, Princeton University, 9-11.03.2006

[12] Womack J. P., Jones D. T.- *Lean thinking. Banish waste and create wealth in your corporation*, The Free Press, New York, 2003

[13] Annual report 2005, Corporate outline 2005, CSR Report 2005, Corporate profile 2006, Seven–Eleven Japan Co.,Ltd, http://www.sej.co.jp (last accessed April 2007)

Improving Service Operation Performance by a Cross-Industry Reference Model

Peter Osadsky[1], Amit Garg[1], Bogdan Nitu[1], Oliver Schneider[2]
and Stefan Schleyer[3]

1 Forschungsinstitut für Rationalisierung,
Pontdriesch 14/16, 52062 Aachen, Germany
incoco@fir.rwth-aachen.de
http://www.fir.de

2 ETH Zurich, Center for Enterprise Sciences (BWI), 8092 Zurich,
Switzerland
OSchneider@ethz.ch
http://www.lim.ethz.ch

3 SKF GmbH, Gunnar-Wester-Str. 12, 97421 Schweinfurt
Stefan.Schleyer@skf.com
http://www.skf.com

Abstract.
The importance of business related services has been growing consequently during the past years. Industrial Services constitute the greatest share of business related service with 30% of their intermediate output. But those kinds of services are facing tremendous challenges in terms of synchronizing their process with the manufacturing processes of their customers. The EU funded project InCoCo-S aims to tackle these challenges by the development of an innovative reference model for the collaboration between Service Providers and Manufacturers. Currently, the development of the structure of the reference model is finished and its validation has already begun. In this scope, SKF, as one of the consortium partners, is currently implementing the results in their new service division – Windmill Condition Monitoring. This paper provides detailed information about conducted surveys within the InCoCo-S project, the Reference Model and its practical application.

1 Introduction

Business related services constitute the largest sector of the economy employing around 55 million persons in 2001 – nearly 55 % of total employment in the EU market economy - and representing around 70% of EU GDP [1]. Since business related services are the dominant part of the European market economy, the

---

Please use the following format when citing this chapter:

Osadsky, P., Garg, A., Nitu, B., Schneider, O., Schleyer, S., 2007, in IFIP International Federation for Information Processing, Volume 246, Advances in Production Management Systems, eds. Olhager, J., Persson, F., (Boston: Springer), pp. 397-404.

European manufacturing industry is highly dependent on external service providers and consumes nearly 30% of the intermediate output of business related services, so called Industrial Services.

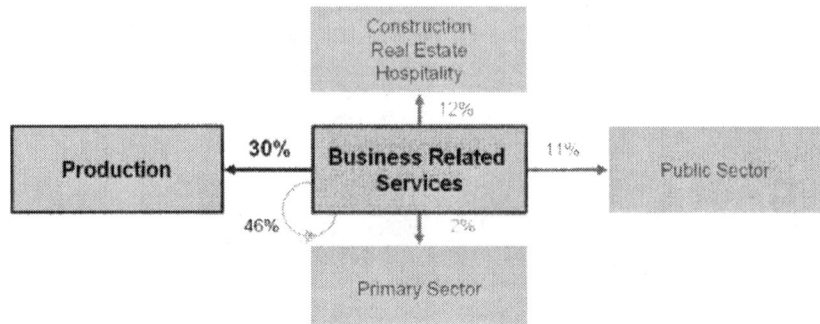

**Figure 1: Key issues to be addressed from manufacturer perspective
and service provider perspective [1]**

Unfortunately, managers of both manufacturing and supporting service organisations are facing massive difficulties when trying to integrate and synchronize their joint activities. The key challenges for the service-supply chain can be categorized as: enabling coordination by synchronisation of business processes and information, developing metrics for measuring the performance of joint activities, advanced support for decision making and finally encapsulation of these developments into standards for the service sector.

The EU funded project "Innovation, Coordination and Collaboration in Service Driven Manufacturing Supply Chains" (InCoCo-S) takes up this challenge by developing an InCoCo-S Reference Model (IRM) for the collaboration between Service Providers and Manufacturers.

1.1 Survey Results

In order to get detailed requirements from the company's perspective, two surveys were designed and carried on involving 81 service providers and 81 manufacturers [2].

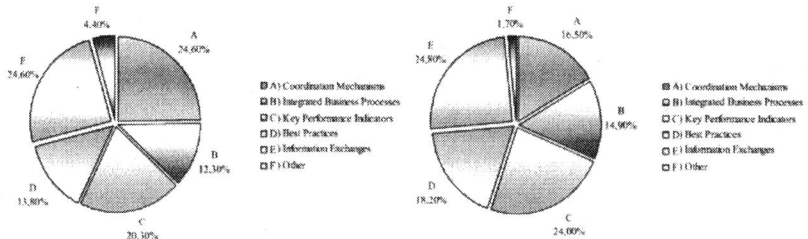

Figure 2: Key issues to be addressed from manufacturer perspective and service provider
perspective [2]

The survey reveals key issues for the IRM. Both Manufacturers and Service Provider state great importance for Coordination Mechanisms, Performance Measurement Systems and Best-Practices.

To consider the stated aspects from the Service Provider and Manufacturer in the IRM, an appropriate framework should be used. Although being used primarily in logistics, the SCOR Model provides such a framework. But Industrial Services are regarded only as a single process element in SCOR so that it cannot be used for detailed service operations. Hence, InCoCo project aims at the development of the IRM specifically for Industrial Services, so that Service Provider and Manufacturer can cope with the aforementioned challenges in their regularly collaboration.

1.2 Approach towards the InCoCo-S Reference Model

The IRM was developed in several workpackages together with the industrial partners of the InCoCo-S consortium. The consortium is composed of strong industrial partners such as SKF, Sigpack-Bosch, SAP, Comau, Adige, Hörmann Industrietechnik, Unitech and Ventana Systems. It covers five major Industrial Services in the following domains:

- Maintenance
- Packaging
- Logistics
- Quality Control
- Retrofit

In the beginning, the aforementioned survey was conducted to gain general requirements from the industry [2]. Detailed requirements were elaborated by a processes analysis of the industrial partner in the consortium. Based on those, To-Be scenarios were derived which reflect optimal service processes [3]. Furthermore, a methodology was developed in order to analyze Service Cost Drivers in the service processes [4]. Resting upon these findings, the design of the IRM was constructed which addresses particularly Best-Practices and Performance Metrics [5].

Since the term Best-Practice is not strictly defined in the scientific community, Coordination Mechanisms and Collaboration Strategies were examined on detailed level in order to compile them to Best-Practices. First, Coordination Mechanisms and Collaboration Strategies were identified and adapted to the Service Domain [6]. Based on these findings, business processes were developed for the service processes and adequate information systems were designed [7]. Finally, they were evaluated quantitatively in order to get detailed information about their current use in practice. In the end, major findings were compiled in more than 100 Best-Practices.

In order to evaluate the service processes, a Performance Measurement System (PMS) was developed with according indicators. The progress of the development started with a comprehensive literature research of existing PMS [8]. The identified indicators were collected and aggregated into an innovative PMS for the service domain which addresses several perspectives such as the satisfaction of service provider and customer, the service object and the service delivery process. Altogether more than 300 indicators were collected that cover quality, time and cost dimensions [9].

The aforementioned Best-Practices and the PMS were developed in close cooperation with the industrial partners of the consortium. To ensure a high quality of the IRM, each of the business processes has to be standardized regularly according to the findings of the workpackages and to industrial experiences. To ensure a durable solution for these tasks, the Technical Steering Committee (TSC) was constituted with representatives of all research partners, SKF, H2O and Comau. In addition to the standardizing activities of the TSC, it is in charge of the acceptance of new processes in the IRM.

Summarizing, the IRM covers more than 100 Best-Practices, 300 indicators and covers service process for maintenance, packaging, logistics, retrofit and quality control. At the moment more than 50 process elements are developed so far for the IRM.

2 The InCoCo-S Reference Model

2.1 Structure

The IRM is structured hierarchically. On the first level, a reference guideline is given for superior activities of services. The second level offers business related adoptions of the Level 1 activities. The 3rd and last level covers each of the Level 1 and Level 2 processes on a more detailed level of the service provider.

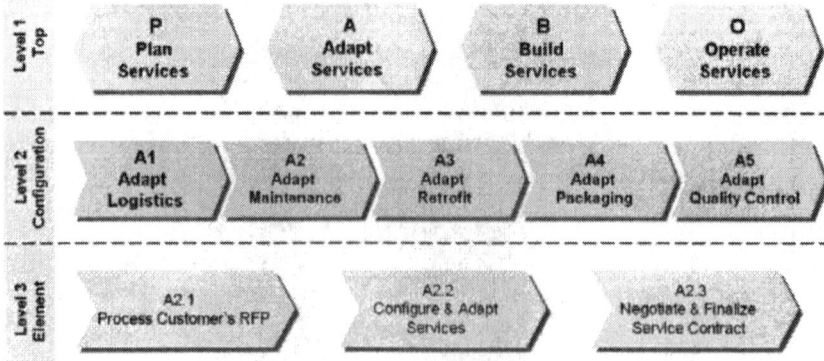

Figure 3: Structure of the IRM

The differentiation between Plan, Adapt, Build and Operate activities on level 1 is a basic feature of the IRM that is continued on each of the three Levels. Plan activities consist of preparatory work for Adapt, Build or Operate activities. Adapt activities address all processes from the customer inquiry to the signed service contract. Processes that have to do with resource allocation in order to prepare the service installation at the customer are defined as Build activities. Finally, Operate activities describe all processes along the service delivery.

2.2 Process Interactions

For each of the Adapt, Build and Operate activities, process interaction diagrams were elaborated which depict the available set of Level 3 processes as shown in figure 4.

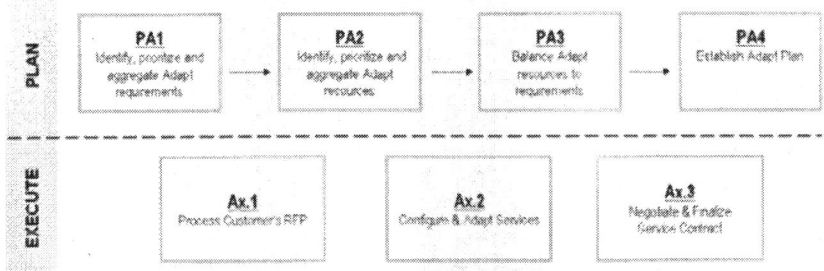

Figure 4: Plan Interaction diagram of the Adapt Phase of the IRM

The interaction diagram describes Level 3 processes for PLAN and EXECUTES activities. The PLAN activities cover all the processes that are necessary to create a service offer. Therefore, the customer requirements and the resources of the service provider are identified and priorized first. Afterwards, they will be matched with one another in order to establish a service plan. This plan is the basis for a service offer that will be given to the customer.

The EXECUTE activities encompassed value-added processes in the service supply chain. In the ADAPT Phase, these ones are defined as the Customer Inquiry, the Configuration of the Services for the Customer and the Negotiation und Finalizing of the Service Contract by both parties.

The Interaction Plan in figure 3 do not encompasses interdependencies of the EXECUTE activities to one another because these processes differ in every service organization. Hence, the IRM provides only a set of processes that have to be linked by every service organization by its own.

2.3 Process Elements

Each of the process elements on Level 3 is described on a more detailed level by using a framework as shown in figure 3.

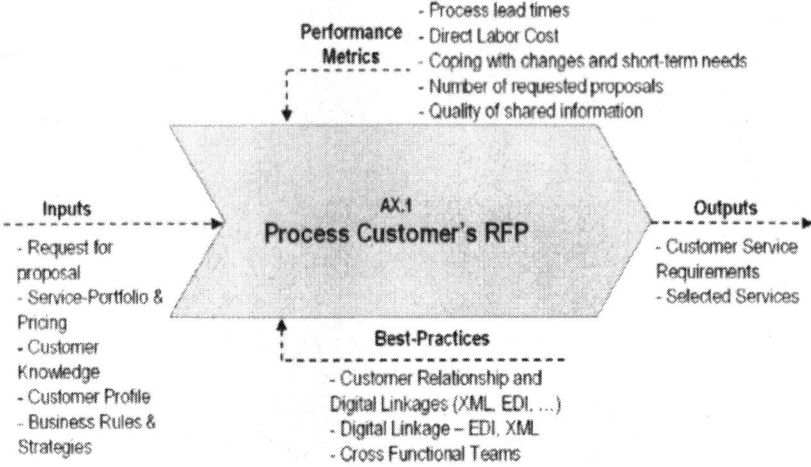

Figure 5: Process Elements on Level 3

The processes are specified by Inputs, Outputs, Performance Metrics and Best-Practices. The analysis of Inputs and Outputs reveals interdependencies to other process elements that enable the Service Provider to reduce inherent complexity and thus to increase its effectively. Performance Metrics cover several indicators for the Service Provider to measure the performance of the services in terms of time, quality and costs. Finally, Best-Practices allow an easy adoption of proven methodologies.

2 Practical Adaptation

SKF is a leading provider of condition monitoring systems for bearings. Currently the company created a new service division to enlarge their activities in a growing market segment. To improve the service effectivity and efficiency of their service processes, SKF is implementing the IRM for their business processes. Figure 3 shows the interactions diagram of the IRM that was used as a starting point.

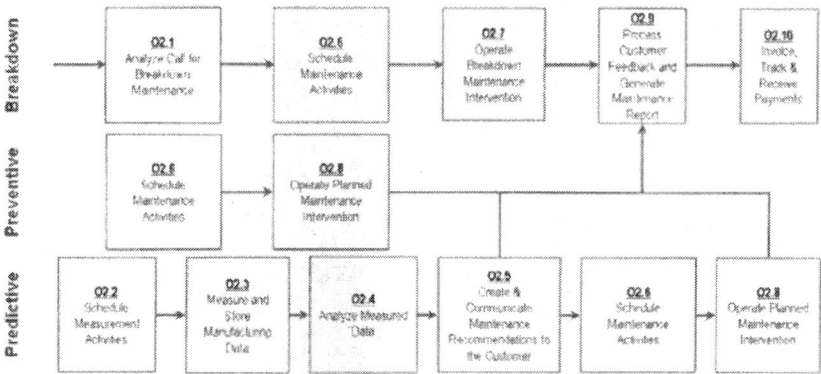

Figure 6: Interaction diagram for the Operate Phase for Maintenance Services

By modeling the business process in the service by using the IRM, SKF gains the possibility to define organization specific processes as shown figure 6. In case of changing customer requirements, these processes can be changed easily with a high level of transparency. New processes can be added, abundant processes can de eliminated and the sequence of the processes can be switched if necessary.

Furthermore, the performance metrics allow a measurement of various dimensions of the service delivery. Therefore a benchmark between each of the conducted orders is possible as well as a focused optimization of the overall service performance.

Finally, the processes model according to the IRM helps a preparation for software implementation by aligning the business process requirements with IT systems. In that way, recent business processes can be standardized the internal & external communication.

Comprising, SKF gains great advantages for their current business field that inherits great potential for further enlargement.

3 Conclusion

The IRM is an innovative approach to model business processes in the service domain. It provides great advantages for service departments in terms of aligned resources and information flows. Based on the principles of a supply chain, the IRM regards various kinds of information such as Inputs and Outputs, Best-Practices and Performance Metrics. A pilot implementation is being done for the new service division of SKF. Thus, SKF could identify further potential to improve the effectivity of their service processes as well as to align the IT Systems with the business process & further plan the personal resources for the new business.

During the next few months, the consortium will focus on developing further case studies with the industry outside the consortium to validate and implement the reference model. There are few ongoing validations where maintenance services are being offered by a collection of service providers. Such validations strengthen the model and make it applicable in generic to providers of industrial services.

Companies operating in the domain of logistics, maintenance, packaging, modernization & quality control are very much invited to participate in this validation exercise. Interested companies who want to join the validation activities should contact the InCoCo-S consortium at incoco@fir.rwth-aachen.de

4 Literature

[1] European Commission: The competitiveness of Business Related Services and their contribution to the performance of European enterprises. Final version released on 4th of December
[2] DL2.1 - Consolidated Supply Chain & Service Providers Survey Results
[3] DL2.4"As Is" Business Use Cases & Requirement Specification in the Service-Supply Chain domain for all the 4 business cases
[4] DL2.6 Design of S-SCOR Framework Structure
[5] DL3.1 Design of coordination mechanisms & collaboration strategies
[6] DL3.2 Development of business processes for service – supply chain
[7] DL4.1 Formulation of the MoP for service – supply chain
[8] DL4.2 Methodology to Measure the Performance of Systems and Development of Tools to quantify added value and to determine market pricing for different services

Integrating Lean and Agile Strategies into the Production Control System for Mixed-model Production Lines

Katsuhiko Takahashi, Kana Yokoyama and Katsumi Morikawa
Department of Artificial Complex Systems Engineering,
Graduate School of Engineering, Hiroshima University
1-4-1, Kagamiyama, Higashi-Hiroshima, 739-8527, Japan.
{takahasi, k4580, mkatsumi}@hiroshima-u.ac.jp
WWW home page: http://www.pel.sys.hiroshima-u.ac.jp

Abstract

For mixed-model production lines, this paper proposes a production control system based on lean and agile strategies for responding to changes in product mix is proposed. Performance of the proposed system under the conditions of unstable changes in product-mix is analyzed by simulation experiments, and it can be claimed that the proposed system can respond to changes in product mix by re-allocating work elements into each work center, and inventories can be decreased without decreasing customer service level.

Keywords

Lean production, Agile production, Production control, Mixed-model line

1 Introduction

Based on the diversification of customers needs, recent production environments become more and more uncertain and competitive. Under the uncertain and competitive production environments, production lines for mass production are requested not only to produce efficiently but also adjust their own systems effectively, that is not only lean but also agile production is requested in the production lines. Then, production lines have to be shifted from single model lines to mixed-model lines, and mass-production has to be shifted to mass-customization. Furthermore, in order to realize mass-customization, mixed-model production lines are requested to have a function to respond to changes in product-mix.

In the previous literature on production control systems for single-model production lines, systems based on lean and agile strategies have been proposed

Please use the following format when citing this chapter:

Takahashi, K., Yokoyama, K., Morikawa, K., 2007, in IFIP International Federation for Information Processing, Volume 246, Advances in Production Management Systems, eds. Olhager, J., Persson, F., (Boston: Springer), pp. 405-412.

(Hopp and Roof [1], Takahashi and Nakamura [2]). In the systems, just-in-time (JIT) ordering based on lean strategy is included, and the function to detect unstable changes in demand and adjust the buffer size, that is based on agile strategy, is also included. However, in the previous literature on mixed-model production lines, line balancing problems have been considered by various researchers only from the view point of lean strategy, and there is no literature on the problems from the view point of not only lean but also agile strategies. In order to integrate agile strategy into the mixed-model production lines, not only adjusting the buffer size but also re-allocating the work elements into each work center are necessary to be considered in responding to changes in product-mix.

Therefore, in this paper, a production control system based on lean and agile strategies is proposed for responding to changes in product mix through re-allocating work elements into each work center, and the performance under the conditions of unstable changes in product-mix is analyzed by simulation experiments.

2   Mixed-model Production Lines

In this section, the mixed-model production line considered in this paper and CONWIP control for the production line are defined.

2.1   Assumptions

The mixed-model production line considered in this paper is assumed as follows:
  (1) The production line produces multiple standard products that can be made to stock.
  (2) The demand for product fluctuates stochastically, and it includes stable and unstable changes. The mean μ_D and the standard deviation σ_D of total demand per time are fixed, but the product-mix, the demand ratio of product j, r_j, fluctuates not only stably but also unstably. Let $d_{i,j}$ be the ith inter-arrival time of product j, the mean $\mu_{i,j}$ fluctuates not only stably but also unstably, but the standard deviation $\sigma_{i,j}$ is fixed. A backorder of demand can be allowed.
  (3) All of the products are produced through a serial production line with N stages, and each stage has a work center called the 1st, 2nd, or Nth stage accordingly, as the process proceeds. The number of stages, N, is fixed.
  (4) The process at each work center is un-paced, and the process time fluctuates stochastically. Work elements are allocated to each work center, and the mean $\mu_P^{(j,n)}$ and standard deviation $\sigma_P^{(j,n)}$ are obtained from the process time of allocated work elements. The allocated work elements are re-allocated in response to unstable changes in product-mix.
  (5) Each stage has an inventory point for stocking the items processed at the stage, and the buffer inventory is stocked at the inventory point of the final stage. The buffer size S_j of item j is controlled dynamically in response to unstable changes in product-mix.

2.2   CONWIP Control

For the production line defined above, a production control system is proposed by integrating lean and agile strategies. In the proposed production control system, constant work-in-process (CONWIP) control [3] is utilized as an alternative to kanban control in just-in-time ordering, that is based on lean strategy, and the CONWIP control is useful to modify the allocation of work elements at each work center (Takahashi and Morikawa [4]).

In the CONWIP control, after releasing the orders of the pre-determined quantities for all products, a new order of a product is released to the first stage just after the product has been completed. Then, at each stage, each order is processed based on first in first out (FIFO) policy.

Based on the CONWIP control, the ith order release time $O_{i,j}^{(n)}$ of product j to stage n can be formulated as follows:

$$O_{i,j}^{(1)} = \max\{D_{i,j}, P_{i-S_j,j}^{(N)}\} \tag{1}$$

$$O_{i,j}^{(n)} = \max\{P_{i,j}^{(n-1)}, \max_{(k,l)\in Q_{i,j}^{(n)}} P_{k,l}^{(n)}\} \quad (n = 2,3,\cdots,N) \tag{2}$$

Here, $D_{i,j}$ is the ith arrival time of product j, that is $D_{i,j} = D_{i-1,j} + d_{i,j}$, and the set of previous order release times $Q_{i,j}^{(n)}$ is formulated as follows:

$$Q_{i,j}^{(1)} = \{(k,l) \mid O_{k,l}^{(1)} \le O_{i,j}^{(1)}\} \tag{3}$$

$$Q_{i,j}^{(n)} = \{(k,l) \mid O_{k,l}^{(n)} \le P_{i,j}^{(n-1)}\} \quad (n = 2,3,\cdots,N) \tag{4}$$

Also, $P_{i,j}^{(n)}$ is the ith completion time of the production of product j at stage n, and it can be formulated as follows:

$$P_{i,j}^{(1)} = \max\{O_{i,j}^{(1)}, \max_{(k,l)\in Q_{i,j}^{(1)}} P_{k,l}^{(1)}\} + p_{i,j}^{(1)} \tag{5}$$

$$P_{i,j}^{(n)} = \max\{O_{i,j}^{(n)}, P_{i,j}^{(n-1)}, \max_{(k,l)\in Q_{i,j}^{(n)}} P_{k,l}^{(n)}\} + p_{i,j}^{(n)} \quad (n = 2,3,\cdots,N) \tag{6}$$

Here, $p_{i,j}^{(n)}$ is the ith process time of the production of product j at stage n.

3   Analyzing the Effects of Product-Mix

The performance of the mixed-model production line defined above is analyzed by simulation experiments, and the effects of product-mix are investigated in this section.

3.1   Experimental Conditions

In this section, the simulation experiments under the following conditions with stable demand are planned.
(1) Two products are produced in the production line.
(2) The inter-arrival time of product demand is assumed to be a normally distributed variable with mean $\mu_D=1.05$ and standard deviation $\sigma_D=0.05$, and the demand ratio between products, $r=0.0, 0.005, 0.010, \ldots, 1.0$.
(3) For producing products, 15 work elements with mean process time and

Table 1. Work elements, immediately preceding elements and mean process time

	element	1	3	4	5	6	8	9	10	13	14	15	
Product 1	immediately preceding	-	1	1	1	1	3	3	4,5 6	8,9	14	13, 14	
	process time	0.6	0.2	0.5	0.7	0.1	0.2	0.6	0.3	0.5	0.5	0.1	
	element	1	2	4	5	6	7	10	11	12	13	14	15
Product 2	immediately preceding	-	-	1	1	1	2	4,5 6,7	7	7	11, 12	10	13, 14
	process time	0.6	0.3	0.5	0.7	0.1	0.4	0.3	0.3	0.2	0.5	0.5	0.4

Table 2. Allocated work elements, ψ, and mean process time $\mu_P^{(j,n)}$ of product j at stage n based on the demand ratio r

Allocation ψ	1			2		
Ratio r	[0.0, 0.20]			(0.20, 0.33]		
Stage n	elements	$\mu_P^{(1,n)}$	$\mu_P^{(2,n)}$	elements	$\mu_P^{(1,n)}$	$\mu_P^{(2,n)}$
1	1,2,6	0.7	1.0	1,2,3,6	0.9	1.0
2	3,4,7,9	1.3	0.9	4,7,9	1.1	0.9
3	5,11	0.7	1.0	5,11	0.7	1.0
4	10,12,14	0.8	1.0	10,12,14	0.8	1.0
5	8,13,15	1.1	0.9	8,13,15	1.1	0.9
Allocation ψ	3			4		
Ratio r	(0.33, 0.42]			(0.42, 0.50]		
Stage n	elements	$\mu_P^{(1,n)}$	$\mu_P^{(2,n)}$	elements	$\mu_P^{(1,n)}$	$\mu_P^{(2,n)}$
1	1,2,3,6	0.9	1.0	1,2,3,8	1.0	0.9
2	4,7,11	0.5	1.2	4,6,7,12	0.6	1.2
3	5,9	1.3	0.7	5,9	1.3	0.7
4	10,12,14	0.8	1.0	10,11,13	0.8	1.1
5	8,13,15	1.1	0.9	14,15	0.9	0.9
Allocation ψ	5			6		
Ratio r	(0.50, 0.64]			(0.64, 0.72]		
Stage n	elements	$\mu_P^{(1,n)}$	$\mu_P^{(2,n)}$	elements	$\mu_P^{(1,n)}$	$\mu_P^{(2,n)}$
1	1,2,3,6	0.9	1.0	1,2,3,6	0.9	1.0
2	4,8,9	1.3	0.5	4,9	1.1	0.5
3	5,7,12	0.7	1.3	7,8,11,12,13	0.7	1.4
4	10,11,13	0.8	1.1	5,10	1.0	1.0
5	14,15	0.9	0.9	14,15	0.9	0.9
Allocation ψ	7			8		
Ratio r	(0.72, 0.83]			(0.83, 1.0]		
Stage n	elements	$\mu_P^{(1,n)}$	$\mu_P^{(2,n)}$	elements	$\mu_P^{(1,n)}$	$\mu_P^{(2,n)}$
1	1,2,8	1.0	0.6	1-3,6	0.9	1.0
2	4,9	1.1	0.5	7,8,9,11,12	0.8	0.9
3	2,5,6,7	0.8	1.5	4,13	1.0	1.0
4	10,11,12,13	0.8	1.3	5,10	1.0	1.0
5	14,15	0.9	0.9	14,15	0.9	0.9

precedence relationship are assumed as shown in Table 1. Also, the process time of each work element is assumed to be a normally distributed variable with the coefficient of variation 0.15 and independent with each other.

(4) Work elements shown in Table 1 are allocated to 5 stages in order to make weighted sum of the mean process times with the demand ratio less than the cycle time 10 by line balancing, and the resulted allocations according to the demand ratio are shown in Table 2.

(5) Buffer size of each product is assumed as $S_j=1, 2, ..., 10, (j=1, 2)$.

(6) A simulation run-length is 100,000 time units, excluding the warm-up run of 5,000 time units, and the number of replications for each simulation run is one.

(7) As performance measures, the mean waiting time of demand, wt, and the total of mean work-in-process inventories, $twip$, are evaluated.

3.2   Results and Implications

Under the experimental conditions shown above, simulation experiments are performed, and the results are shown in Fig. 1.

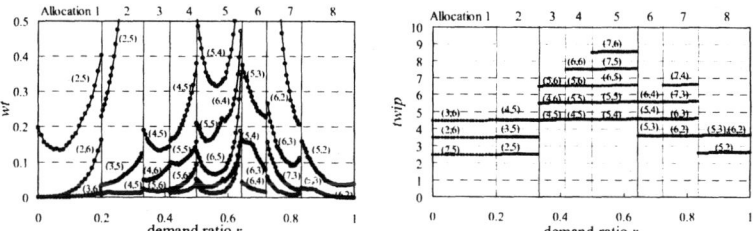

Fig. 1. The effects of the demand ratio, r, the allocation of work elements, ψ, and buffer sizes, (S_1, S_2), on the two performance measures, wt and $twip$

With Fig. 1, it can be seen that wt increases as r increases from 0 to 0.5 or decreases from 0.5 to 1.0 if the allocation of work elements and buffer size are fixed at a certain level. At the same time, $twip$ does not change even if r changes. It can be stated that, if r changes as an unstable change and the allocation of work elements and buffer size are fixed, wt suffers from a significant influence of the unstable change. However, with Fig. 1, it can be seen that wt decreases whereas $twip$ increases, as the allocation of work elements changes or the buffer size increases. Therefore, in order to minimize $twip$ while maintaining wt at less than a certain level, the allocation of work elements and/or the buffer size should be altered in response to the changed r. In responding to the changed r, altering only the buffer size is insufficient, but the allocation of work elements should be also altered. For example, in order to minimize $twip$ while maintaining $wt<0.20$, the buffer size should be altered from (2,5) to (2,6) in response to changes in r from [0,1.4) to [1.4, 0.2] under allocation 1. In order to respond to changes in r as $r>0.2$, it is difficult to maintain $wt<0.2$ even if the buffer size is altered. In the case, the allocation of work elements should be altered from allocation 1 to 2, and the buffer size is also altered in response to changes in r.

4   Proposed Production Control System

Based on the effects of product-mix analyzed above, a lean and agile system that can detect unstable changes in product-mix and control the allocation of work elements and buffer size as a reaction to the detected unstable changes is proposed in this section.

4.1   Detecting Unstable Changes

In order to react to unstable changes in product-mix, the time series of the inter-arrival time of demand is monitored, and only unstable changes must be detected from the time series data. For the purpose, control charts are utilized in process control. Various control charts have been developed, and the EWMA chart [5] is utilized to detect small changes quickly. In this paper, the EWMA chart is utilized to detect unstable changes in demand like Takahashi and Nakamura [2]. At first, the monitored time series data $d_{i,j}$ are filtered by exponential smoothing, a kind of low-pass filter, and the EWMA $H_{i,j}$ are calculated as follows;

$$H_{i,j} = \alpha_j \, d_{i,j} + (1-\alpha_j) \, H_{i-1,j} \tag{7}$$

where α_j is the smoothing constant.

Based on the filtered data of inter-arrival time, the demand ratio can be calculated as $r=H_{i,2}/H_{i,1}$. However, in this paper, the EWMA $H_{i,j}$ are compared with the upper and lower control limits, and unstable changes are detected whenever any of the data is out of these limits. Based on the steady-state mean and variance, the upper and lower control limits, $LCL_{c,j}$, $UCL_{c,j}$ to detect the unstable change in demand ratio for which the current allocation of work elements ψ_c and buffer size $S_c =(S_j)$ should be altered can be formulated as follows;

$$UCL_{c,j} = \overline{\mu}_c^{(j)} + \gamma_j \sqrt{\frac{\alpha_j}{2-\alpha_j}} \sigma_D^{(j)} \tag{8}$$

$$UCL_{c,j} = \overline{\mu}_c^{(j)} - \gamma_j \sqrt{\frac{\alpha_j}{2-\alpha_j}} \sigma_D^{(j)} \tag{9}$$

where γ_j is a parameter of allowance. Also, $\overline{\mu}_c^{(j)}$ and $\underline{\mu}_c^{(j)}$ are the upper and lower limits of the interval of demand of product j for which the current allocation ψ_c and the buffer size S_c under the condition of the allocation are appropriate, and the limits are obtained from the simulation results under the stable-demand conditions above.

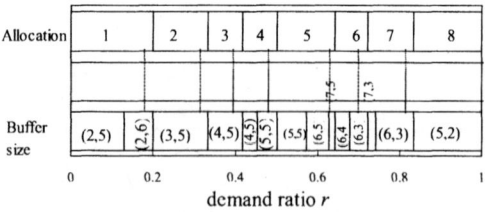

Fig. 2. The allocation of work elements and buffer size (ψ,S) to minimize *twip* while maintaining *wt*<0.2 in response to the demand ratio r

For example, the allocation of work elements and buffer size (ψ, S) to minimize *twip* while maintaining *wt*<0.2 in response to the demand ratio are obtained as shown in Fig. 2. Obtained interval of demand ratio can be translated into the interval of demand inter-arrival time of each product.

4.2   Adjusting WIP Level and Re-allocating Work Elements

If an unstable change is detected, the current allocation of work elements and/or buffer size must be altered in response to the detected change. For the purpose, the control rule for modifying the allocation of work elements and buffer sizes must be developed. The relationship between demand ratio r and the appropriate allocation of work elements and buffer size shown in Fig. 1 can be utilized, and Fig. 2 shows the allocation and buffer sizes to minimize *twip* while maintaining *wt*<0.2.

Based on the relationship in Fig. 2, the allocation of work elements and buffer sizes can be controlled in response to unstable changes in demand ratio r. When an unstable change in demand ratio r is detected by the limits, the allocation of work elements and buffer size are altered from the current ones (ψ_c, S_c) as follows:

$$(\psi, S) = \begin{cases} (\psi_c, S_c) & (\forall j, LCL_j < H_{i,j} < UCL_j) \\ (\psi_m, S_m) & (otherwise) \end{cases} \qquad (10)$$

Here, the modified allocation and buffer size (ψ_m, S_m) is obtained from the interval of the demand ratio in which the demand ratio calculated from the exceeded EWMA $H_{i,j}$ is included.

5   Investigating Effectiveness of the Proposed System

Effectiveness of the proposed system is investigated and compared with those of the alternative system based on only lean strategy. In the investigation, no delay in detecting unstable changes in demand and modifying the allocation of work elements and buffer sizes is assumed. Under the assumption, performance of the proposed system can be expected from the simulation results under stable demand condition shown in section 3, and the obtained results are shown in Fig. 3.

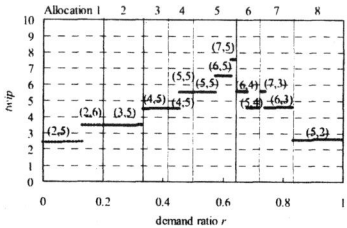

Fig. 3. The expected *twip* of the proposed system for maintaining *wt*<0.2

With Fig. 3, we can calculate the average of the expected *twip* for demand ratio $0.0 < r < 1.0$. The obtained average is 4.16, and it is about 8.4% less than that of the fixed allocation (allocation 8) and fixed buffer sizes (5, 4), that is the best in utilizing only lean strategy. The simulation results show that the proposed production control system based on lean and agile strategies can decrease the total of mean work-in-process inventory without increasing the mean waiting time of demand, and it can be claimed that the proposed production control system can respond to unstable changes in product-mix effectively.

6   Conclusions

Under the recent uncertain and competitive production environments, lean and agile production lines with the production control system proposed in this paper are valuable for achieving a practical goal to decrease inventories without decreasing customer service level. Then, it can be claimed that the proposed production control system is practically valuable.

In the mixed-product production line in this paper, the allocation of work elements to each work center is assumed to be able to modify in response to changes in product-mix. Therefore, multi-skilled workers or highly functional work centers are assumed implicitly. However, the fixed mean process time of each work element is assumed, and the learning effects of process time have not yet considered. Also, the mean and variance of the total demand are assumed to be fixed and changes in the mean and variance of demand have never been considered in this paper. If changes not only in product-mix but also in the mean and variance of demand are considered, the number of work centers as well as the re-allocation of work elements should be changed in response to the detected changes.

Acknowledgements
The work in this paper was partially supported by the Grant-in-Aid for scientific research in Japan Society for the Promotion of Science in 2006-2008.

References
1. W. J. Hopp and M. L., Roof, Setting WIP Levels with Statistical Throughput Control (STC) in CONWIP Production Lines, *IJPR*, 36(4), 867-882, (1998).
2. K. Takahashi and N. Nakamura, Reacting JIT Ordering Systems to the Unstable Changes in Demand, *IJPR*, 37(10), 2293-2313, (1999).
3. M. L. Spearman, D. L. Woodruff, and W. L. Hopp, CONWIP: A pull alternative to kanban, *IJPR*, 28(5), 879-894, (1990).
4. K. Takahashi and K. Morikawa, A System for Monitoring and Supervising Reconfigurable Manufacturing Systems, *Proc. 7th Int'l Conf. Monitoring & Automatic Supervision in Manufacturing*, 1-8, (2004).
5. J. M. Lucas and M. S. Saccucci, Exponentially Weighted Moving Average Control Schemes: Properties and Enhancements, *Technometrics*, 32(1), 1-12, (1990).

The Role of Culture in Implementing Lean Production System

Meiling Wong
Department of Industrial Engineering & Enterprise Information,
Tunghai University, Taichung, Taiwan, R.O.C.
Department of Industrial Engineering & Management, National Chinyi
Institute of Technology, Taipin, Taichung, Taiwan, R.O.C.
12F,#13, Jun-Ho St., Pai-Tun, Taichung, Taiwan, R.O.C.
email:mlwong@ncut.edu.tw Tel:886-1-4-24362803

Abstract.
Culture is a powerful, latent, and often unconscious set of forces that determine both of our individual and collective behavior, ways of perceiving, thought patterns, and values. Organizational culture in particular matters because cultural elements determine strategy, goals, and modes of operating. Many international managerial theories or production methods work well locally, but can not receive expected result once they are practiced cross nationally. Although SDWT can be seen as an example of cultural adaptation of lean manufacturing system, yet little is known about the inefficiency caused by the cultural differences. The academic community has remained primarily dedicated to single culture and comparative research which is no more sufficient. Cultures are patterns of interacting elements. To decipher that pattern, we propose an analytical framework based upon the investigation on how Taiwanese enterprises cope with the cultural resistance to achieve expected goals.

Keywords
SDWT, lean production, organization culture

1 Introduction

Organizational culture, contained of common values, symbols, beliefs, and behaviors, comes down to a common way of thinking, which drives a common way of acting on the job or producing a product in a factory. These shared assumptions,

Please use the following format when citing this chapter:

Wong, M., 2007, in IFIP International Federation for Information Processing, Volume 246, Advances in Production Management Systems, eds. Olhager, J., Persson, F., (Boston: Springer), pp. 413-422.

beliefs, and values are implicit, yet they can make the difference (functional or dysfunctional) between a company that wins and loses (Goffee and Jones, 1998, p.14). Many international managerial theories or production methods work well locally, but can not receive effect as expected once they are cross nationally practiced.

In last twenty years, the process of Lean Production System globally learned and imported, originated from Japanese Toyota Motor, is a very good example. In order to have Lean Production System put through thoroughly, the SDWT (Self-directed Work Team) implemented on shop floor in North America is a typical case of culturally adaptive feature.

To date, the academic community, by itself, has remained primarily dedicated to single culture and comparative research (Hofstede, 1982; 1991; 1991a; Trompenaars, 1984; Hampden-Turner and Trompenaars, 1993; Joynt and Warner, 1996) which, while still necessary, is no more sufficient (Bartholomew and Adler, 1996). Culture conceived as essence and difference is a massive wheel reinventing itself (Holden, 2002, p.53), we can not simply generalize the entire world from one another. Cultures are patterns of interacting elements. If we don't have a way of deciphering that pattern then we may not understand the cultures at all (Schein, 1999). Thus constructing an analytical framework based upon the culturally adaptive features in investigating how enterprises cope with cultural resistance by adjusting their organization and methods to achieve expected goal is an important research direction.

The purpose of our research is to clarify the culturally adaptive features on site in Taiwanese enterprises with Lean Production implementation. We start out with reviewing the previous research of Lean Production and culture. And using Hofstede's national culture dimensions and study of Japanese enterprises in Taiwanese to derive out the sense of problem of how Taiwanese enterprises cope with cultural difference with their organizational mechanism in Lean Production System implementation. Then on top of this base, we apply Schein's theory of three levels of corporate culture to set up our analytical framework for culturally adaptive features on site in Taiwanese enterprises with lean production system implemented, also have our case study of two Taiwanese enterprises carried on. At the end of our paper, we conclude with important findings and managerial implication from the research.

2   Lean Production System and Culture

Leading the global learning fever of lean production, the book of "The Machine That Changed the World" by Womack, Jones and Ross in 1990, was the first to use the name of Lean Production System. It introduced ideas such as Just In Time, Kanban, Jidoka, quick-switching in modes and lines, and maintenance, etc. Moreover in recent years, concepts like "lean manufacturing", "lean thinking", and "The Toyota Way" in North America keeps fermenting among nations (Womack and Jones, 1996; Shah and Ward, 2003; Liker, 2004).

Lean Production System has been one of the competitive advantages for Japanese enterprises, its cultural element underneath (Recht and Wilderom, 1998), well-known by academy and practice fields which may not exist in other countries or enterprises. While during the process of being "lean", due to their different national or organizational cultures, enterprises of nations who intend to put through Lean Production system in their organizations appear to have various culturally adaptive mechanisms SDWT in North America is seen as the adaptive system accordingly as needed (Wilson & Grey-Taylor, 1995; Janz, 1999; Rafferty & Tapsell, 2001). SDWT works in organizations as to bring out the advantage by strengthening collective power and avoiding disadvantage of individualism. Nevertheless its effect is still limited to the cultural constraint.

National culture plays an important role in constructing the corporate culture (Adler and Jelinek, 1986; Doktor, 1990; Hofstede, 1991; O'Conner, 1995), where its impact is reflected in response to the limit on the organization from external environment and appeared on the human resource management with respect to the employees' psychological level and preference.   In cross national business or cooperation, the inconsistence of national cultures increases the difference of the organizational cultures (Oudenhoven, 2001), which indirectly hinders the transfer of managerial modes or production system (Yoshiaki, T; M. Hayashi, and K. Hidaka, 2000).

Hofstede and Bond (1988, p.10-16) revealed that national cultures differ mainly along four dimensions: Power distance, Individualism/Collectivism, Masculinity/ Femininity, and Uncertainly avoidance.   Further a fifth dimension of Confucian Dynamism (later is revised to long term versus short term orientation in Hofstede, 1994) to distinguish countries of the teaching of Confucius.   Also in their national culture comparison, Taiwan and Japan have relatively closer national culture features, especially in the dimensions of power distance and Confucian Dynamism (Hofstede and Bond, 1988, p.12-13, exhibit.2). This may be explained by their historical connection.

However Japanese enterprises in Taiwan including small group activities have very strong features of Japanese production system on site, which seem obviously different in comparison with North America and China (Liu, 1996, 2000).   We therefore would like to propose that Taiwanese enterprises have the 'cultural interface' and in turn develop out the agilely adaptive Lean Production System which is totally different from SDWT, and have its mechanism verified through out our case study.

3   The Analytical Framework of Culturally Adaptive Features

In this article, we define culturally adaptive features as: the phenomenon appeared or adaptive behavior adopted during implementation of managerial styles or production systems by the organization in coping with cultural difference in terms of value points or ways of thinking.

For decades, many researches contribute the economy success of East Asia countries to their common background of Confucianism (Hofstede & Bond, 1988; Ruttan, 1995; Yeung & Tung, 1996; Fang, 2001). In Hostede's research (1994), the results point out that long term orientation is correlated with national economic growth showing that what led to the economic success of the East Asian economies during the past 25 years is their populations' cultural stress on the future-oriented values of thrift and perseverance. Jones and Davis (2000, p.28) also claim that those characteristics normally associated with the positive pole of the Confucian dynamism dimension, including the focus on hard work and perseverance should be the concepts of 'face' and reciprocation, and concerns for traditions and fulfilling social obligations (Hofstede 1994, Hofstede and Bond, 1988). It seems this long term versus short term orientation lies underneath the motives is what makes it most different between Japanese and Taiwanese behavioral modes.

We apply Schein's (1992) three levels of organizational cultural theory to elaborate our analysis. Schein believes that there are three levels of organizational culture, they are: Level one--- artifacts: they are visible organizational structures and processes, such as the décor, and the climate, but hard to decipher. Level two---espoused values: strategies, goals, philosophies, which provide deeper level of thoughts and perceptions that drive the overt behavior. And Level three---shared tacit assumptions: they are unconscious, taken-for-granted beliefs, perceptions, thought and feelings, where the value system is from (Schein, 1999, p.15-20).

Horsley and Buckley (1990) describe the Confucian 'work ethic', introduced to feudal Japan through Chinese and Korean models, as a comprehensive social code governing relationships, respect for authority and conformity to the rule of law. People of these countries thus regard work not as a hardship but as a positive life-asserting activity. They all are diligent, and value collective harmony in general, yet there is difference in their behavioral modes in terms of long-term /short-term perspectives.

The Tanaka formula (Tanaka, 1992) for adding value to its people, products and processes, based on the traditional Japanese cultural trait of continuous improvement or Kaizen, is a model for achieving business excellence. When Kaizen is adopted as a personal philosophy, it becomes a lifetime process of continuous personal development (Cartwright, 1999, p.14). In contrast with this long-term perspective, Taiwanese core values are deeply influenced by Chinese culture, which are inclined to characteristics of self-centered, think highly of "guanxi" and "face", ignore accuracy, and care for benefit at the moment, etc. (Smith, 2000, p.7, 31, 51, 75). In such high-context culture country, trust and commitment to another is secured by the potential damage to one's social position or face, which may result from failing to honor exchange obligations (Yi and Ellis, 2000). Guanxi may bring personal gains to individuals, vital resources and cost savings to the organizations. But these benefits are often achieved at the expense of other individuals (Fan, 2002).

Overall speaking, Taiwanese enterprises have the cultural approach of short-term vision that drives individual self conservation value in contract with Japanese enterprises' long-term vision that drives group's self realization value. Japanese and Taiwanese enterprises flaunt different value system. While implement Lean Production System, Taiwanese enterprises need to conduct the cultural adaptation by taking an alternative way and organization to cope with the difference, which is the basic assumption of our research.

4   Case Study

Both of the enterprises as our case study with listed companies are globally well-known and leading companies of Taiwanese optical and bicycle industries. They actively promote and implement Lean Production System in recently years and receive quite achievement. The Asia Optical was established in 1981, located at Taichung, with employees of 1,146 people. Their main products are optical lenses, microscope, aiming device, laser measurement of distance and components and parts. etc. Merida was established in 1972, located at Chanhwa, with employees of 888 people. Their main products are high class bicycle, superb bikes, exercise bikes, auto bikes and magnesium-alloyed computer shells, etc. Our research consisted of in-depth person-to-person interviews of 2-3 hours duration each with two informants per enterprise (thus, a total of 4 interviews).

In its 25- year business, the Asia optical continuously focuses on trading with Japanese enterprises as their major customer, and has been strongly influenced by Japanese style. Those strongholds co-invested with Japanese enterprises in China are particularly worldwide known (Liu, 2003). The most noticeable reform of Japanese Canon in last ten years has also had direct impact on the Asia Optical. In June 2002, the Asia Optical announced to promote Toyota Production System with determination in hiring the former executive of Japan Canon, Mr. Nishimae as the consultant and the term of "production renovation" used which was also originally from Japan Canon. The first year, they started from the important cities of production where Chinese Business grouped and received great achievement after fifteen months. One after another, the activities were continuously taken place in Taiwan on Sept. of 2003, Philippine on Feb. of 2004, and in Burma on July of 2005. Whenever and wherever there is activity being held, the President, Mr. Yi-Ren, Lai always comes forward to take the lead and starts out from conscious revolution. President Lai as the leader, along with his crew, gives spiritual lecture to express their determination of carrying out through to the end. In order to put into effect, the Asia Optical establishes the general office with sub units as the regional offices who are in charge of promotion. All regional office, being supportive for each business promotion team, holds competition activities and renovation conference periodically. To create the climate for inspiring improvement thoughts, there is no hierarchy but treating each other as classmates. At the meantime, the role of middle management in the promoting teams switches from a traditional supervisor to a helper to assist the employees in putting the improvement thoughts into practice.

Merida has also tried to improve their production though they did not have much history with Japan. For the sake of leading bicycle industry and grouping A-Team, they implement TPS in March of 2003 under the instruction of the president of Kuozui Motors, Takehiko Harada. The shop floor improvement is planned in accordance with the present condition of A-Team member and its whole future requirement, which is discussed and decided by A-Team office that also does plans for the yearly performance target. The improvement project contains educational training courses, monthly instruction from Kuozui's TPS consultant team, recruitment of factory seed talent personnel, interactive learning forum, and result exhibition, etc. Either internally or externally, Merida actively participate A-Team's semi-year result presentation, started from one Team and later expanded into six Teams. In last three years, time required of material input to bike frames has reduced from 6 days to 2 days, material stock has been cut down to 1/5, and online stock has been reduced to one third, which in turn has brought up 30% of the production efficiency. Mr. Suon-Tzu Tsen, the president of Merida who is also the vice president of A-Team, has expressed to both inside and outside with his strong vision and will in putting though to the end. He spoke with satisfactory: it was easy to make progress from grades of 60 to 90. But it gets tougher when we intend to move forward. Nevertheless we believe that improvement on site should never end.

Merida requires a whole crew participation, which mainly contains: evaluating with performance award to encourage employees' learning motivation, combining promotion channels while take a long-term foster and training to those who are creative and active, and for monthly performance awards that belonged to individuals and groups, Merida thinks it is necessary to promote multi-skilled operators and process production. When targets of various stages achieved, Merida begins to pay more attention on nurturing the organizational environment and less emphasis on immediate result and effect of improvement, and incline to a gradual decrease for individual incentive.

During Lean Production System implementation in these two enterprises, in contrast with the basic assumption and analytical framework raised by our research, there are at least two significant differences. One is the short-term individual or team incentive that exists in Merida and its necessity is emphasized. The other one is the incentive system and organizational goal are gradually adjusted as needed. The proportion of short-term individual or group incentive is progressively decreased when Lean Production System has been put into effect and making improvement.

Without lifetime employment, seniority-based wages or other long-term human resource incentive system, the short-term individual incentive system is prevailing in lots of Taiwanese enterprises (Liu, 1996). In the beginning stage of Lean Production System implementation, it is understandable to have systems like short-term incentive system. And Merida's adjustment of incentive system and organizational goals may be seen as an indicator of the degree of value system change. Obviously, being long-term influenced by Japanese enterprises, the employees of Asia Optical has already gone through this stage. It seems that having the organizational climate of long-term prospect and vision is quite enough to achieve the goals. This kind of effort devoted shows the value system change in thinking highly of groups and self-realization of the operators driven by long-term vision. This change is helpful for fermenting the manufacturing site with long-term vision to carry on continuous improvement for long-term incentive effect, which not only enables continuous learning, but also indirectly create the atmosphere needed for mutual assistance between shop floor teams. Being relatively passive, Taiwanese enterprise shop floor cares more for "face", with the supporting mechanism of comparing and learning, the external experts and high rank of unit in charge are able to achieve the goals of improving mutual assistance ability within teams or collaborative capability cross departments.

In conclusion, there exists degree of difference among Japanese enterprises, Asia Optical, and Merida in flaunting their values of motives, diligent, and think highly of group harmony. The culturally adaptive organization and method in enabling self-realization on site seems also cause the quality change of the flaunted value system. The top manager's participation in person, help from external experts, and higher rank of unit in charge and so on do have important effect. At the meantime, if Taiwanese enterprises could turn themselves toward to the flaunted value system driven by long-term vision, short-term individual incentive system will gradually fade away.

5   Conclusion and Implication

The characteristic of Lean Production System is that it contains not only visible skills or methods (task side), but also the organizational collaborative factor between people (human side), which explains why simply duplicate it cannot receive the same effect. There is the need for cultural adaptation. Our study shows that since Japanese and Taiwanese enterprises have different flaunted value system, it is required to have cultural adaptation conducted while implementing Lean Production System. In other words, it is necessary to take a different organization and method to cope with it so as to reserve the spirit of Lean Production System, have it put through thoroughly, thus in turn the competitive advantage with positive cycle can be derived.

Specifically speaking, there are four culturally adaptive features for the manufacturing site in Taiwan when implementing the Lean Production System:

1)   The top manager participates in person to expresses his strong ambition and vision. Top managers who take part personally and express their determination in solid action to set up good environment or atmosphere for competition can break up the conception of "I" of being selfness and help to drive the change of value system with long-term vision.

2)   Help from external experts and raise higher rank of unit in charge seem to be the common organization type and method, in combining with the atmosphere driven by top managers (discussed above) and assorted mechanism of both competition and inspecting and learning can protect it from being perfunctory, which is an indispensable organizational competency in refining the Lean Production System implementation.

3)   Short-term individual or group incentive is still necessary. When there is no either lifetime employment, seniority-based wages or any other long-term human resource incentive system, the short-term incentive system is still needed, especially at the beginning stage.

4)   The proportion of short-term individual or group incentive may be gradually decreasing or fading away while Lean Production System gets refiner and more practicable. This can be seen as an indicator for the degree of value system change.

This research found that national culture has significant impact during the process of Lean Production System implementation in Taiwanese enterprises. Yet there are two more subjects required more attention.   One is the generalization problem of this article. Do all Taiwanese enterprises who have successfully implemented Lean Production System have similar criterions? Another one is that when a complete Lean Production System across product development and collaborative system is implemented, more organizational and wider cultural levels will be involved. Will culturally adaptive features still be the same in that situation? We shall be delighted to share and work with scholars who are interested in further relevant research.

References

Adler, N.J. et al (1986). "From the Atlantic to the Pacific Century: Cross-Cultural Management Reviewed". *Journal of Management*, 12(2), pp.295-318.

Bartholomew, S. and Adler, N. (1996). Building networks and crossing borders: the dynamics of knowledge generation in a transnational world. In: Joynt, P. and Warner, M. *Managing across cultures: Issues and perspectives*. London: International Thompson, pp.7-32.

Cartwright, J. (1999). *Cultural Transformation, nine factors for improving the soul of your business*. London: Pearson Education LTD.

Deal, T and Kennedy, A. (1982). *Corporate Cultures*. Addison-Wesley, Reading, MA.

Doktor, R H. (1990). "Asian and American CEOs: A Comparative Study". *Organizational Dynamics*, Vol. 18, No.3, pp.46-56.

Fan, Y. (2002). "Questioning guanxi: definition, classification and implications". *International Business Review*, 11, pp.543-561.

Fang, T. (2001). "Culture as a driving force of interfirm adaptation: A Chinese case". Industrial Marketing Management, 30, pp.51-63.

Goffee, R. and Jones, G.. (1998). *The character of a corporation*. 2nd ed. London: Profile Books LTD.

Hampden-Turner, C. and Trompenaars, F. (1993). *The seven cultures of capitalism: value systems for creating wealth in the United States, Britain, Japan, Germany, France, Sweden and the Netherlands*. London: Judy Piatkus.

Hofstede, G.. (1980/1991). *Culture's consequences: International differences in work-related values*. Beverly Hills, Ca: Sage Publications.

Hofstede, G. (1982). "Intercultural cooperation in organizations". *Management Decision*, 20. pp.53-67

Hofstede, G.. (1991a; 2003). *Culture and organizations: Intercultural cooperation and its   importance for survival – software of the mind*. London: McGraw-Hill.

Hofstede, G. (1994). "The business of international business is culture". *International Business Review*, Vo. 3, No. 1, pp.1-14.

Holden, N. J. (2002). *Cross-cultural management, a knowledge management perspective*. London: Pearson Education Ltd., pp.53.

Horsley, W. and Buckley, R. (1990). *Nippon*. London : BBC Books.

Janz, B. (1999). "Self-directed Teams in IS: Correlates for Improved Systems Development Work Outcomes". *Information & Management,* 35, pp.171-192.

Joynt, P. and Warner, M. (1996). *Managing across cultures: Issues and perspectives*. London: International Thompson.

Jones, G.K., and Davis, H. (2000). "National culture and innovation: implications for locating global R&D operations". *Management International Review*, vol.40, 00.11-39.

Liker, J. K. (2004). *The Toyota Way, 14 management principles.* New York, U.S.: McGraw-Hill Books Company.

Liu Ren-Jye. (1996). "A Study of Japanese Management System of the Japanese Affiliated Companies in Taiwan and China," *Journal of Economics & Business Administration* (in Japanese), Edited by Kobe University, Vol.174 No.1, pp.37-52.

Liu, Ren-Jye. (2000). "International Division of Labor and Japanese Production System," in Munakata, M. Nuki, T. and Sakamoto K., eds. *Contemporary Production System* (in Japanese), Kyoto: Minerva, pp.218-237.

Liu, Ren-Jye.(2003). "An Empirical Study of Strategic Alliances between Taiwanese and Japanese Enterprises in Mainland China." *Journal of Asian Business.* 19(3): 71-94.

Martin J. and Meyerson, D. (1988). *Organizational Cultures and the Denial, Chenneling and Acknowledgement of Ambiguity and Change*, pp.93-125. John Wiley & Sons, Chichester.

Nathan, J. (1999). *Sony: The private life.* London: HarperCollins.

O'Connor, N.G. (1995). "The Influence of Organizational Culture on the Usefulness of Budget Participation by Singaporean-Chinese Managers". *Accounting, Organizations and Society*, Vol. 20, No. 5, pp.383-403.

Oudenhoven, J. P. (2001). "Do Organizations Reflect National Cultures? A 10-nation study". *International Journal of International Relations*, 25, pp.89-107.

Rafferty, J and Tapsell J. (2001). "Self-Managed Work Teams and Manufacturing Strategies: Cultural Influences in the Search for Team Effectiveness and Competitive Advantage". *Human Factors and Ergonomics in Manufacturing*, Vol. 11 (1), pp.19-34.

Recht .R. and Wilderom, C. (1998). "Kaizen and Culture:on the Transferability of Japanese Suggestion Systems". *International Business Review*, 7, pp.7-22.

Shah, Rachna and Ward, Peter T. (2003). "Lean manufacturing: context, practice bundles, and performance," *Journal of Operations Management*, Vol.21, Issue2, pp.129-149.

Schein, E.H. (1992). *Organizational Culture and Leadership*, 2nd ed. C.A.: Jossey-Bass, San Francisco, pp15-20.

Schein, E.H. (1999). *The Corporate Culture Survival Guide.* C.A.: Jossey-Bass, San Francisco.

Smith, A.H. (2000). *Chinese character.* H.K.: Joint Publishing Co.

Tanaka, K. (1992). *Building a New Japan.* Tokyo: Simul Press,.

Trompenaars, F. (1984). The organization of meaning and the meaning of organization – a comparative study on the conceptions and organizational structure in different cultures. PhD thesis, University of Pennsylvania.

Wilkins, A. L. and Ouchi, W. G. (1983). "Efficient Cultures: Exploring the Relationship Between Culture and Organizational Performance". *Administrative Science Quarterly*, Vol. 28, pp.468-481.

Wilson.J.R and Grey-Taylor, S.M. (1995). "Simultaneous Engineering for Self-directed Work Teams Implementation: A Case Study in the Electronics Industry". *International Journal of Industrial Ergonomics*, 16, pp.353-365.

Womack, James, Daniel Jones and Daniel Ross, (1990). *The Machine That Changed the World.* N.Y.: Rawson Associates, Macmillan.

Womack, James, and Daniel Jones, (1996). *Lean Thinking.* N.Y.: Simon and Shuster.

Yeung, I. Y.M., and Tung, R.L. (1996) "Achieving Business Success in Confucian Societies: the importance of guanxi (connections)". *Organizational Dynamics*, Autumn, pp.54-65.

Yi, L.M., and Ellis, P. (2000). "Insider-outsider perspectives of guanxi". *Business Horizons*, Jan-Feb., pp.25-30.

Yoshiaki, T., M. Hayashi, and K. Hidaka. (2000). *International Transfer of Management Style* (in Japanese), Tokyo: Chuuo University Press.

Printed in the United States of America